Springer Series in Geomechanics and Geoengineering

Series Editors

Wei Wu
Ronaldo I. Borja

For further volumes:
http://www.springer.com/8069

Springer Series in Geomechanics and Geoengineering

For further volumes:

http://www.springer.com/series/8069

Stéphane Bonelli • Cristian Dascalu
François Nicot
Editors

Advances in Bifurcation and Degradation in Geomaterials

Proceedings of the 9th International Workshop on Bifurcation and Degradation in Geomaterials

Springer

Editors
Stéphane Bonelli
Cemagref
Unité de Recherche Ouvrages Hydrauliques
Route de Cezanne, CS 40061 3275
13182 Aix-en-Provence Cedex 5
France
stephane.bonelli@cemagref.fr

Cristian Dascalu
Université J. Fourier
Laboratoire Sols Solides
Structures - Risques
PO Box 53
Grenoble Cedex 9
France
cristian.dascalu@hmg.inpg.fr

François Nicot
Cemagref, Geomechanics Group
Unité de Recherche Erosion
Torrentielle
PO Box BP 76
Saint Martin d'Hères
France
francois.nicot@cemagref.fr

ISBN 978-94-007-3779-2 ISBN 978-94-007-1421-2 (eBook)
DOI 10.1007/978-94-007-1421-2
Springer Dordrecht Heidelberg London New York

Cover design: eStudio Calamar S.L.

Printed on acid-free paper

Springer is part of Springer Science+Business Media (www.springer.com)

Preface

Geomaterials exhibit intriguing mechanical behaviours with a variety of failure modes ranging from diffuse to localized deformations depending on stress, density, microstructure and loading conditions. For instance, diffuse failure in a soil mass under constant gravity-driven load occurs with a sudden phase change triggering catastrophic large fluid-like movements. Another remarkable example of related phenomena is degradation arising from cyclic loading, ageing, weathering, chemical attack and capillary effects, among others. Today, it is well-recognized that the various manifestations alluded to in the above are a result of material and/or geometric instabilities that can be studied within the framework of micromechanics and bifurcation theory.

Interests in localization and related instabilities in the field of geomechanics date back to the early 1980s when the first International Workshop on Localization of Soils was organized in Karslrule, Germany, February 1988. This aroused so much enthusiasm and interest in the fundamental aspects of bifurcation theory for soils that the second workshop followed right after in Gdansk, Poland, September 1989. The topic was then extended to rock mechanics at the third international workshop in Aussois, France, September, 1993. Interests grew steadily and the scope was expanded to instabilities and degradations in geomaterials at the fourth, fifth, sixth, seventh and eighth workshops that were held in Gifu, Japan, September 1997; Perth, Australia, November 1999; Minnesota, USA, June 2002; Crete, Greece, June 2005; and Lake Louise, Canada, May 2008.

The ninth international workshop continued the central theme of bifurcations and degradations in geomaterials, with further extensions into geo-environmental applications. The workshop venue, Porquerolles Island in Southern France, was an idyllic setting for discussing the science and engineering in the natural beauty of the Mediterranean region.

IWBDG 2011 was attended by 90 participants representing 12 countries, with 55 presentations given over three days. In particular, this meeting fostered the participation of young researchers, including doctoral students and postdoctoral fellows. As a major co-sponsor, CATERPILLAR Inc., USA provided generous financial support to IWBDG 2011, similar to the previous Workshop in Lake Louise. Additional support was provided by (in alphabetical order), CEMAGREF, CNRS, EdF, GeophyConsult, Provence Alpes-Cotes d'Azur Province, Université Joseph Fourier.

This proceeding contains 46 peer reviewed full papers that are a sampling of presentations given at IWBDG 2011, capturing the state-of-the-art in the specialized field of geomechanics and contemporary approaches to solving the central issue of failure.

We would like to dedicate this Workshop to Professor Ioannis Vardoulakis who tragically passed away in September, 2009. He played an eminent role as one of the founders of this series of workshops. His substantial contributions, both technical and as a mentor to young researchers were pivotal to the success of the IWBDG series. We trust this is a fitting tribute to an outstanding man and scientist.

The members of the Advisory Board,

Ronaldo Borja
Félix Darve
Jacques Desrues
Hans Muhlhaus
Richard Wan

Contents

Part III Numerical Modeling of Failure and Localized Deformation in Geomaterials

Part IV Failure, Localization and Multi-physical Couplings

Contributors

Sara Abedi Department of Civil & Environmental Engineering, University of Southern California, Los Angeles, CA 90089-2531, USA, abedimas@usc.edu

R. Affes LMGC, CNRS-Université Montpellier II, Place Eugène Bataillon, 34095 Montpellier cedex, France and Laboratoire MIST, IRSN-CNRS-Université Montpellier 2, France

A. Aït-Mokhtar University of La Rochelle, avenue Michel Crépeau, 17042 La Rochelle cedex, France

Mustafa Alsaleh Virtual Product Development Technology, Caterpillar Inc.; Mossville, IL 61616 USA, alsaleh_mustafa_i@cat.com

M. Aris CEMAGREF d'Aix en Provence, 3275, Route de Cézanne CS, 40061 13182 Aix en Provence Cedex 5, France

E. Azéma LMGC, CNRS – Université Montpellier 2, Place Eugène Bataillon, 34095 Montpellier cedex 05, France, azema@lmgc.univ-montp2.fr

M. Badiane CEMAGREF d'Aix en Provence, 3275, Route de Cézanne CS 40061, 13182 Aix en Provence Cedex 5, France, mdbadiane@yahoo.fr

Erich Bauer Institute of Applied Mechanics, Graz University of Technology, Graz, Austria, erich.bauer@tugraz.at

Thomas Baumgartl Center for Mined Land Rehabilitation, Sustainable Minerals Institute, The University of Queensland, St. Lucia, QLD 4072, Australia

R. Beguin Laboratoire d'étude des Transferts en Hydrologie et Environnement, Université de Grenoble, BP 53, 38041 Grenoble Cedex 09, France

EDF-Centre d'Ingénierie Hydraulique (CIH), 73373 Le Bourget du Lac Cedex, France, remi.beguin@ujf-grenoble.fr

N. Benahmed CEMAGREF d'Aix en Provence, 3275, Route de Cézanne CS, 40061 13182 Aix en Provence Cedex 5, France, nadia.benahmed@cemagref.fr

P. Bésuelle CNRS – UJF – Grenoble INP, Laboratoire 3SR, BP53, 38041 Grenoble Cedex 9, France, Pierre.Besuelle@hmg.inpg.fr

Ronaldo I. Borja Department of Civil and Environmental Engineering, Stanford University, Stanford, CA 94305, USA, borja@stanford.edu

K. Bourbatache LTI, 48 rue Ostende 02100 Saint-Quentin

Nicolas Brantut Laboratoire de Géologie, CNRS UMR 8538, Ecole Normale Supérieure, Paris, France

D. Caillerie Laboratoire Sols Solides Structures – Risques, Grenoble Université & CNRS, BP 53, 38041 Grenoble Cedex, France, Denis.Caillerie@inpg.fr

G. Chambon Cemagref, BP 76, 38402 St Martin d'Heres, Montpellier, France

R. Chambon Laboratoire 3S-R, University of Grenoble, BP 53X, 38041 Grenoble Cedex, France, Rene.Chambon@hmg.inpg.fr

C. S. Chang University of Massachusetts, Amherst, MA 01003, USA

Robert Charlier Université de Liege, ArGEnCo – *Geo*3, Bat B52/3 chemin des chevreuils 1, 4000 Liège, Belgium, robert.charlier@ulg.ac.be

K.T. Chau Department of Civil and Structural Engineering, The Hong Kong Polytechnic University, Kowloon, Hong Kong, China, cektchau@polyu.edu.hk

J.S. Chen Civil & Environmental Engineering, University of California, Los Angeles, CA, USA, jschen@seas.ucla.edu

S.W. Chi Civil & Environmental Engineering, University of California, Los Angeles, CA, USA

Gäel Combe CNRS UMR 5521, Grenoble-INP, UJF-Grenoble 1, 3SR-Lab, Grenoble F-38041, France

Frédéric Collin Université de Liège, ArGEnCo – *Geo*3, Bat B52/3 chemin des 132 chevreuils 1, 4000 Liège, Belgium, f.collin@ulg.ac.be

Cristian Dascalu CNRS UMR 5521, Grenoble-INP, UJF-Grenoble 1, 3SR-Lab, Grenoble F-38041, France, cristian.dascalu@hmg.inpg.fr

Jean-Yves Delenne LMGC UMR 5508 CNRS-Université Montpellier 2/MIST, IRSN DPAM-CNRS, cc. 048, Pl. E., Bataillon, 34095 Montpellier cedex 5, France, jean-yves.delenne@univ-montp2.fr

Jacques Desrues CNRS UMR 5521, Grenoble-INP, UJF-Grenoble 1, 3SR-Lab, Grenoble F-38041, France, Jacques.Desrues@grenoble.cnrs.fr

Thomas A. Dewers Sandia National Laboratories, PO Box 5800, Albuquerque, NM 87185–0751 USA, tderwers@sandia.gov

B. Ducoin LAMSID (EDF/R et D), 1 Avenue du Gnral De gaulle, 92140 Clamart, France

A. V. Dyskin School of Civil and Resource Engineering, University of Western Australia, Perth, WA, Australia, arcady@civil.uwa.edu.au

N. Eckert Cemagref, BP 76, 38402 St Martin d'Heres, Montpellier, France

Itai Einav School of Civil Engineering, The University of Sydney, Sydney, NSW 2006, Australia, itai.einav@sydney.edu.au

Y.-H. Faure Laboratoire d'étude des Transferts en Hydrologie et Environnement, Université de Grenoble, BP 53, 38041 Grenoble Cedex 09, France, yhfaure@ujf-grenoble.fr

R. Fernandes LAMSID (EDF/R et D), 1 Avenue du Gnral De gaulle, 92140 Clamart, France, romeo.fernandes@edf.fr

A. Foucault Laboratoire MSSMat, CNRS UMR 8579, École Centrale Paris, Grande voie des vignes, F-92290 Châtenay-Malabry, France, alexandre.foucault@edf.fr

Laboratoire LaMSID, UMR EDF/CNRS 2832, 1, avenue du Général de Gaulle, F-92141 Clamart, France

Département Analyses Mécanique et Acoustique, EDF Recherche et Développement, 1, avenue du Général de Gaulle, F-92141 Clamart, France

Zhongzhi Fu Institute of Hydraulic Engineering, Hohai University, Nanjing, China, fuzhongzhi@yahoo.com

T. Fukuda Department of Civil and Earth Resources Engineering, Kyoto University, Kyoto 615-8540 Japan

E. Garcia Department of Civil and Earth Resources Engineering, Kyoto University, Kyoto, Japan

On leave from the Department of Civil Engineering, University of Antioquia, Colombia

J. Gaume Cemagref, BP 76, 38402 St Martin d'Heres, Montpellier, France, johan.gaume@cemagref.fr

Lutz Gross Earth Systems Science Computational Center, The University of Queensland, St. Lucia, QLD 4072, Australia

P. C. Guan Civil & Environmental Engineering, University of California, Los Angeles, CA, USA

Alice Guest Department of Geoscience, University Dr., Calgary, AB, T2N 1N4, alice.m.guest@gmail.com

N. Guo Department of Civil and Environmental Engineering, Hong Kong University of Science and Technology, Clearwater Bay, Kowloon, Hong Kong, ceguo@ust.hk

Peijun Guo Department of Civil Engineering, McMaster University, Hamilton, ON, Canada, L8S 4L7, guop@mcmaster.ca

Marte Gutierrez Division of Engineering, Colorado School of Mines, Golden, CO, 80401, USA, mgutierr@mines.edu

S.A. Hall CNRS - UJF - Grenoble INP, Laboratoire 3SR, BP53, 38041 Grenoble Cedex 9, France

P.-Y. Hicher GeM, Ecole Centrale de Nantes, Nantes, France, pierre-yves.hicher@ec-nantes.fr

Y. Higo Department of Civil and Earth Resources Engineering, Kyoto University, C1 Bld. C cluster, Kyotodaigaku-Katsura 4, Nishikyo-ku, Kyoto, 615–8540, Japan, higo.yohsuke.5z@kyoto-u.ac.jp

David J. Holcomb Sandia National Laboratories, PO Box 5800, Albuquerque, NM 87185–0751 USA, djholco@comcast.net

Mathew D. Ingraham Clarkson University, 8 Clarkson Ave #5725, Potsdam, NY 13699, USA, ingrahmd@clarkson.edu

Kathleen A. Issen Clarkson University, 8 Clarkson Ave #5725, Potsdam, NY 13699, USA, issenka@clarkson.edu

H. Iwai Department of Civil and Earth Resources Engineering, Kyoto University, Kyoto 615–8540 Japan

S. Kimoto Department of Civil and Earth Resources Engineering, Kyoto University, C1 Bld. C cluster, Kyotodaigaku-Katsura 4, Nishikyo-ku, Kyoto, 615–8540, Japan, kimoto.sayuri.6u@kyoto-u.ac.jp

Djimédo Kondo Université de Sciences et Technologies de Lille, Laboratoire de Mécanique de Lille – Cité Scien-tifique, 59655 Villeneuve d'Ascq Cedex, France, djimedo.kondo@univ-lille1.fr

Poul V. Lade Department of Civil Engineering, The Catholic University of America, Washington, DC 20064, USA, Lade@cua.edu

Danuta Lesniewska Institute of Hydro-Engineering, Polish Academy of Sciences, Geomechanics Division, 80–328 Gdańsk, Koscierska 7, Poland, d.lesniewska@ibwpan.gda.pl

Koszalin Technical University, 75–453 Koszalin, Sniadeckich 2, Poland

Séverine Levasseur Université de Liege, ArGEnCo – *Geo*3, Bat B52/3 chemin des chevreuils 1, 4000 Liège, Belgium, severine.levasseur@ulg.ac.be

P. Lin Department of Hydraulic Engineering, Tsinghua University, Beijing, China, celinpe@tsinghua.edu.cn

Fushen Liu ExxonMobil Research and Engineering, 1545 Route 22 East, Annandale, NJ 08801, USA, fushen.liu@exxonmobil.com

M. Mansouri Département de génie civil, université Ferhat-bbas, 19000, Setif, Algeria

Benjy Marks School of Civil Engineering, The University of Sydney, Sydney, NSW 2006, Australia, benjy.marks@sydney.edu.au

F. Marnelli Laboratoire 3S-R, Grenoble Université, Université Joseph Fourier BP 53 X Grenoble cedex 9

Y. Matsuhima Department of Civil and Earth Resources Engineering, Kyoto University, C1 Bld. C cluster, Kyotodaigaku-Katsura 4, Nishikyo-ku, Kyoto, 615–8540, Japan, yoshiki@handballer.mbox.media.kyoto-u.ac.jp

Y. Miki Department of Civil and Earth Resources Engineering, Kyoto University, Kyoto 615–8540 Japan

O. Millet University of La Rochelle, avenue Michel Crépeau, 17042 La Rochelle cedex, France, olivier.millet@univ-lr.fr

A. Modaressi Laboratoire MSSMat, CNRS UMR 8579, École Centrale Paris, Grande voie des vignes, F-92290 Châtenay-Malabry, France

Arash Mohajeri Earth Systems Science Computational Center, The University of Queensland, St. Lucia, QLD 4072, Australia, arash.mohajeri@uqconnect.edu.au

Abdalsalam Mohamed Division of Engineering, Colorado School of Mines, Golden, CO, 80401, USA

Y. Monerie IRSN, DPAM, CE Cadarache, Bat. 702, 13115 St. Paul-lez-Durance Cedex, France and Laboratoire MIST, IRSN-CNRS-Universit'e Montpellier 2, France

Hans Muhlhaus Earth Systems Science Computational Center, The University of Queensland, St. Lucia, QLD 4072, Australia

David Muir Wood University of Dundee, Fulton Building, Dundee DD1 4HN, United Kingdom, d.muirwood@dundee.ac.uk

M. Naaim Cemagref, BP 76, 38402 St Martin d'Heres, Montpellier, France

Micha Nitka CNRS UMR 5521, Grenoble-INP, UJF-Grenoble 1, 3SR-Lab, Grenoble F-38041, France, micnitka@pg.gda.pl

F. Oka Department of Civil and Earth Resources Engineering, Kyoto University, C1 Bld. C cluster, Kyotodaigaku-Katsura 4, Nishikyo-ku, Kyoto, Japan, oka.fusao.2s@kyoto-u.ac.jp

Euripides Papamichos Department of Civil Engineering, Aristotle University of Thessaloniki, Thessaloniki GR-54124, Greece

SINTEF Petroleum Research, Trondheim N-7465, Norway, epapamic@civil.auth.gr

E. Pasternak School of Mechanical Engineering, University of Western Australia, Perth, WA, Australia, elena@mech.uwa.edu.au

Dunja Perić Kansas State University, Manhattan, KS 66506–5000, USA, peric@ksu.edu

P. Philippe CEMAGREF d'Aix en Provence, 3275, Route de Cézanne CS 40061, 13182 Aix en Provence Cedex 5, France, pierre.philippe@cemagref.fr

M. Pinheiro Thurber Engineering Ltd., Calgary, Alberta, Canada

F. Radjaï LMGC UMR 5508 CNRS-Université Montpellier 2 / MIST, IRSN DPAM-CNRS, cc. 048, Pl. E. Bataillon, 34095 Montpellier cedex 5, France, radjai@lmgc.univ-montp2.fr

Helia Rahmani Department of Civil and Environmental Engineering, Stanford University, Stanford, CA 94305, USA, rahmani@stanford.edu

Amy L. Rechenmacher Department of Civil & Environmental Engineering, University of Southern California, Los Angeles, CA 90089-2531, USA, arechenm@usc.edu

Richard A. Regueiro Department of Civil, Environmental, and Architectural Engineering, University of Colorado at Boulder, 1111 Engineering Dr., Boulder, CO 80309, USA, regueiro@colorado.edu

X. Ren Civil & Environmental Engineering, University of California, Los Angeles, CA, USA

Pierre Rognon School of Civil Engineering, The University of Sydney, Sydney, NSW 2006, Australia, pierre.rognon@sydney.edu.au

M. J. Roth U. S. Army Engineer Research and Development Center, Vicksburg, MS, USA

T. Sanagawa Foundation and Geotechnical Engineering Structures Technology Division, Railway Technical Research Institute, 2-8-38 Hikari-cho, Kokubunji-shi, Tokyo, Japan, sanagawa@rtri.or.jp

A. Seridi Laboratoire de mécanique des solides et systémes (LM2S), université M'hamed-Bougara, Boumerdes, Algeria

Antonin Settari Department of Chemical and Petroleum Engineering, The Schoolich School of Engineering, University of Calgary, 2500 University Dr., Calgary, AB, Canada, T2N 1N4, asettari@ucalgary.ca

Y. Sieffert Laboratoire 3S-R, Grenoble Université, Université Joseph Fourier BP 53 X Grenoble cedex 9, yannick.sieffert@ujf-grenoble.fr

T. R. Slawson U. S. Army Engineer Research and Development Center, Vicksburg, MS, USA

Jean Sulem CERMES UR Navier, CNRS UMR 8205, Ecole des Ponts ParisTech, Marne-la-Vallée, France, jean.sulem@enpc.fr

Jacek Tejchman Gdańsk University of Technology, Faculty for Civil and Environmental Engineering, 80–233 Gdańsk-Wrzeszcz, Narutowicza 11/12, Poland, tejchmk@pg.gda.pl

V. Topin IRSN, DPAM, CE Cadarache, Bat. 702, 13115 St. Paul-lez-Durance Cedex, France and Laboratoire MIST, IRSN-CNRS-Universit'e Montpellier 2, France

Antoinette Tordesillas Department of Mathematics & Statistics, University of Melbourne, Parkville, VIC 3010, Australia, atordesi@ms.unimelb.edu.au

F. Voldoire Laboratoire LaMSID, UMR EDF/CNRS 2832, 1, avenue du Général de Gaulle, F-92141 Clamart, France

Département Analyses Mécanique et Acoustique, EDF Recherche et Développement, 1, avenue du Général de Gaulle, F-92141 Clamart, France

David M. Walker Department of Mathematics & Statistics, University of Melbourne, Parkville, VIC 3010, Australia, dmwalker@unimelb.edu.au

Richard Wan Department of Civil Engineering, University of Calgary, Calgary Alberta, Canada, T2N 1N4, wan@ucalgary.ca

Y. Wang Institut für Geotechnik, Universität für Bodenkultur, Vienna 1180, Austria

M. Wawra Institut für Geotechnik, Universität für Bodenkultur, Vienna 1180, Austria

W. Wu Institut für Geotechnik, Universität für Bodenkultur, Vienna 1180, Austria

Beichuan Yan Community Surface Dynamics Modeling System, University of Colorado at Boulder, 3100 Marine, St., Boulder, CO 80303, USA, beichuan.yan@colorado.edu

Z.-Y. Yin Shanghai Jiao Tong University, Shanghai 200240, PR, China

M.S. El Youssoufi LMGC UMR 5508 CNRS-Université Montpellier 2 / MIST, IRSN DPAM-CNRS, cc. 048, Pl. E. Bataillon, 34095 Montpellier cedex 5, France

J. Zhao Department of Civil and Environmental Engineering, Hong Kong University of Science and Technology, Clearwater Bay, Kowloon, Hong Kong, jzhao@ust.hk

Instability and Bifurcation for Inelastic Geomaterials

R. Chambon

Abstract During the past issues of the workshop on localization and bifurcation in soils and rocks, the two words instability and bifurcation are used by many participants and induce many strong discussions. I am convinced that most of these discussions are only semantic ones and that indeed this is due to two main things. In the best case discussion is triggered by the fact that different people have different definitions of these words, in the worst case some people have no explicit definition of the word used. We try in this paper once more to put forward the intrinsic difficulties of true instability and bifurcation studies for geomaterials and see how numerical tools can finally help us.

Keywords Instability • Bifurcation • Numerical computation • Controllability • Inelasticity

1 Introduction

Difficulty about distinguishing stability from bifurcation is in our opinion a consequence of the fact that mainly for elastic situations, loss of uniqueness is related with the so called exchange of stability which means that as soon as several solutions are available, the "old" one becomes unstable whereas the new ones are stable (this is typical for elastic buckling for instance). Consequently, in this case, it is sufficient to study loss of uniqueness to deduce loss of stability. Unfortunately geomaterials behave mostly as plastic materials. This means that cohesion and/or friction sliding is a main feature of their behavior, inducing energy dissipation independently of the

R. Chambon (✉)
Laboratoire 3S-R, University of Grenoble, BP 53X, 38041 Grenoble Cedex, France
e-mail: Rene.Chambon@hmg.inpg.fr

S. Bonelli et al. (eds.), *Advances in Bifurcation and Degradation in Geomaterials*, Springer Series in Geomechanics and Geoengineering 11,
DOI 10.1007/978-94-007-1421-2_1, © Springer Science+Business Media B.V. 2011

velocity. An other consequence of the plastic behavior is the strong non linearity of the problem studied. Non linearity is indeed not a main problem provided it is possible to study a linearized problem, like in large hyperelasticity. Unfortunately the non linearity of the geomaterials is an incremental non linearity and this renders many method used to study instability by considering bifurcation useless. In this paper we start by recalling a useful check list of questions. Then looking at this check list allows us first to recall a clear distinction between stability and uniqueness of the quasi static solution. Then the distinction between element and structural studies is emphasized. In a following Sect. 5 linearization is studied. Finally results using numerical tools are recalled and the limitations of these tools are put forward.

2 A Useful Check List

During the 2002 workshop, I already proposed to clarify the discussions by proposing a check list. This was not published in the proceedings of the workshop but later on (see Chambon et al. 2004). Here is the conclusion of that paper.

What are you studying?

- stability
- uniqueness

In either case, what is the system that you are studying?

- a complete system (a structure)
- a single element – in which case your study is a material study

In either case, are you using:

- a justified linearization
- a partial (unjustified) linearization
- the complete non linear model

In either case,

- specify the interaction of the system with the *outside*

If you then end up with a criterion, is it:

- a sufficient condition
- a necessary condition
- both necessary and sufficient

If you are studying uniqueness,

- do you assume a specific mode
- do you restrict your study to a class of modes
- do you perform a complete study

If you are studying stability, then you must define:

- which perturbation (input) is being considered, and a measure of its magnitude
- which effect of the perturbation (output) is being considered, and a measure of its magnitude

3 Stability Versus Uniqueness

As mentioned in the introduction both notions are closely related for (hyper)elastic (even non linear) problems. However contrary to viscosity, cohesion and friction which are the main physical basis of geomaterial inelasticity induce a lot of difficulties. The simple problem of a material point put on a frictional surface shows clearly that such a problem can have an infinite number of solutions and that the most part of these solutions are stable in the Lyapunov (1992) sense. Consequently a quasi static analysis can only gives us result on uniqueness, and dynamic studies are necessary to study instability.

It is possible in some cases to study uniqueness of the solution (or the loss of uniqueness which means bifurcation). But it is necessary to well define the studied problem. Usually the problems we are interested in are initial boundary value problems. This means that we know an initial state, a history of boundary conditions is given and we are looking for the time evolution of the studied domain. Unfortunately in the most cases we are only able to study the rate problem. This means that we know a state; velocities of the boundary conditions are given ad we are looking for the velocities of any point of the studied domain. Some general results have already been obtained in Caillerie and Chambon (2004) and Chambon and Caillerie (1999). Clearly if uniqueness is lost for the rate problem for a state corresponding to a position of a solution of the initial boundary value problem, it is also lost for the initial boundary value problem.

Stability study requires a dynamic analysis. Some of these studies can be done by studying wave propagations, but we will see in the following that this induces some difficulties related to the linearization of the dynamic equations.

4 Element (Material) Versus Structural Studies

The complete initial boundary value problem is very difficult to tackle. In some case however it is possible to study a very particular problem where the studied domain is homogeneous and the boundary conditions allows a homogeneous solution. This problem is also important because it corresponds to laboratory experiments like triaxial, biaxial, true triaxial, oedometers etc... which are called element tests. Since in such studies the constitutive equations play a crucial role they are often called material studies. Mainly they are called material instability studies whereas they should be called material bifurcation studies (Cf. Sect. 3 for the difference between the two notions).

When alternative solutions are also homogeneous ones, the problem is then the controllability problem as defined in the pioneering work of Nova given during the second workshop (Nova 1989). There is now a general result about such material bifurcation. It has been established that the annulment of the so called second order work implies this specific material bifurcation, namely loss of controllability (Chambon 2005), other results are discussed in this paper but we have not sufficient place to recall them here.

An other material bifurcation study is the so called localization study sometimes improperly referred as shear band analysis. In this case an alternative solution involving a band is searched. It has to be noticed first that compaction band or shear band are triggered by the same mechanism and their threshold are obtained using the same equations where the constitutive equations play a crucial role. The main difference comes when the persistent (or non persistent) nature of the bands are studied. If the strength of the material decreases in the band (which is often the case in shear band), then the band is persistent. On the contrary when the strength of the material increases in the band (which is often the case in compaction bands) then the band is not persistent and the localized zone moves with respect to the materials. Let us finally emphasize that the criterion involving the acoustic tensor often presented as the localization criterion is available only in few cases clearly detailed in the pioneering work of Rice (1976) but often forgotten nowadays. For a complete discussion of this problem and some complete analysis (without linearization as defined in Sect. 5) see our already published work (Chambon et al. 2000). In fact the first (to our knowledge) first non linear localization analysis based on our work (Chambon and Desrues 1985) was given also during the second workshop in Gdansk.

Even if an element analysis has been able to be performed, then a question arises: what is the link with the complete initial boundary value problem or rate problem? This question has no general answer but it is clear that the fact that a material bifurcation is detected in a point of a structure does not imply that the uniqueness is lost at the structural level. Loss of uniqueness at the element level is related to some boundary condition at this element level. It is not sure that this element will undergo the same "boundary conditions" coming from the remain of the structure. This point is particular clear for localization analysis for which potential orientation of localized zones can be inhibited by the boundary conditions of the problem. Element analysis is useful for laboratory tests and is only a warning for a complete structural problem.

A final remark has to be done. If localization can be found in a structural analysis then it is necessary to use enhanced model to obtain an objective description of the post localized states. The Karlsruhe workshop (the first one) was a good opportunity to disseminate this idea (see Vardoulakis and Aifantis 1989; Vardoulakis 1989; Muhlhaus 1989).

5 Linearization

With respect to bifurcation or stability analysis, it is clear that it is easier to deal with linear problem which turns out to be only a spectrum analysis of a linear operator, or in the case of a finite number of degrees of freedom, to be a study of eigenvalues of some matrix.

Once more the problem for geomaterials is more difficult than for metals for instance. In the latter case since these materials are associative, the powerful notion of linear comparison solid can be used (Hill 1978) and there is some clear link between the results of bifurcation for the linearized analysis and the ones for the underlying true non linear problem. For non associative materials the only result is the one of Raniecki and Bruhns (1981), but it is very difficult to use it. Keeping these restriction in mind, linearization of the numerical problem and eigenvalue analysis of corresponding matrical equations can be used in order to detect bifurcation points and associated modes of the linearized problem, which in some case gives interesting results (de Borst 1989).

The problem is more crucial for stability problems. Deriving a linear dynamic problem by using a linearization of the constitutive equation can yield some difficulties. This is particularly the case when studying the so called flutter instability analysis. In this case the solutions exhibited are more or less periodic which means loading (for which the linearization has been done) followed by (elastic) unloading. Physically flutter instability is due to energy supply inside the studied system (see Chambon et al. (2004) for a simple example) due the non conservative properties of some forces acting on the system. It is clear that the non associativeness of the constitutive equation cannot be at the origin of this energy since on the contrary plasticity is dissipative. This has been clearly demonstrated by the numerical computation of dynamic problems by Simoes and Martins (2005) who proved that unless if strong non conservative forces are prescribed on the boundary the true non linear problem never exhibits the instabilities predict by a linear analysis. In other word flutter instability studied are in most cases meaningless.

6 The Powerful of Numerical Analysis

As seen above very few general results are known concerning bifurcation and stability. Simplified analysis for element problem as well as (even not justified) linearization can give us some warning about these problems. There is also a complementary tool, namely numerical computations either for the initial boundary value problem or for the rate problem. The standard way to deal with computation of quasi static initial boundary value problems is based on a discretization of the time, and for each time step on the solution of a non linear problem based on a full Newton-Raphson method. If a time step tends towards zero then the problem becomes the rate problem. It is well known even that if there are solutions, it is possible that the Newton-Raphson does not to converge. In the case of multiple solutions, if the method converges the found solution can depend on the initial guess necessary to start the method.

These remarks allowed us to propose a method to search the loss of uniqueness of the solution of the studied problem. The initial idea was to start the Newton-Raphson method with a randomly chosen initial guess whereas the standard way is to use an initial guess coming from the converged solution of the previous time step. This idea allows us to find bifurcation points (Chambon et al. 2001b). So provided we look for alternative solutions in some cases we are able to find them. Clearly if solutions involve localized band then it is necessary to use enhanced models (Matsushima et al. 2002 for instance) but the method can be generalized without difficulty in this case (see Chambon and Moullet 2004; Bésuelle et al. 2006; Sieffert et al. 2009). Let us emphasize that changing the initial guess changes neither the underlying mathematical problem nor the approached numerical one. In fact it has been experienced that other numerical input which do not enter in the definition of the problem such as the value of some tolerance, size of the time step are also able to trigger bifurcations. Such a method similarly to others is not completely satisfactory: the inability to find other solutions does not imply uniqueness.

It is quite clear that numerical methods can be useful also for stability problem as already mentioned in Sect. 5 (see Simoes and Martins 2005).

One of the advantages of the numerical method is that it is possible to deal with true non linear problem, however we have to keep in mind that we are solving discrete approximation of the true problem. This can biased the results. However, when we have for the same problem theoretical results and numerical ones the consistency between both results are quite good (Crochepeyre 1998).

7 Conclusion

It is clear that since the Karlsruhe conference, we have improved a lot our knowledge about bifurcation and stability in geomaterials. The important point now is to use all this knowledge in practical problems keeping in mind that this is due to the non associativeness and to the decreasing of the strength of the geomaterials, and that uniqueness and/or stability studies cannot be discarded for our materials of interest. Our workshops have in the past contributed a lot in new discussions and presentations of pioneering works.

References

P. Bésuelle, R. Chambon, F. Collin, Switching deformation modes in post-localization solutions with a quasibrittle material. J. Mech. Mater. Struct. **1**, 1115–1134 (2006)

D. Caillerie, R. Chambon, Existence and uniqueness for rate problems of geomechanics – application to elastoplastic and hypoplastic (CLoE) models. Revue Francaise de Génie Civil **8**, 537–562 (2004)

R. Chambon, Some theoretical results about second order work, uniqueness, existence and controllability independent of the constitutive equation. J. Eng. Math. **52**, 53–61 (2005)
R. Chambon, D. Caillerie, Existence and uniqueness theorems for boundary value problems involving incrementally non linear models. Int. J. Solids Struct. **36**, 5089–5099 (1999)
R. Chambon, J. Desrues, Bifurcation par localisation et non linéarité incrémentale: un exemple heuristique d'analyse complète, in *Plastic Instability* (E.N.P.C., Paris, 1985), pp. 101–119
R. Chambon, J. Moullet, Uniqueness studies in boundary value problem involving some second gradient models. Comput. Meth. Appl. Mech. Eng. **193**, 2771–2796 (2004)
R. Chambon, S. Crochepeyre, J. Desrues, Localization criteria for non-linear constitutive equations of geomaterials. Mech. Cohes. Frict. Mater. **5**, 61–82 (2000)
R. Chambon, S. Crochepeyre, R. Charlier, An algorithm and a method to search bifurcation points in non linear problems. Int. J. Num. Meth. Eng. **51**, 315–322 (2001)
R. Chambon, D. Caillerie, G. Viggiani, Loss of uniqueness and bifurcation vs instability: some remarks. Revue Francaise de Genie Civil **8**, 517–535 (2004)
S. Crochepeyre, Contribution à la modélisation numérique et théorique de la localisation et de la post localisation dans les géomatériaux. Ph.D. Thesis, Grenoble, 1998
R. de Borst, Numerical methods for bifurcation analysis in geomechanics. Ing. Arch. **59**, 160–174 (1989)
R. Hill, Aspect of invariance in solids mechanics. Adv. Appl. Mech. **18**, 1–75 (1978)
A.M. Lyapunov, *The General Problem of Stability of Motion*, English translation (1892) (Taylor & Francis, London, 1992)
T. Matsushima, R. Chambon, D. Caillerie, Large strain finite element analysis of local second gradient models, application to localization. Int. J. Num. Meth. Eng. **54**, 499–521 (2002)
H.-B. Muhlhaus, Application of Cosserat theory in numerical solutions of limit load problems. Ing. Arch. **59**, 124–137 (1989)
R. Nova, Liquefaction, stability, bifurcations of soil via strainhardening plasticity, in *Second International Workshop on Localisation and Bifurcation in Soils and Rocks*, eds. by E. Dembicki, G. Gudehus, Z. Sikora (Technical University of Gdansk, Gdansk, 1989), pp. 117–131
B. Raniecki, O.T. Bruhns, Bounds to bifurcation stresses in solids with non associated plastic flow rule at finite strain. J. Mech. Phys. Solids **29**, 153–172 (1981)
J. Rice, The localization of plastic deformation, in *International Congress of Theoretical and Applied Mechanics*, ed. by W.D. Koiter (North Holland Publishing Comp, Amsterdam, 1976)
Y. Sieffert, S. Al-Holo, R. Chambon, Loss of uniqueness of solutions of the borehole problem modeled with enhanced media. Int. J. Solids Struct. **46**, 3173–3197 (2009)
F.M.F. Simoes, J.A.C. Martins, Flutter instability in a non associative elastic-plastic layer: analytical versus finite element results. Int. J. Eng. Sci. **43**, 189–208 (2005)
I. Vardoulakis, Shear-banding and liquefaction ingranula materials on the basis of a Cosserat continuum theory. Ing. Arch. **59**, 106–113 (1989)
I. Vardoulakis, E.C. Aifantis, Gradient dependent dilatancy and its implications in shear banding. Ing. Arch. **59**, 197–208 (1989)

Part I
Micromechanical Origins and Multiscale Analysis of Instabilities in Geomaterials

Deciphering D'Alemberts Dream: New Tools for Uncovering Rules for Self-Organized Pattern Formation in Geomaterials

Antoinette Tordesillas and David M. Walker

Abstract Techniques from Complex Networks are used to study the evolving topology and functional connectivity in various granular systems in both two and three dimensions. A generic process of self-organization in all samples is realized and is characterized by the co-evolutionary synergy between force chains and 3-cycles. Three-cycles provide force chains a twofold benefit: (i) they prop-up force chains in the way that a counterfort or buttress supports a wall, and (ii) they frustrate rolling at contacts which is a critical mechanism for buckling. All samples reflect an inherent structural hierarchy where cyclic (minimal cycles) and linear (force chains) motifs of various length scales serve as basic building blocks for self-organization.

Keywords Complex network • Minimal cycles • Granular material • Force chains • Self-organization

1 Introduction

The rheological behavior of a granular geomaterial, in particular, its shear strength, is heavily influenced by the way the constituent grains are packed. A key issue is to understand how the fine structure of such packings on the mesoscopic scale – a grain and its contacting and surrounding neighbours – influence the bulk load-carrying capacity of the material. Here we employ the techniques and principles of Complex Networks (Newman 2003) to characterize the evolving contact topology of packed grains to uncover rules for their self-organization in dense cohesionless granular materials. In particular, we investigate the reciprocity between the minimal cycles

A. Tordesillas (✉) · D.M. Walker
Department of Mathematics & Statistics, University of Melbourne, Parkville, VIC 3010, Australia
e-mail: atordesi@ms.unimelb.edu.au; dmwalker@unimelb.edu.au

S. Bonelli et al. (eds.), *Advances in Bifurcation and Degradation in Geomaterials*, Springer Series in Geomechanics and Geoengineering 11,
DOI 10.1007/978-94-007-1421-2_2,

within the contact network and the important load-bearing force chain structures. Force chains are quasi-linear arrangements of three or more contacting particles which each bear above average load in the sense of the maximum eigenvalue of the local stress tensor calculated at the particles contacts (Muthuswamy and Tordesillas 2006).

A complex network is summarized by a set of nodes and a set of links connecting nodes. A granular contact network defines the nodes as individual particles and a link exists between nodes if the corresponding particles are in physical contact. The resulting contact network can be described by its degree distribution, clustering coefficient and a vast arsenal of other measures usefully encapsulated in a feature vector (Walker and Tordesillas 2010). These measures provide a global overview of the entire assembly but can also be specified on a per particle/node basis.

An important aspect of a complex network is its minimal cycle basis (Tordesillas et al. 2010c). The growth of high-order cycles in a contact network is indicative of the dilatation within the material but of most import are the low-order cycles. In particular the 3-cycles and 4-cycles have a crucial interpretation for a granular assembly. Three-cycles are associated with frustrated rotations whereas 4-cycles behave as roller bearings allowing rotation between participating grains. We pay particular attention to the strain evolution of these cyclic motifs and identify the role played by these cycles on the shear strength of the material by quantifying the density of 3-cycles around the primary load-bearing structures of force chains.

2 Methodology

In this investigation we explore bulk as well as mesoscopic patterns of behaviour and dynamics in the networks of distinct polydisperse assemblies of cohesionless particles in 2D and 3D DEM simulations and an experimental 2D system subject to a variety of loading conditions (Tordesillas 2007; Ben-Nun et al. 2010; Darve et al. 2007; Zhang et al. 2010). Our aim is to characterize the evolution of functional connectivity using complex network feature vectors, minimal cycles and their interplay with the important load bearing force chain structures. An overall goal is to discover generic attributes of this interplay and differences across a variety of loading conditions, material properties and dimensionality.

We have studied eight samples and the loading conditions and salient features are summarized in Table 1. The first four systems consider biaxial compression 2D DEM simulations with differences in boundary conditions (constant confining pressure versus constant volume) and material properties (high versus low rolling resistance, Tordesillas 2007). The macroscopic stress response of these four samples exhibit similar behaviour. We also consider an experimental system of 2D bidisperse photoelastic disk subject to cyclic shear, quasistatically deformed, to preserve constant volume (Zhang et al. 2010). In this sample, failure by strain localization is

Table 1 Salient features and loading conditions of the eight samples

Sample	Dimension	Loading	Salient aspects of system
BCP1	2D Simulation	Biaxial, constant confining pressure	Low rolling resistance; Spherical particles moving along a plane (Walker and Tordesillas 2010; Tordesillas 2007)
BCP2	2D Simulation	Biaxial, constant confining pressure	High rolling resistance; Spherical particles moving along a plane (Tordesillas 2007; Tordesillas et al. 2010a)
BCV1	2D Simulation	Biaxial, constant volume	Low rolling resistance; Spherical particles moving along a plane (Tordesillas 2007; Tordesillas et al. 2010a)
BCV2	2D Simulation	Biaxial, constant volume	High rolling resistance; Spherical particles moving along a plane (Tordesillas 2007)
BCYS	2D Experiment	Cyclic shear, constant volume	Bidisperse circular disks (Zhang et al. 2010)
TCP1	3D Simulation	Triaxial, constant confining pressure	Spherical particles, fails in the absence of localization (Darve et al. 2007)
TCP2	3D Simulation	Triaxial, constant confining pressure	Polyellipsoidal particles, fails in the absence of localization (Tordesillas et al. 2010b)
UCC	2D Simulation	Confined comminution	Particle breakage to an ultimate state; particles are circular disks (Ben-Nun et al. 2010)

evident but due to the cyclic nature of the shear no persistent shear band or critical state is observed. It is more challenging to study the response of a real, experimental or simulated assembly of granular materials in 3D but less of an undertaking for simulations. We consider two 3D DEM simulations both subject to triaxial compression under constant confining pressure distinguished mainly by particle shape (spherical versus polyellipsoidal) as well as differing material properties (Darve et al. 2007; Tordesillas et al. 2010b). Both triaxial compressed 3D systems fail in the absence of localization with TCP1 in particular resulting in an assembly with zero strength (Darve et al. 2007). The final system, comprising circular disks, included an additional failure mechanism through comminution although the effect of particle breakage actually strengthens the assembly (Ben-Nun et al. 2010).

A way to characterize the self-organization of an assembly of grains in response to quasistatic loading is by examining the properties of its contact network for each equilibrium state in the loading history (Walker and Tordesillas 2010; Tordesillas et al. 2010c). Similarly if we wish to understand how this packing distribution evolves in response to the applied load, then we may examine the changes in the properties of this contact network from one equilibrium state to the next. For each equilibrium state a complex network is used to identify possible building blocks for self-organization, in particular, linear and cyclic motifs.

2.1 Minimal Cycle Basis

An n-cycle in a network is a self-avoiding closed path of length n which starts and ends at a given node. A cycle basis is a collection of cycles such that all nodes and links of the network are traversed. A minimum cycle basis is a cycle basis with minimum path length, i.e., the total length of all cycles in the basis is a minimum. Minimum cycle bases are not unique, however, the distribution of cycle size is the same for all minimum cycle bases which is of most important to us here. We calculate a minimum cycle basis for the contact networks of all our samples at the observed strain states using an algorithm presented in Mehlhorn and Michail (2006).

The use of cycles to characterize the hierarchical structure of Complex Networks is commonplace. For example, small world networks can be characterized by the connectedness of neighbouring nodes, in particular linked neighbours corresponding to a triangle (3-cycles) quantified by the clustering coefficient (Newman 2003). Three-cycles and 4-cycles have a special interpretation in the context of granular contact networks due to them frustrating or permitting rotations. The low-order cycles have also been observed to play a role in phase transition behaviour of evolving networks through L-percolations (Costa 2004). Higher order cycles are also critical to the understanding of the deformation within a granular material as they indicate the extent of dilatation and emblematic of void space in the material.

Here we use minimal cycles to succinctly summarize the contact topology around self-organized load-bearing structures, i.e., force chains. Minimal cycle bases for all of our test systems across all of their strain stages have been calculated. In Figs. 1 and 2 we summarize this information by showing the evolving cycle distribution with respect to size as well as the total population of the important 3-cycle and 4-cycle population together with the accumulated population of higher order odd and even length cycles. In all samples, barring comminution, we see a decrease in the dominant 3-cycle and 4-cycle populations as the loading increases and causes failure. A notable difference is apparent in the three constant volume examples, i.e., BCV1, BCV2 and BCYS. In BCV1 and BCV2 we observe an upturn in the population of 3-cycles before failure of the material (Tordesillas et al. 2010a). For the experimental BCYS system, we observe the oscillatory response of the contact networks to the cyclic nature of the shear. Also noteworthy is the apparent invariance of cycle population once the material has undergone failure and from this point we see the appearance of high order n-cycles in the material indicative of the process of dilatation within shear bands when they nucleate. For the TCP1 sample which fails completely due to diffuse localization we see the cycle population captures this failure with the population of cycles of any length collapsing entirely. Of interest for this sample is the contrary pattern of 4-cycles dominating over 3-cycles but the population of higher odd-cycles is the highest before failure. Thus, in this case, configurations which frustrate rotations are drawn from 5-cycles and higher order odd-cycles more so than the 3-cycle population.

The system with the most contrarian behaviour is the comminution system where the latter stages see an explosion in the 3-cycle population. This results in the sample

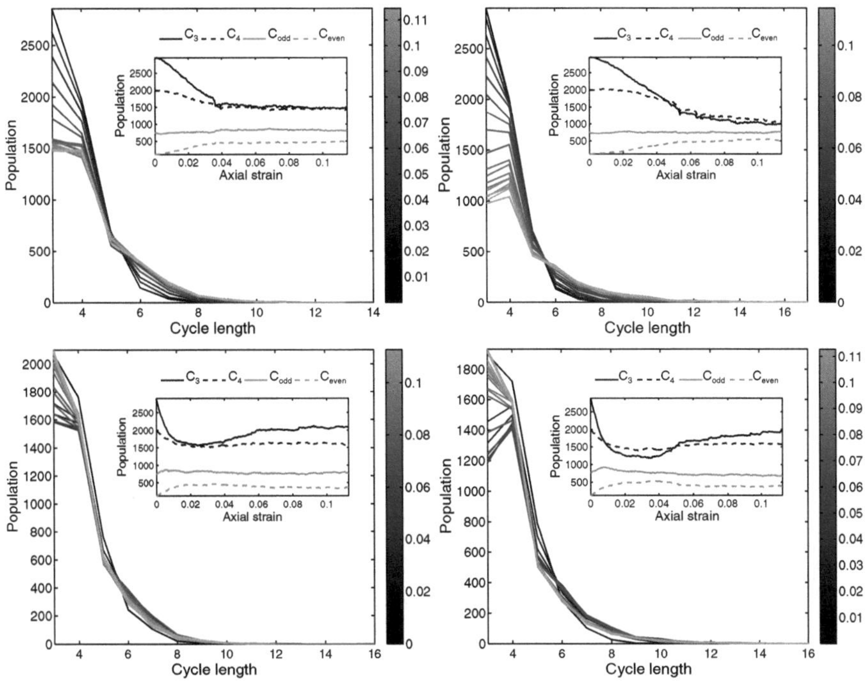

Fig. 1 Clockwise from *top left*, the cycle basis population: BCP1, BCP2, BCV2 and BCV1

becoming stronger and stronger in terms of the shear strength it can withstand. The 3-cycle population explosion is aided by the fracture rule replacing one particle with three and also the asymptotic scale free distribution of grain size allows the particles to pack tighter and tighter together forming the lowest length 3-cycles.

As we can see a view of the evolving distribution of n-cycles or straightforward count of their population is suggestive of the behaviour within the material as it responds to loading, however, more insight can be gained by studying measures of connectedness at the particle scale.

2.2 *Feature Vectors and Force Chains*

Granular assemblies respond to increasing compression through rearrangements of its particles. This rearrangement is reflected in an evolving contact network, which can be summarized succinctly by a feature vector. Feature vectors are n-tuples of network measurements (Costa et al. 2007) defining a feature space. The trajectory of feature vectors in feature space gives valuable insights into the response of the material. One of the most fundamental properties of a network is its degree distribution. The degree of a node is simply its total number of links, or the total

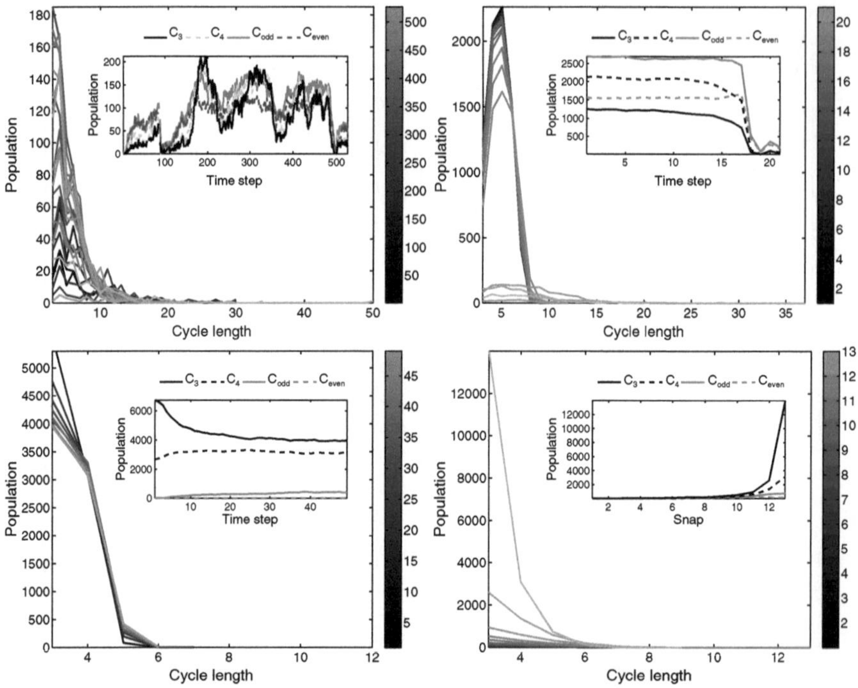

Fig. 2 Clockwise from *top left*, cycle basis population: BCYS, TCP1, UCC and TCP2

number of contacts, i.e., it is equivalent to the coordination number of a particle. The first component of our feature vector will be the average degree of a particle.

A second component of our feature vector will be the average value of the clustering coefficient of a particle (Newman 2003). The clustering coefficient can distinguish heterogeneities in the local arrangements of particles, i.e., contact anisotropy in local particle neighbourhoods. The clustering coefficient is a measure of the number and density of 3-cycles in a network. The clustering coefficient for each particle can be calculated by considering (Newman 2003)

$$C_i = \frac{\text{number of triangles connected to node } i}{\text{number of triples centred on node } i} \tag{1}$$

where a triple is a node with links connecting an unordered pair of other nodes. C_i measures the fraction of triples that have their third link filled in to complete the triangle. We note that in order for a particle with a higher number of contacts to maintain or increase its clustering coefficient compared to a particle with a lower number of contacts it must be more densely packed since the denominator in the ratio for higher degree particle is necessarily larger. We also remark that in all of our average calculations rattlers (particles with zero or only one contact) are discounted.

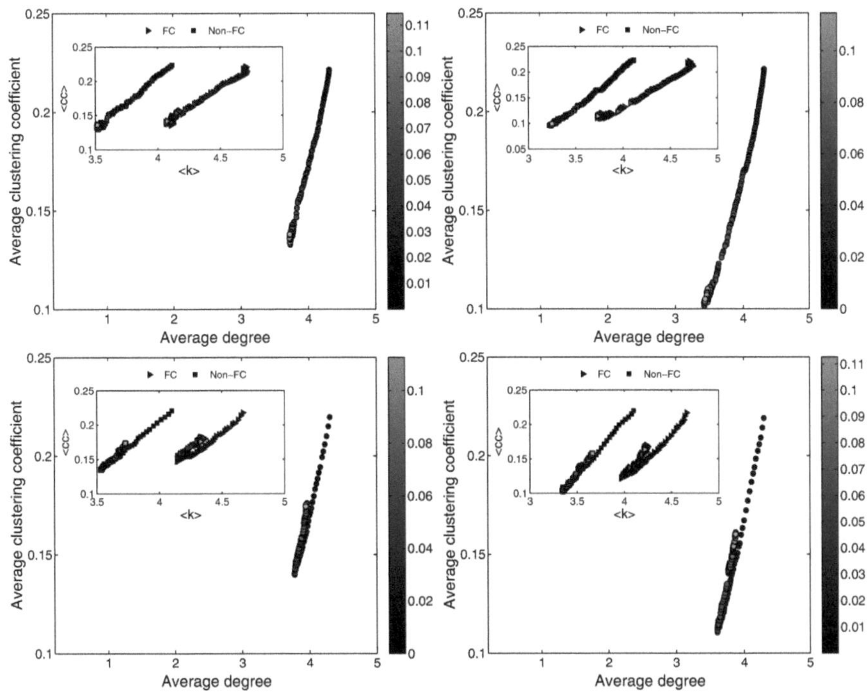

Fig. 3 Clockwise from *top left*, feature space by force chain: BCP1, BCP2, BCV2 and BCV1

In Figs. 3 and 4 we show the trajectory in feature space of all systems throughout loading. The main plot displays averages for all particles (global) whereas the insets show separate averages for force chain particles and non-force chain particles. We see that the global trajectories reflect the materials response to loading. In the 2D DEM simulations we see a steady progression towards decreasing average degree and decreasing clustering coefficient before a jump to an almost steady state coinciding with failure of the material. The "loop-back" of the constant volume samples is indicative of the 3-cycle population rise seen above and appears to be typical of constant volume boundary conditions. The case of BCV1 has been discussed extensively in Tordesillas et al. (2010a).

In all samples we see that force chain particles have a relatively higher number of contacts than non-force chain particles and also have a higher or comparable value of clustering coefficient. This is also confirmed by a count of the average number of 3-cycles per force chain particle in all samples being higher. This shows that force chains reside in highly connected local environments. In particular, around force chains the granular material tends to form 3-cycles to provide not only lateral support by "propping-up" chains but also to frustrate a major mechanism for failure, i.e., force chain buckling. (In Tordesillas et al. 2010c we have quantified this support mechanism in great detail.)

We can thus give an explanation to the possible co-evolution between 3-cycles and force chains. We know from previous studies that the 3-cycle population

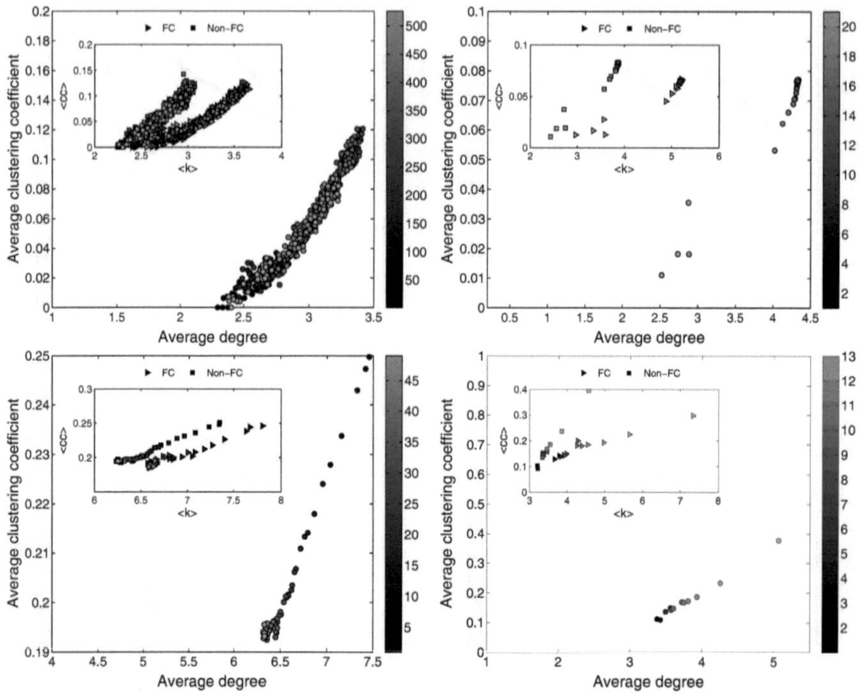

Fig. 4 Clockwise from *top left*, feature space by force chain: BCYS, TCP1, UCC and TCP2

signified by the clustering coefficient plays an important role in frustrating rotations between constituent particles (Tordesillas et al. 2010c). These 3-cycles when conjoined with particles from load bearing force chains effectively act as trusses. Since the constituent force particles have higher connectivity and higher clustering coefficient most force chains receive dual support from 3-cycle trusses: (i) rolling between force chains particles and their neighbours is reduced impeding buckling, and (ii) lateral support. The benefits of this dual support is available and, as expected, manifest themselves to varying degrees in all samples.

We also note that the sample collapse seen in the global feature space plots at failure is mirrored when considering only force chain particles. But the collapse is not as dramatic and force chain particles still typically have higher number of contacts and local connectivity than non-force chain particles. Thus, even in the regime of failure, new and old force chains still marginally benefit from 3-cycle support. The dynamics of shear bands is governed by the continual "birth" and "death" of force chains and drives the material to (locally) explore configurational state space towards states where its force chain network has increased ability to sustain greater load. In particular, the dilatation that necessarily accompanies buckling of force chains (note the drop in degree and clustering coefficient in feature space and the increase in higher order cycles within the minimum cycle basis) drives the degree of non-force chain particles to the edge of isostaticity (towards degree equal to 3 for 2D) while attempting to invoke new, increasingly harder, force chains to form in low void regions by conjoining with 3-cycles. It is important to remember that through continued and increased loading we systematically thwart this attempt of the material to resist increasing shear strength.

3 Conclusions

We have shown through the use of complex network analysis that granular materials exploit the benefits topology can offer functional connectivity to the maximum effect. In response to loading the material sends its stabilizing agents, 3-cycles, to support the emerging load bearing force chain network (for example, in BCP1 we found >95% of 3-cycles are conjoined with force chains (Tordesillas et al. 2010c)). The feature vector analysis show that force chains reside in highly connected local environments. This is evident in the first instance in the relatively high number of contacts on average for those particles in force chains (high average degree) and secondly force chain particles prefer to make contact with its laterally supporting neighbors in 3-cycle formations (force chain particles have a clustering coefficient higher than or on a par with weaker particles). This provides force chains a twofold benefit as not only can they be propped up in the way that a counterfort or buttress supports a wall but also so that rolling at contacts (a critical mechanism for buckling) is frustrated. This appears to be a generic process of granular materials subject to loading as all eight samples considered here reflect an inherent structural hierarchy where minimal cycles serve as one class of basic building blocks for self-organization.

Acknowledgements We thank our collaborators for permission to include data from their simulations and experiment. This work was supported by US Army Research Office (W911NF-07-1-0370) and the Australian Research Council (DP0986876 and DP0772409). We also thank the Victorian Partnership for Advanced Computing for computing resources.

References

O. Ben-Nun, I. Einav, A. Tordesillas, Force attractor in confined comminution of granular material. Phys. Rev. Lett. **104**, 108001 (2010)
L.F. da Costa, L-percolations of complex networks. Phys. Rev. E **70**, 056106 (2004)
L.F. da Costa, F.A. Rodrigues, G. Travieso, R.R. Villas Boas, Characterization of complex networks: a survey of measurements. Adv. Phys. **56**, 167–242 (2007)
F. Darve, L. Sibille, A. Daouadji, F. Nicot, Bifurcations in granular media: macro- and micro-mechanics approaches. C.R. Mecanique **335**, 496–515 (2007)
K. Mehlhorn, D. Michail, Implementing minimum cycle basis algorithms. ACM J. Exp. Algorithmics **11**(2.5), 1–14 (2006)

M. Muthuswamy, A. Tordesillas, How do interparticle contact friction, packing density and degree of polydispersity affect force propagation in particulate assemblies? J. Stat. Mech. Theory Exp. P09003 (2006)

M.E.J. Newman, The structure and function of complex networks. SIAM Rev. **45**, 167–256 (2003)

A. Tordesillas, Force chain buckling, unjamming transitions and shear banding in dense granular assemblies. Philos. Mag. **87**, 4987–5016 (2007)

A. Tordesillas, P. O'Sullivan, D.M. Walker, Paramitha, Evolution of functional connectivity in contact and force chain networks: feature vectors, k-cores and minimal cycles. C.R. Mecanique **338**, 556–569 (2010a). doi:10.1016/j.crme.2010.09.004

A. Tordesillas, S. Pucilowski, J.P. Peters, D.M. Walker, M. Hopkins, Network analysis of evolving three-dimensional granular structures. Dyn. Contin., Discr. Impuls. Syst.-B (accepted) (2010b)

A. Tordesillas, D.M. Walker, Q. Lin, Force chains and force cycles. Phys. Rev. E **81**, 011302 (2010c)

D.M. Walker, A. Tordesillas, Topological evolution in dense granular materials: a complex networks perspective. Int. J. Solids Struct. **47**, 624–639 (2010)

J. Zhang, T.S. Majmudar, A. Tordesillas, R.P. Behringer, Statistical properties of a 2D granular materials subjected to cyclic shear. Granul. Matter **12**, 159–172 (2010)

Microscopic Origins of Shear Strength in Packings Composed of Elongated Particles

E. Azéma and F. Radjaï

Abstract We investigate the rheology, force transmission and texture of granular materials composed of elongated particles by means of contact dynamics simulations. The particles have a rounded-cap rectangular (RCR) shape described by a single elongation parameter varying from 0 for a circular particle to 1 for an infinitely thin or long particle. We study the quasi-static behavior, structural and force anisotropies as a function of the elongation parameter for packings submitted to biaxial compression. The shear strength is found to increase linearly with this parameter whereas the solid fraction both at the initial isotropic state and in the critical state is nonmonotonous. We show that for these elongated particles a harmonic decomposition of the stress tensor provides a fairly good approximation of the internal state. Our data suggest that the increase of shear strength with reflects both enhanced friction mobilization and anisotropic particle orientation as the elongation of the particles increases.

Keywords Elongated particles • Fabric properties • Force transmission • Harmonic decomposition

1 Introduction

A wide variety of particle shapes can be found in nature and industry: elongated and platy shapes, angular and facetted shapes, and nonconvex shapes. The issue of shape effect opens actually the door to a vast and substantial scientific domain given a multitude of potential particle morphologies. The effect of particle shape is mediated by the specific *granular texture* induced by each particle shape. For example,

E. Azéma (✉) · F. Radjaï
LMGC, CNRS – Université Montpellier 2, Place Eugène Bataillon,
34095 Montpellier cedex 05, France
e-mail: azema@lmgc.univ-montp2.fr; radjai@lmgc.univ-montp2.fr

S. Bonelli et al. (eds.), *Advances in Bifurcation and Degradation in Geomaterials*, Springer Series in Geomechanics and Geoengineering 11,
DOI 10.1007/978-94-007-1421-2_3,

it is found that hard ellipses can be jammed even though they are underconstrained (Donev et al. 2007). In the case of anisometric or elongated particle shapes, such as spheroids and sphero-cylinders, the particles tend to develop orientational order affecting force transmission and frictional behavior (Ouadfel and Rothenburg 2001; Hidalgo et al. 2009). This "nematic" ordering occurs while the particles interact only via contact and friction.

In this paper we focus particularly on the connectivity of particles and we show, depending on the elongation, how each contact type contributes to the internal angle of friction (Azéma and Radjai 2010). We first introduce our numerical approach in Sect. 2. Then, in Sect. 3, the stress-strain behavior is presented for different values of η. The microstructure is analyzed in Sect. 4.

2 Numerical Procedures

The CD method is based on implicit time integration of the equations of motion and a nonsmooth formulation of mutual exclusion and dry friction between particles (Moreau 1994; Radjaï and Richefeu 2009). This method requires no elastic repulsive potential and no smoothing of the Coulomb friction law for the determination of forces. For this reason, the simulations can be performed with large time steps compared to molecular dynamics simulations. We used LMGC90 which is a multipurpose software developed in our laboratory, capable of modeling a collection of deformable or undeformable particles of various shapes by different algorithms (Dubois and Jean 2006).

The particles are modeled as a juxtaposition of two half-disks of radius R' with one rectangle of length L and width $2R'$ to which we will refer as Rounded-Cap Rectangular (RCR) particles, see Fig. 1. The RCR shape can be characterized by a single aspect ratio α or by an *elongation* parameter $\eta = (R - R')/R = (\alpha - 1)/\alpha$ varying from $\eta = 0$, for a circle, to 1 corresponding to a line.

We prepared eight different packings of 13,000 RCR particles with η varying from 0 to 0.7 by steps of 0.1. The radius $R \in [R_{min}, 3R_{min}]$ of the circumscribing circle defines the size of a RCR particle. All samples were compacted by isotropic

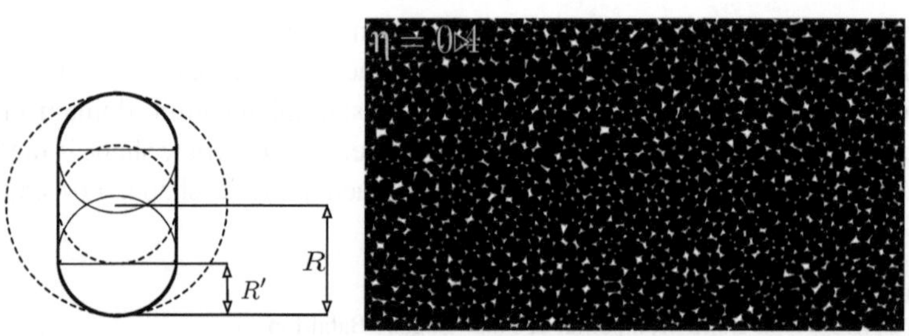

Fig. 1 Shape of a rounded-cap rectangle (RCR) (*left*). Snapshot of the portion of the packing for $\eta = 0.4$ (*right*)

compression inside a rectangular frame of dimensions $l_0 \times h_0$ in which the left and bottom walls are fixed, and the right and top walls are subjected to the same compressive stress σ_0. The gravity was set to zero in order to avoid force gradients in the samples. The coefficient of friction was set to 0 between grains and walls during the isotropic compression. Thus, at equilibrium, all samples were in isotropic stress state. Figure 1 displays snapshot of the packings for $\eta = 0.4$ at the end of isotropic compaction. The isotropic samples are then subjected to vertical compression by downward displacement of the top wall at a constant velocity v_y for a constant confining stress σ_0 acting on the lateral walls. The friction coefficient μ between particles is set to 0.5 and to zero with the walls.

3 Stress-Strain Behavior

The stress tensor $\boldsymbol{\sigma}$ can be evaluated from the simulation data as an average over the dyadic product of contact force $\boldsymbol{f}^c$ and branch vector $\boldsymbol{\ell}^c$: $\sigma_{\alpha\beta} = n_c \langle f^c_\alpha \ell^c_\beta \rangle_c$ (Moreau 1994), where n_c is the number density of contacts c, and the average $\langle \ldots \rangle_c$ run over all contacts in a control volume. We can extract the mean stress $p = (\sigma_1 + \sigma_2)/2$ as well as the stress deviator $q = (\sigma_1 - \sigma_2)/2$ where σ_1 and σ_2 are the principal stresses values. The principal strain values are $\varepsilon_1 = \int_{h_0}^{h} dh'/h'$ and $\varepsilon_2 = \int_{l_0}^{l} dl'/l'$. The control variable is the shear strain given by $\varepsilon_p = \varepsilon_1 + \varepsilon_2$.

During shear, the shear stress jumps initially to a high value before decreasing to a nearly constant value in the steady state. The steady-state shear stress $(q/p)^*$ characterizes the shear strength of the material. According to the Mohr-Coulomb model, the internal angle of friction, representing the shear strength of the material, is defined by $\sin\varphi^* = (q/p)^*$ (Mitchell and Soga 2005). Interestingly, as shown by Fig. 2b, the shear strength varies linearly with the elongation parameter.

Figure 2 displays the solid fraction ϕ as a function of η at different levels of shear deformation ε_q. It is remarkable that, at all levels of deformation, the solid fraction increases with η, reaches a maximum at $\eta \simeq 0.4$ and then declines as η further increases. We note that solid fractions as large as 0.90 are reached for $\eta = 0.4$ in the initial state. A similar nonmonotonous behavior was observed for

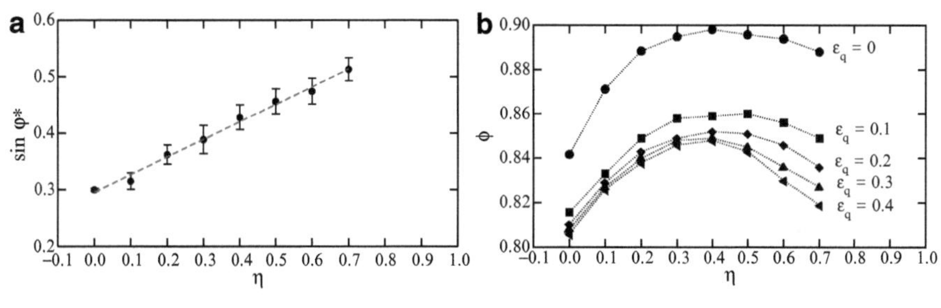

Fig. 2 (**a**) Internal angle of friction φ^* as a function elongation η. (**b**) Solid fraction as a function of particle shape parameter η at different levels of shear strain

packings of ellipses or ellipsoidal particles (Donev et al. 2007). This is somewhat a counterintuitive finding as the shear strength (a monotonous function of η) does not follow the trend of solid fraction (nonmonotonous).

4 Granular Texture and Force Transmission

Remembering that RCR particles are clumps of two disks with one rectangle, a major effect of elongation is then to allow for multiple contacts between two touching particles: cap-to-cap (cc), cap-to-side (cs) and side-to-side (ss) contacts in each packing. The side-to-side or side-to-cap contacts are able to accommodate force lines that are usually unsustainable by cap-to-cap contacts. Figure 3 (left) shows the proportions k_{cc}, k_{cs} and k_{ss} of cc, cs and ss contacts averaged over the residual state as a function of η. We see that k_{cc} declines with η from 1 (for disks) to 0.2 for $\eta = 0.7$. At the same time, k_{cs} and k_{ss} increase from 0 to 0.6 and to 0.2, respectively. Interestingly, $k_{cs} \simeq k_{cc}$ for $\eta \simeq 0.4$, and $k_{ss} \simeq k_{cs}$ for $\eta = 0.7$. In this way, as the particle elongation increases, the packing passes from a contact network dominated by cc contacts to a contact network dominated by the *complex* contacts cs and ss.

An additive decomposition of the stress tensor can be performed by grouping the contacts according to their types: $\boldsymbol{\sigma} = \boldsymbol{\sigma}_{cc} + \boldsymbol{\sigma}_{cs} + \boldsymbol{\sigma}_{ss}$, where $\boldsymbol{\sigma}_{cc}$, $\boldsymbol{\sigma}_{cs}$ and $\boldsymbol{\sigma}_{ss}$ are obtained from the expression of the stress tensor (see Sect. 3) by restricting the summation to cc, cs and ss contacts, respectively. The corresponding stress deviators q_{cc}, q_{cs} and q_{ss} are then calculated and normalized by the mean pressure p. Note that, by construction we have $q/p = (q_{cc} + q_{cs} + q_{ss})/p$. Figure 3 (right) shows q_{cc}/p, q_{cs}/p and q_{ss}/p averaged in the residual state as a function of η. We see clearly that q_{cc}/p follows a trend opposite to that of q_{cs}/p. For $\eta < 0.3$, $(q/p)^*$ is dominated by cc contacts. For $\eta \simeq 0.3$, cc and cs contacts participate equally to the shear stress, and for $\eta > 0.3$, the cs contacts dominate $(q/p)^*$. Remarkably, $q_{ss}/p \simeq 0$ for $\eta < 0.4$. In this way, the growth of the number of cs and ss contacts is clearly at the origin of a gradual consolidation of the packings as η increases.

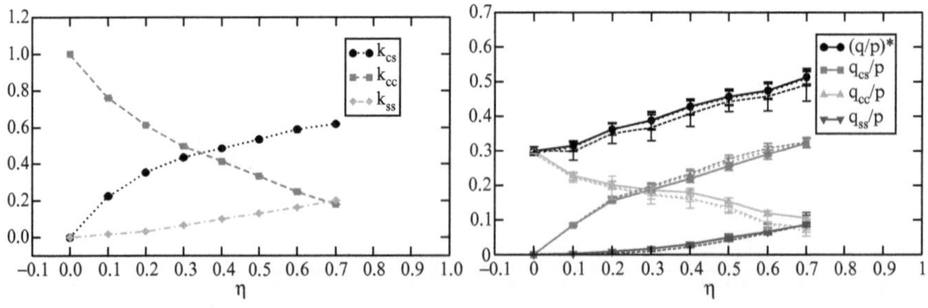

Fig. 3 (*left*) Proportions of side-to-side (ss), cap-to-side (cs) and cap-to-cap (cc) contacts and (*right*) shear strength $(q/p)^*$ (*circle*) for cs (*square*), ss (*triangle up*) and cc (*triangle down*) contacts as a function of η, together with the harmonic approximation fits (see below)

The shear strength of a granular material reflects its ability to develop force and fabric anisotropies (Radjai et al. 1998; Ouadfel and Rothenburg 2001; Azéma et al. 2009; Azéma and Radjai 2010). The fabric anisotropy $a_{c\gamma}$, where γ stands alternatively for $\{cc, cs, ss\}$, is the anisotropy of the distribution $P_\gamma(\theta)$ of contact orientation θ. At the lowest order, we have

$$P_\gamma(\theta) = \{1 + a_{c\gamma} \cos 2(\theta - \theta_\sigma)\}/2\pi, \tag{1}$$

where θ_σ is the major principal stress direction ($\theta_\sigma = \pi/2$). Note that, by construction the whole contact anisotropy is given by $a_c = a_{css} + a_{ccs} + a_{ccc}$. In the same way, the normal and tangential branch anisotropies ($a_{ln\gamma}$, $a_{lt\gamma}$) and forces anisotropies ($a_{fn\gamma}$, $a_{ft\gamma}$) can be defined from the distributions of the angular average of normal and tangential branch length and forces, respectively, by:

$$\begin{cases} \langle \ell_{n\gamma} \rangle(\theta) = \langle \ell \rangle \{1 + a_{ln\gamma} \cos 2(\theta - \theta_\sigma)\}, \\ \langle \ell_{t\gamma} \rangle(\theta) = \langle \ell \rangle a_{lt\gamma} \sin 2(\theta - \theta_\sigma), \\ \langle f_{n\gamma} \rangle(\theta) = \langle f \rangle \{1 + a_{fn\gamma} \cos 2(\theta - \theta_\sigma)\} \\ \langle f_{t\gamma} \rangle(\theta) = \langle f \rangle a_{ft\gamma} \sin 2(\theta - \theta_\sigma), \end{cases} \tag{2}$$

where $\langle f \rangle$ and $\langle \ell \rangle$ are the mean force and the mean branch length. By construction, the total normal and tangential branch and force anisotropies are given by the sum of the corresponding partial anisotropies.

Using the above Fourier approximations together with the expression of the stress tensor it can be shown that (Ouadfel and Rothenburg 2001; Azéma et al. 2009; Azéma and Radjai 2010):

$$\frac{q_\gamma}{p} \simeq \frac{1}{2}(a_{c\gamma} + a_{ln\gamma} + a_{lt\gamma} + a_{fn\gamma} + a_{ft\gamma}) \tag{3}$$

We can see in Fig. 3 (right) that this decomposition is nicely verified by our numerical data both for the partial shear stress q_γ/p as well as for the whole shear stress $(q/p)^*$. The contribution of the normal and tangential branch anisotropies is negligible. Figure 4 shows only the partial critical-state anisotropies $a_{c\gamma}$, $a_{fn\gamma}$,

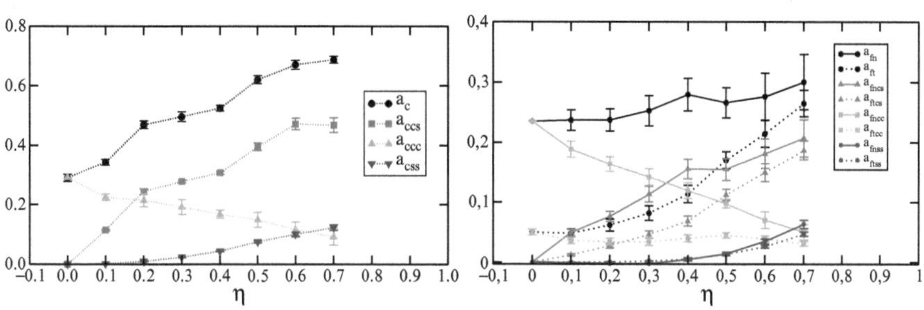

Fig. 4 (*left*) Partial contact orientation anisotropies a_{ccc}, a_{ccs} and a_{css} as a function of η in the critical state. (*right*) Partial normal (*line*) and tangential (*dashed line*) force anisotropies a_{fncc}, a_{fncs} and a_{fnss}, and a_{ftcc}, a_{ftcs} and a_{ftss} as a function of η in the critical state

and $a_{ft\gamma}$, together with the total critical-state anisotropies a_c, a_{fn} and a_{ft}. The anisotropy a_{css} of ss contacts increases slowly with η from 0 to 0.18. At the same time, a_{ccc} decreases and at $\eta = 0.7$ we have $a_{css} = a_{ccc}$. Hence, although the ss contacts represent at $\eta = 0.7$ nearly 20% of contacts, their contribution to the contact anisotropy remains modest and of the same order as cc contacts. The variation of the contact anisotropy a_c is thus largely governed by that of a_{ccs}. We see also that, the cs contacts carry most of the normal and tangential force anisotropies. The ss contacts contribute modestly to the global force anisotropies only for $\eta \geq 0.4$. The anisotropy a_{fncc} declines with η, mainly due to their low number, and a_{ftcc} stays nearly constant. The increase of a_c with η and a hight value of a_{ft} for the most elongated particle reveal that (1) particles tend to be aligned orthogonally to the compressive direction and (2) a strong mobilization of friction due the increase of the side length of particles (Hidalgo et al. 2009; Azéma et al. 2009; Azéma and Radjai 2010).

5 Conclusions

In this paper, we applied the contact dynamics method to simulate large samples of elongated particles. It was shown that the shear strength is an increasing linear function of elongation whereas the solid fraction first increases and then declines. We find that both force and texture anisotropies contribute to the increase of shear strength. The increasing mobilization of friction force (and the associated anisotropy) as well as a local nematic ordering of the particles (reveled by contact anisotropy), seem to be the key effect of particle elongation. Currently, we work to extend these results to 3D systems composed with rounded-cap-cylinders (spherocylinders).

References

E. Azéma, F. Radjai, Stress-strain behavior and geometrical properties of packings of elongated particles. Phys. Rev. E **81**, 051304 (2010)

E. Azéma, F. Radjai, R. Peyroux, G. Saussine, Quasistatic rheology, force transmission and fabric properties of a packing of irregular polyhedral particles. Mech. Mater. **41**, 729–741 (2009)

A. Donev, R. Connelly, F.H. Stillinger, S. Torquato, Underconstrained jammed packings of nonspherical hard particles: ellipses and ellipsoids. Phys. Rev. E **75**, 051304 (2007)

F. Dubois, M. Jean, The non smooth contact dynamic method: recent LMGC90 software developments and application, in *Analysis and Simulation of Contact Problems*, vol. 27 (Springer, Berlin/Heidelberg, 2006)

R.C. Hidalgo, I. Zuriguel, D. Maza, I. Pagonabarraga, Role of particle shape on the stress propagation in granular packings. Phys. Rev. Lett. **103**, 118001 (2009)

J.K. Mitchell, K. Soga, *Fundamentals of Soil Behavior* (Wiley, New York, 2005)

J.J. Moreau, Some numerical methods in multibody dynamics: application to granular. Eur. J. Mech. A/Solids **13**, 93–114 (1994)

H. Ouadfel, L. Rothenburg, 'Stress-force-fabric' relationship for assemblies of ellipsoids. Mech. Mater. **33**, 201–221 (2001)
F. Radjaï, V. Richefeu, Contact dynamics as a nonsmooth discrete element method. Mech. Mater. **41**, 715–728 (2009)
F. Radjai, D.E. Wolf, M. Jean, J.J. Moreau, Bimodal character of stress transmission in granular packings. Phys. Rev. Lett. **80**, 61–64 (1998)

Multi-scale Analysis of Instability in Sand

C.S. Chang, Z.-Y. Yin, and P.-Y. Hicher

Abstract Instability of sand can occur under drained or undrained loading conditions in loose sand or dense sand. A micromechanics approach is used for the analysis of local instabilities of inter-particle contacts and their relations to the global instability of assembly. The comparisons between experimental and predicted results on Toyoura Sand show the capability of the model in capturing the modes of instability at the assembly level. Analysis at inter-particle contact level for loose sand under an undrained triaxial loading condition show that the number of unstable inter-particle planes increases continuously, while the assembly remains stable. The assembly becomes unstable when the sum of all local second-order work becomes zero. After this point, the overall shear stress begins to decrease during a strain controlled test, and progressively, more inter-particle contact planes become instable.

Keywords Instability • Second-order work • Sand • Micromechanics • Local behavior

1 Introduction

Instability of granular materials is an important topic in geotechnical engineering because it may lead to catastrophic events such as gross collapse of earth structures. It is well known that granular soil can become unstable even before the stress state

C.S. Chang
University of Massachusetts, Amherst, MA 01003, USA

Z.-Y. Yin
Shanghai Jiao Tong University, Shanghai 200240, PR, China

P.-Y. Hicher (✉)
GeM, Ecole Centrale de Nantes, Nantes, France
e-mail: pierre-yves.hicher@ec-nantes.fr

S. Bonelli et al. (eds.), *Advances in Bifurcation and Degradation in Geomaterials*, Springer Series in Geomechanics and Geoengineering 11,
DOI 10.1007/978-94-007-1421-2_4, © Springer Science+Business Media B.V. 2011

reaches the Mohr-Coulomb envelop. Analyses regarding material instability have been extensively studied at the constitutive level for a soil element (Nova 1994, 2004; Darve et al. 2004; Nicot and Darve 2007). Instability of a granular assembly is evidently a consequence of the instability behaviour at the level of inter-particle. In the present paper, we study the material instability using a micromechanics based model in order to relate instability at the macroscopic level to local instabilities at grain contact level.

2 Instability Analysis

The micromechanical model used here is based on a model for sand proposed by Chang and Hicher (2005). In this model, the stress-strain relationship for an assembly can be determined by integrating the behaviour of the inter-particle contacts in all orientations using a static hypothesis, which relates the average stress of the granular assembly to a mean field of particle contact forces.

The model needs a number of input parameters such as mean particle size, particle stiffness, inter-particle friction, initial porosity, and parameters defining the critical state of the sand. The laboratory test results for Toyoura sand by Verdugo and Ishihara (1996) are used here for comparison with model prediction. All parameters were determined based on isotropic compression and undrained triaxial tests, summarized in Table 1 (definition of parameters see Chang and Hicher 2005).

2.1 Instability of Sand in Triaxial Tests

Figure 1 presents the numerical predictions for undrained triaxial tests using the set of parameters given in Table 1 for Toyoura sand samples with initial void ratio equal to 0.907, 0.833, and 0.735, isotropically consolidated to a confining pressure of 2000 kPa. The predictions are in reasonable agreement with the experimental results.

To study the change of the second-order work with the applied load, three typical curves in Fig. 1 are selected, and replotted in Fig. 2a. In order to examine the inception of instability, the second-order work is plotted against shear strain $\varepsilon_d = 2(\varepsilon_1 - \varepsilon_3)/3$ in Fig. 2b. The second-order work for the conventional triaxial tests can be expressed as,

$$d^2W = dp\, d\varepsilon_v + dq\, d\varepsilon_d \tag{1}$$

For undrained conditions ($d\varepsilon_v = 0$), the second-order work is reduced to $d^2W = dq d\varepsilon_d$. Since the deviatoric strain increases continually ($d\varepsilon_d > 0$) in the conventional triaxial tests, the second-order work can become non-positive if and only if $dq < 0$ (i.e., decrease in q). Fig. 2b shows that the instability begins at the peak of shear stress where $d^2W = 0$ for $e_0 = 0.833$ and 0.907.

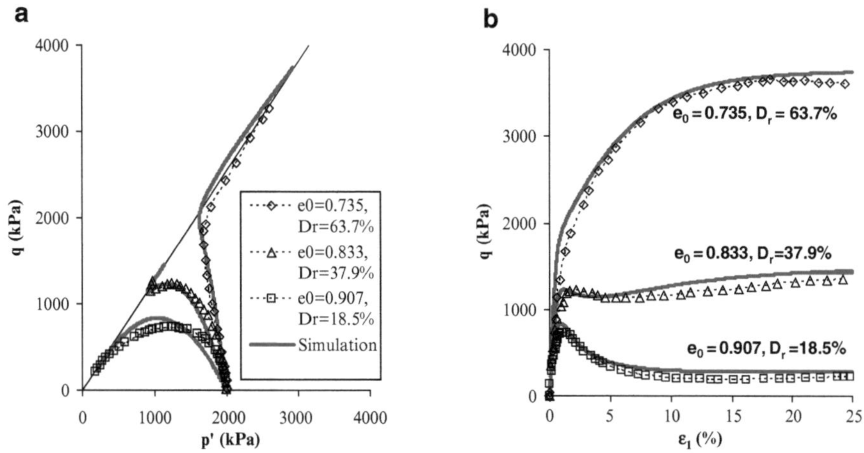

Fig. 1 Experimental data and simulations for undrained triaxial tests on Toyoura sand

Table 1 Model parameters for Toyoura Sand

e_{ref}	ζ	λ	k_{pR}	k_{rR}	D	ϕ_μ (°)	m
0.934	0.7	0.019	0.23	0.5	1	33	1

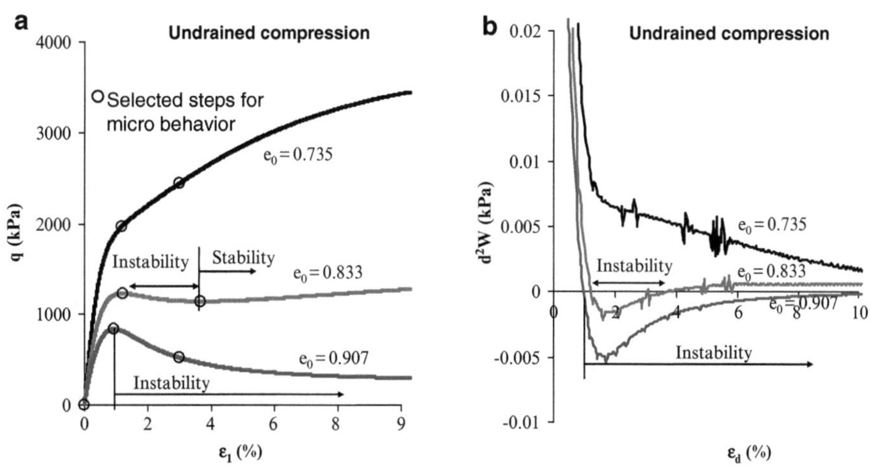

Fig. 2 Predicted stress-strain curves and second-order work for undrained triaxial tests

For loose sand, the second-order work is negative after the peak stress. At the peak, the stress has not reached its plastic limit (see Fig. 1a). At large shear strain, the stress state approaches the plastic limit and the negative second-order work approaches back to zero. For the medium sand ($e_0 = 0.883$), the second-order work is also equal to zero at the peak stress, and negative after the peak stress.

Then, the second-order work returns towards zero and becomes positive due to soil dilatancy. Thus, the soil experience temporary instability, before becoming stable again at large strains. For dense sand, the second-order work is always positive, thus the soil remains stable.

The above analyses are focused on the instability behaviour of the granular assembly. It is of interest to examine the instabilities of inter-particle contacts, in relation with questions such as: (1) would the overall instability occur when inter-particle contact planes in one of the orientations has reached instability? If not, then (2) how many contact planes would cause the overall instability of the system? This will be discussed in the next section.

2.2 *Instability of Contact Planes Between Grains*

In order to study the behaviour at grain level during the deformation of the assembly, we examined the inter-particle contact planes oriented in several directions. The direction of a contact plane is defined by an angle θ, which is measured between the outward vector n on the contact plane and the vertical axis. A larger value of θ indicates a more inclined contact plane. We define a local normal stress $\sigma^\alpha = f_n^\alpha Nl/3V$ and a local shear stress $\tau^\alpha = f_r^\alpha Nl/3V$, where l is the branch length of two particles, representing the distance between the centers of two neighboring particles, and N/V is the total number of contacts per unit volume. The corresponding local normal strain is defined as $\varepsilon^\alpha = \delta_n^\alpha / l$ and local shear strain as $\gamma^\alpha = \delta_r^\alpha / l$.

The response in local contact planes are plotted for the undrained triaxial test of loose sand (Fig. 3), for seven selected contact planes. In each curve, three circles are marked, corresponding to the three different loading stages shown in Fig. 2a: the initial stage, the peak and the residual stage at assembly level. Because of the static hypothesis used in this model, the local stresses on each contact plane can be determined from the applied stress. For the example shown in Fig. 3a, the maximum local shear stress is on the 45° inclined plane; the 0° plane carries only normal forces.

The shear deformation in the contact plane is plotted in Fig. 3b. The points corresponding to the peak stress at global level are also on the peaks of local shear stress-strain curves. This indicates that all contact planes reach the peak and begin to soften simultaneously. It is of interest to see, whether this indicates that all contact planes would experience instability at the same time.

In order to answer this question, the local second-order works for all contact planes are computed. The second-order work for the αth contact plane, expressed in terms of local stress and local strain, is given by

$$d^2 w^\alpha = df_i^\alpha d\delta_i^\alpha = \frac{3V}{N}(d\sigma^\alpha d\varepsilon^\alpha + d\tau^\alpha d\gamma^\alpha) \qquad (2)$$

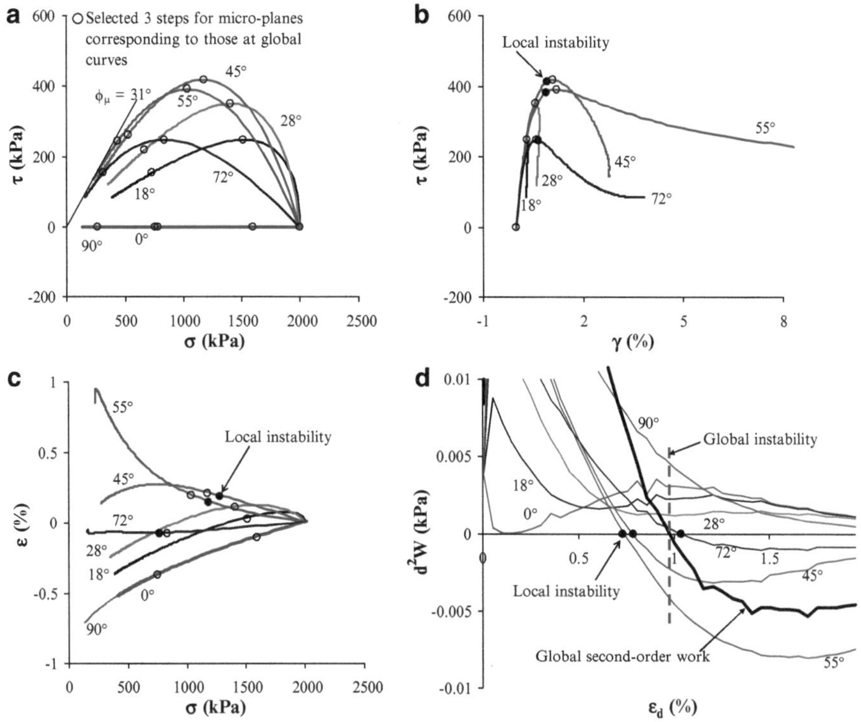

Fig. 3 Local stress-strain behaviour in inter-particle contact planes for loose sand

Note that in an undrained test, the overall volume change for sand is zero (i.e., the summation of all contact normal strains are zero $\Sigma\varepsilon^{\alpha} = 0$), but in each contact plane, the local normal strain ε^{α} does not need to be restricted (see Fig. 3c).

The second-order work curves for the six contact planes are plotted in Fig. 3d. For the less-inclined contact planes (0°, 18° and 28°), the second-order work is always positive, thus the contacts remain stable during the entire undrained triaxial test. The softening for these contacts is merely due to elastic unloading (see Fig. 3b). For the more-inclined contact planes (45°, 55° and 72°), the second-order works becomes negative indicating that the instability is predominately caused by the decrease of τ (see Fig. 3b). It is noted that on the 55° contact plane, before the peak stress of the assembly, the local normal strain contracts ($d\varepsilon^{\alpha} > 0$) as the normal stress is released ($d\sigma^{\alpha} < 0$), producing large negative second-order work $d\sigma^{\alpha}d\varepsilon^{\alpha} < 0$. Although the second-order work due to shear component is positive before peak stress ($d\tau^{\alpha}d\gamma^{\alpha} > 0$), the second-order work due to both normal and shear components becomes zero before peak, as indicated by the solid circle in Fig. 3b.

As shown in Fig. 3d, the contacts do not become simultaneously unstable. The points crossing the zero axis are marked by solid circles and labeled as local

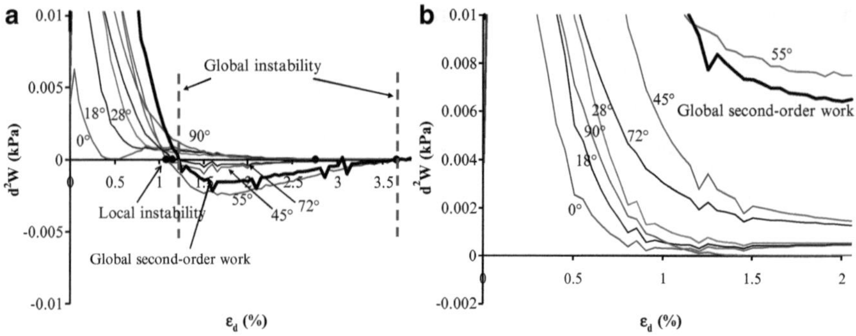

Fig. 4 Local stress-strain behaviour for medium loose and dense sand

instability. The zero second-order work for the overall sample (marked as global instability) is also plotted here for comparison. It can be seen that, even though the local shear stress-strain curves for all contact planes display shear softening simultaneously, only part of the contact planes experience instability, and the instability on each plane occurs at different times, some before the global instability and some after (see the solid circles marked in Fig. 3). The second-order work for the overall assembly is equal to the summation of the local second-order work for all planes.

$$d^2W = \sum_{\alpha} d^2w^{\alpha} = \frac{1}{V}\sum_{\alpha} df_i^{\alpha} d\delta_i^{\alpha} = \frac{3}{N}\sum_{\alpha} (d\sigma^{\alpha} d\varepsilon^{\alpha} + d\tau^{\alpha} d\gamma^{\alpha}) \qquad (3)$$

The summation becomes zero when more than one half of the inter-particle contact planes are still stable at this time.

The local second-order work for the medium sand is shown in Fig. 4a. Most local behaviours are similar to those of loose sand in Fig. 3. The main difference is that the assembly, after a short-term softening, regains its stability due to dilation. At the point of assembly dilation, all local stress paths in the contact planes reverse their direction simultaneously with $d\sigma^{\alpha} > 0$ and $d\tau^{\alpha} > 0$. This action reverses gradually the evolution of the local second-order works from being negative to positive, indicating that each contact plane regain its stability.

The dense sand behaviour is shown in Fig. 4b. All planes display positive second-order work. The sample is unconditionally stable.

3 Summary and Conclusion

A micromechanics approach was used to analyse local instabilities of inter-particle contacts and their relations to the global instability of the assembly. The inter-particle contacts are assumed to have an elasto-plastic behaviour with non-associated flow rule. The analysis is based on Toyoura Sand behaviour under undrained triaxial loading. The comparisons between experimental and predicted results showed the capability of the model in capturing the instability at the assembly level. Analysis at inter-particle contact level for loose sand under an undrained triaxial loading condition showed that the number of unstable inter-particle planes increases continuously during loading, the assembly remaining stable as long as the sum of all local second-order works remains positive. Afterwards, a reduction of the overall shear stress begins with shear strain, and progressively, more inter-particle contact planes become instable.

References

C.S. Chang, P.-Y. Hicher, An elastic-plastic model for granular materials with microstructural consideration. Int. J. Solids Struct. **42**, 4258–4277 (2005)

F. Darve, G. Servant, F. Laouafa, H.D.V. Khoa, Failure in geomaterials: continuous and discrete analyses. Comput. Methods Appl. Mech. Eng. **193**, 3057–3085 (2004)

F. Nicot, F. Darve, A micro-mechanical investigation of bifurcation in granular materials. Int. J. Solids Struct. **44**, 6630–6652 (2007)

R. Nova, Controllability of the incremental response of soil specimens subjected arbitrary loading programmes. J. Mech. Behav. Mater. **5**(2), 193–201 (1994)

R. Nova, The role of non-normality in soil mechanics and some of its mathematical consequences. Comput. Geotech. **31**, 185–191 (2004)

R. Verdugo, K. Ishihara, The steady state of sandy soils. Soils Found. **36**(2), 81–91 (1996)

Observation of Microstructural Changes and Strain Localization of Unsaturated Sands Using Microfocus X-ray CT

Y. Higo, F. Oka, S. Kimoto, T. Sanagawa, and Y. Matsuhima

Abstract It is known that unsaturated soil exhibits more brittle failure due to the collapse of the water meniscus, caused by shearing or the infiltration of water, than saturated soil. The aim of this paper is to observe the strain localization behavior and the microstructural changes in partially saturated soil during the deformation process using microfocus X-ray CT. The strain localization of fully saturated, partially saturated, and air-dried sand specimens during triaxial compression tests is observed and discussed. In addition, the microstructures in the shear bands of partially saturated specimen are discussed.

Keywords Unsaturated sand • Strain localization • Microstructural change • μX-ray CT

1 Introduction

It is well known that the pore water of partially saturated soils exists as a meniscus with a suction force that behaves as a capillary force between the soil particles. The water meniscus strengthens the soil through this capillary force, while the collapse of the water meniscus causes a drastic loss in strength by shearing or by the infiltration of water. Corresponding to the loss in strength, partially saturated soil

Y. Higo (✉) · F. Oka · S. Kimoto · Y. Matsuhima
Department of Civil and Earth Resources Engineering, Kyoto University, C1 Bld.
C cluster, Kyotodaigaku-Katsura 4, Nishikyo-ku, Kyoto, 615-8540, Japan
e-mail: higo.yohsuke.5z@kyoto-u.ac.jp; oka.fusao.2s@kyoto-u.ac.jp;
kimoto.sayuri.6u@kyoto-u.ac.jp; yoshiki@handballer.mbox.media.kyoto-u.ac.jp

T. Sanagawa
Foundation and Geotechnical Engineering Structures Technology Division, Railway Technical Research Institute, 2-8-38 Hikari-cho, Kokubunji-shi, Tokyo, Japan
e-mail: sanagawa@rtri.or.jp

S. Bonelli et al. (eds.), *Advances in Bifurcation and Degradation in Geomaterials*, Springer Series in Geomechanics and Geoengineering 11,
DOI 10.1007/978-94-007-1421-2_5, © Springer Science+Business Media B.V. 2011

exhibits a more brittle mode of failure than fully saturated soil, especially in cases of lower confining pressure (e.g., Cunningham et al. 2003).

Investigations of strain localization using X-ray computed tomography (CT) have produced various results (e.g., Desrues et al. 1996; Otani et al. 2000; Alshibli et al. 2000), by which the strain localization behavior inside geomaterials has been observed in a nondestructive manner. Recently, microfocus X-ray CT (hereinafter referred to as "μX-ray CT") and X-ray micro CT using synchrotron radiation were able to accomplish a very high spatial resolution of less than 10 μm, which makes it possible to study microstructural changes in the fine-grained sand of shear bands (e.g., Oda et al. 2004; Matsushima et al. 2006; Hall et al. 2010). To the author's knowledge, however, the published studies are mainly for dry sand, although most natural geomaterials are fully saturated or unsaturated porous media.

In the present paper, strain localization and the microstructural changes during triaxial compression tests for air-dried, partially saturated, and fully saturated Toyoura sand specimens using μX-ray CT are presented. The development of strain localization is observed by the X-ray CT scanning of the entire specimen during the triaxial compression tests. Through the obtained results, the strain localization behavior in partially saturated sand is discussed by comparing the results with those for fully saturated sand and air-dried sand specimens. Furthermore, the microstructure of partially saturated sand is viewed in a partial CT scan, namely, a part of the specimen of interest is scanned with high magnification. In particular, focus is placed on the microstructures in shear bands.

2 Material and Methods

The test sample used in this study is Toyoura sand. Toyoura sand is semi-angular in shape with an average diameter D_{50} of 0.185 mm.

The μX-ray CT system used in this study is KYOTO-GEO μXCT (TOSCANER-32250 μHDK), which was assembled by TOSHIBA IT & Control Systems Corporation and installed in Kyoto University (Higo et al. 2010). The focus size of the microfocus X-ray tube is very small, 4 μm, which provides the very high spatial resolution of 5 μm. Since the triaxial cell can be mounted on a rotation table, specimens can be scanned during the triaxial tests.

We used the conventional triaxial test procedure in which the moist-tamping method was applied to prepare the partially saturated specimen. Air-dried specimen was also prepared by compacting method. For the fully saturated specimen, the water pluviation method was used; thus, sample saturation is ensured. The height of every specimen is 70 mm and the diameter is 35 mm. A confining pressure of 50 kPa was isotropically applied by air pressure without the use of any back pressure. The axial load was applied through a displacement control system with a constant axial strain rate of 0.5%/min under drained conditions for air and water. During the triaxial shearing, three-dimensional CT images of the specimen were taken by the cone beam technique every axial strain of 2%. The volume of the specimen has been calculated by the integration of the voxels of the specimen. Note that the voxel size

Table 1 Test cases

Case no.	Case U	Case S	Case D
Initial void ratio	0.673	0.659	0.652
Relative density (%)	84	88	89
Water content (%)	15.00	25.06	0.16
Initial degree of saturation (%)	58.62	100	0.66
Voxel size[a] of CT image (μm)	$53^2 \times 90$	$51^2 \times 80$	$53^2 \times 90$

[a]Voxel size: (spatial resolution)^2 (thickness) of one CT image

can be precisely calibrated by CT scanning of the specimen with known size, by which the accuracy of the proposed method is ensured.

The triaxial compression tests performed in this study are listed in Table 1. We prepared the dense specimens for partially saturated, fully saturated and air-dried sand.

3 Stress-Strain Relations, Dilatancy and Strain Localization in Dry, Unsaturated, and Saturated Sand Specimens

The stress-strain relations for the triaxial compression tests are shown in Fig. 1. It can be seen that the peak stress level of the partially saturated specimen is higher than that of the fully saturated specimen, and that the peak deviator stress of the air-dried specimen is lower than that of the fully saturated specimen. For the axial strain at the peak deviator stress level, the strain at the peak stress of the partially saturated specimen is the smallest and that of the air-dried specimen is the largest. These results suggest that partially saturated sand is more brittle than the other types of sand. The partially saturated specimen exhibits higher peak stress than the other specimens due to the capillary force among the soil particles. However, the degradation of this capillary force during the shearing results in softening behavior. It is seen that the peak stress for the fully saturated specimen is higher than that for the air-dried specimen. This is possibly because of the fact that the shear strength of the specimens prepared by the water pluviation is higher than that by the air pluviation (e.g., Zlatović and Ishihara 1997). In addition, it is worth noting that the residual stress levels at an axial strain of 20% are almost the same for all three cases. This means that the steady states of the sand specimens are not influenced by the degree of saturation.

The relation between the volumetric strain and the axial strain is also shown in Fig. 1. A low level of volume compression can be observed in the early stages of loading for the partially saturated specimen. After that, all the specimens exhibit volume expansion and the volume expansion of the partially saturated specimen is larger than that in the other two cases. Finally, the volumetric strain levels at an axial strain of 20% are almost the same for all three cases.

The locations of the cross sections are illustrated in Fig. 2. CT images at an axial strain of 20% are shown in Fig. 3. Several lower density regions denoted by dark

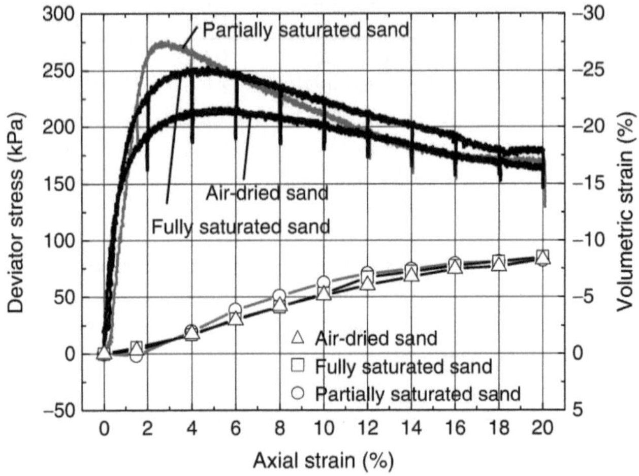

Fig. 1 Deviator stress-axial strain relations and volumetric strain-axial strain relations for triaxial compression tests on partially saturated sand, fully saturated sand, and air-dried sand (solid lines: deviator stress, open marks with lines: volumetric strain, compression is positive)

Fig. 2 Locations of the cross sections

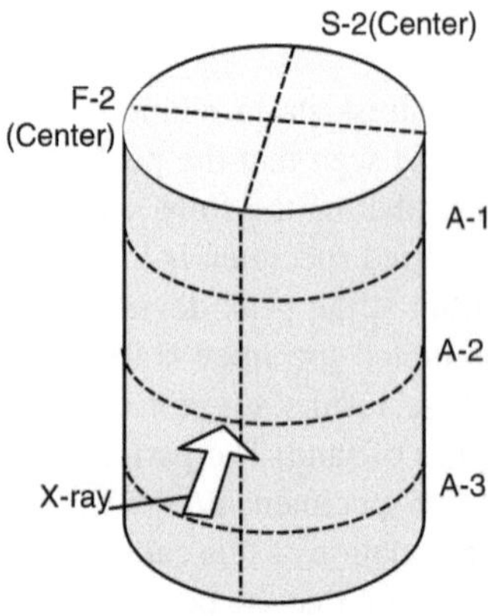

grey or black are clearly seen in the CT images. Volume expansion partly due to positive dilatancy probably occurs in the lower density regions. Thus, we can assume the lower density regions as shear bands. In the partially saturated specimen, the cone type of shear bands develop from the top and the bottom of the specimen, while radial shear bands appear in the middle of the specimen. The localization mode is similar to that in the results, e.g., Desrues et al. (1996). The shear bands that developed in the partially saturated specimen are the clearest among these three cases. In the fully saturated case, it is seen that the shear bands developed from the top of the specimen; however, the number of shear bands is fewer than that in the partially saturated case. The strain localization for the air-dried case is not so clear. This suggests that the strain localization behavior of the partially saturated sand specimen is significant because of the strain softening brought about by the degradation of the capillary force.

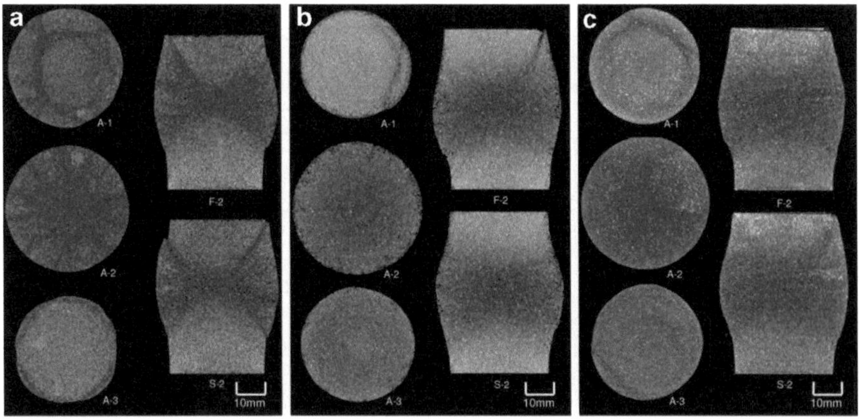

Fig. 3 CT images for partially saturated, fully saturated and air-dried triaxial sand specimens at an axial strain of 20%. (**a**) Partially saturated sand, (**b**) Fully saturated sand, (**c**) Air-dried sand

4 Microstructures in the Shear Bands of Partially Saturated Sand

Partial CT scan has been performed near the shear band for the partially saturated specimen (case U). The partial CT image obtained is shown in Fig. 4. The voxel size of the partial CT image is $6.86^2 \times 11.0$ (μm). The region of shear band indicated between the two dashed lines corresponds to the lower density region observed in the full CT image. The lower density region indicates the volume expansion partly due to dilatancy, while in the partial CT image with the higher spatial resolution, it is seen that the void areas in the shear band are larger than those outside of the shear band. In Fig. 4(b), chain-like structures in the shear band, loosely connected particles, are depicted. The structures cross the shear band with small contact areas between the particles. This structure is similar to the "Columnar structure" observed in the shear bands of the dry sand specimens after the plane strain compression tests by Oda et al. (2004). The columnar structure developed in the shear band results in a large void ratio and sustains the axial force. The thickness of the shear band is 1.8 mm which corresponds to about ten times the average diameter of Toyoura sand. This finding is consistent with that obtained for dry Toyoura sand by Oda et al. (2004), namely, ten times the particle diameter.

As mentioned above, the microstructures of partially saturated sand and air-dried sand are similar in their shear bands. In addition, the residual states of fully saturated sand, partially saturated sand, and air-dried sand are almost the same. At residual states, in which soils reach the critical state with development of clear shear bands, it can be assumed that most of deformation due to external loading occurs in the shear bands; hence, the stress-strain relation for the whole specimen is governed mainly by the behavior in the shear bands. This suggests that the behavior at the critical

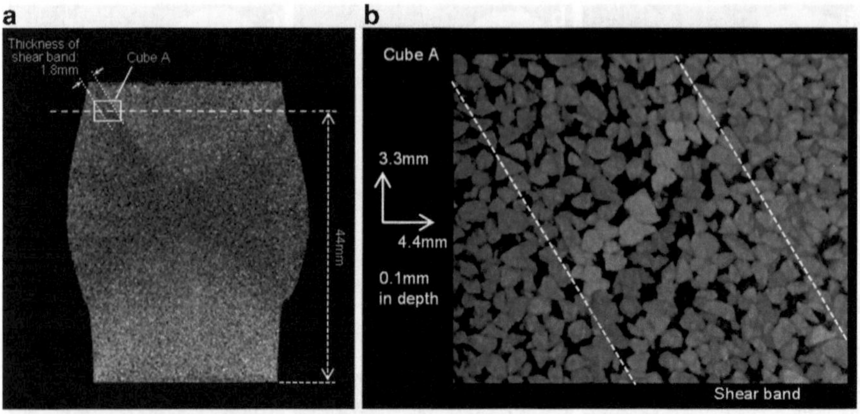

Fig. 4 Microstructure in the shear band of the partially saturated sand specimen (Case U). (**a**) Full CT image (F-2 vertical section), (**b**) Partial CT image (vertical section)

state of sand is unique and is almost independent of the initial degree of saturation. It is, however, possible to say that the water meniscus remaining in the shear bands increases the shear strength of the partially saturated sand. Further study is required on this point.

5 Conclusions

In this paper, the strain localization behavior and microstructural changes in partially saturated Toyoura sand during and after triaxial compression under drained conditions for air and water have been studied using microfocus X-ray CT through a comparison of the drained triaxial test results for fully saturated sand and for air-dried sand specimens. Shear bands were seen to have developed more clearly and the stress-strain relations exhibited stronger softening behavior in the partially saturated sand than in the fully saturated or air-dried sand. The stress levels and the volumetric strain levels at the residual state of fully saturated sand, air-dried sand, and partially saturated sand were found to be almost the same, and the microstructures of partially saturated sand and air-dried specimens were similar. This means that the behavior of the sand in the critical state seems to be uniquely independent of the initial degree of saturation. Further studies on localized zones from the microscopic point of view are necessary to confirm this conclusion.

References

K.A. Alshibli, S. Sture, N.C. Costes, M.L. Frank, M.R. Lankton, S.N. Batiste, R.A. Swanson, Assessment of localized deformations in sand using X-ray computed tomography. Geotech. Test. J. **23**(3), 274–299 (2000)

M.R. Cunningham, A.M. Ridley, K. Dineen, J.B. Burland, The mechanical behaviour of a reconstituted unsaturated silty clay. Géotechnique **53**(2), 183–194 (2003)

J. Desrues, R. Chambon, M. Mokni, F. Mazerolle, Void ratio evolution inside shear bands in triaxial sand specimens studied by computed tomography. Géotechnique **46**(3), 539–546 (1996)

S.A. Hall, M. Bornert, J. Desrues, Y. Pannier, N. Lenoir, G. Viggiani, P. Bésuelle, Discrete and continuum analysis of localised deformation in sand using X-ray μCT and volumetric digital image correlation. Géotechnique **60**(5), 315–322 (2010)

Y. Higo, F. Oka, S. Kimoto, T. Sanagawa, M. Sawada, T. Sato, Y. Matsushima, Visualization of strain localization and microstructures in soils during deformation using microfocus X-ray CT, in *Proceedings of the 3rd International Conference on X-ray CT for Geomaterials, GeoX2010*, New Orleans, Louisiana, USA, March 1–3, 2010, ed. by K.A. Alshibli, A.H. Reed (ISTE Ltd., London and John Wiley & Sons, Inc., Hoboken, 2010), pp. 43–51

T. Matsushima, K. Uesugi, T. Nakano, A. Tsuchiyama, Visualization of grain motion inside a triaxial specimen by micro X-ray CT at SPring-8, Advances in X-ray tomography for geomaterials, in *Proceedings of the Second International Workshop on X-ray CT for Geomaterials, GeoX 2006*, Aussois, France, ed. by J. Desrues, G. Viggiani, P. Bésuelle (ISTE Ltd., London 2006), pp. 255–261

M. Oda, T. Takemura, M. Takahashi, Microstructure in shear band observed by microfocus X-ray computed tomography. Géotechnique **54**(8), 539–542 (2004)

J. Otani, T. Mukunoki, Y. Obara, Characterization of failure in sand under triaxial compression using an industrial X-ray CT scanner. Soils Found. **40**(2), 111–118 (2000)

S. Zlatovi, K. Ishihara, Normalized behavior of very loose non-plastic soils: effect of fabric. Soils Found. **37**(4), 47–56 (1997)

Signature of Anisotropy in Liquefiable Sand Under Undrained Shear

Jidong Zhao and Ning Guo

Abstract This paper presents a study on the anisotropic behavior of liqufiable sand subjected to undrained shear, by using a 3D Discrete Element Method with two different approaches describing particle rolling. By using a sliding and free-rolling model, the force network in relation to anisotropy in medium-loose or dense samples presents a clear bimodal character, while the liquefiable loose specimen behaves differently. Appreciable degree of anisotropy is found developed in the weak force network when the sample tends to liquefy. When the rolling resistance is considered, all samples show marked increases in anisotropy in both the weak and strong force networks as well as the overall shear strengths, as compared with the free-rolling case. The loose sample tends also to be more resistant to liquefaction in the latter case than in the free rolling case under otherwise similar conditions.

Keywords Granular sand • Liquefaction • Anisotropy • Discrete element method • Rotational resistance

1 Introduction

Granular anisotropy has been found playing a key role in the complex behavior of granular sand in response to external loads, such as strain localization and liquefaction failure, along with many other intriguing phenomena observed in sand. In particular, the problem of liquefaction in sand has attracted much attention due to its theoretical and practical significance. A proper understanding of the anisotropic behavior in liquefiable sand may shed light on deciphering the numerous puzzles concerning liquefaction such as its occurrence mechanism and post-liquefaction

J. Zhao (✉) · N. Guo
Department of Civil and Environmental Engineering, Hong Kong University of Science and Technology, Clearwater Bay, Kowloon, Hong Kong
e-mail: jzhao@ust.hk; ceguo@ust.hk

S. Bonelli et al. (eds.), *Advances in Bifurcation and Degradation in Geomaterials*, Springer Series in Geomechanics and Geoengineering 11,
DOI 10.1007/978-94-007-1421-2_6,

behavior (Jefferies and Been 2006). Available experimental tools, however, are practically limited in various ways to identify granular anisotropy and its evolution (Mitchell and Soga 2005). To this end, Discrete Element Method (DEM) offers a unique and convenient way for the study of granular anisotropy in sand from the grain level (Oda and Iwashita 1999), and will be adopted for the study. In this paper, a three-dimensional Discrete Element Method is employed to investigate the behavior of anisotropy in granular sand which develops deformation towards liquefaction. Two different approaches on particle rolling are compared. In particular, we examine the bimodal characteristic of force network in relation to granular anisotropy observed in (Azéma et al. 2007; Radjaï et al. 1998) for dense sand under drained shear. In the bimodal theory, it is found that the interparticle forces can be largely split into two networks, one with contacts of normal force greater than the average value and the other smaller than the average force. The former is called strong force network and the latter weak network. The two complimentary network differ from each other in a number of important ways in supporting the external load (for detail see Azéma et al. (2007) and Radjaï et al. (1998)).

2 Approach and Formulation

In the 3D DEM code used here, the interparticle sliding behavior is described by a linear force-displacement contact law, with the normal and tangential stiffnesses being set equal: $k_n = k_s = 10^5 N/m$ following (Oda and Iwashita 1999). The interparticle sliding is assumed to be governed by Coulomb's law and the coefficient of friction is set to $\mu = 0.5$. In addition to particle sliding, particle rolling has been found important in dictating the steric rearrangement of particles, and hence influences the macroscopic granular responses such as peak strength, dilation and liquefaction. In this study, we present two approaches handling the rolling case. The first considers free rotation of the particle body, i.e. the rolling is taken as a direct consequence of the inter-particle friction. The second approach takes into account particle rolling resistance by considering contact moment M_r (Ishihara and Oda 1998; Estrada et al. 2008). The interparticle contact is assumed to be plane-to-plane with a triangular distribution of normal force along the contact plane. A threshold moment is defined $M_{max} = \mu_r f_n l$ such that $|M_r| \leq M_{max}$, where μ_r denotes the coefficient of rolling resistance, and f_n is the normal force and l is a characteristic length related to the particle radii and the overlapping depth between particles (Guo and Zhao 2010). An assembly of 31,769 sphere particles with polydisperse radii ranged from 0.2 mm to 0.6 mm are randomly generated in a cube with rigid and frictionless bounding walls. It is then consolidated isotropically to reach different initial void ratio e_0 as shown in Table 1. These samples are then sheared under undrained condition which is achieved by adjusting the horizontal strain to keep the total volume of the assemblage constant with the major compression applied in the vertical direction. Note that no strain localization has been observed in the tested specimens. The stress tensor is defined as $\sigma_{ij} = \sum_{N_c} f_i^c d_j^c / V$, where V is the

Table 1 Initial properties and descriptions of samples for testing

Test	Label	e_0	Description
Series I	UL	0.644	Loose sample under undrained shear
	UM	0.634	Medium loose sample under undrained shear
	UD	0.612	Dense sample under undrained shear
Series II	UR0	0.644	Undrained shear with $\mu_r = 0.0$
	UR1	0.647	Undrained shear with $\mu_r = 0.1$
	UR3	0.648	Undrained shear with $\mu_r = 0.3$
	UR5	0.649	Undrained shear with $\mu_r = 0.5$

total volume of the assemblage, N_c is the total number of contacts, $\mathbf{f}^c$ is the contact force and $\mathbf{d}^c$ is the branch vector joining the centers of the two contact particles (Christoffersen et al. 1981). Stress quantities commonly used in triaxial tests are then obtained: $p' = \sigma_{ii}/3$, $q = \sqrt{3\sigma'_{ij}\sigma'_{ij}/2}$, where $\sigma'_{ij} = \text{dev}(\sigma_{ij})$. Log-form strains are defined by the boundary displacements $\varepsilon_1 = \ln H_0/H$, $\varepsilon_v = \varepsilon_1 + \varepsilon_2 + \varepsilon_3 = \ln V_0/V$, where H_0 and V_0 are the initial height and volume of the assembly respectively (compression is taken as positive).

Two sources of anisotropy are considered: the geometrical anisotropy and the mechanical anisotropy (Satake 1982; Oda 1982; Ouadfel and Rothenburg 2001; Sitharam et al. 2009). The geometrical anisotropy is expressed in terms of the distribution of contact normals (and branch vectors if non-spherical particles are used), which is defined by the following fabric tensor:

$$\phi_{ij} = \int_{\Theta} E(\Theta) n_i n_j d\Theta, \quad E(\Theta) = \frac{1}{4\pi}\left[1 + a^c_{ij} n_i n_j\right] \tag{1}$$

where $\mathbf{n}$ is the unit contact normal vector, Θ is the solid angle in the spherical coordination system. Note that $a^c_{ij} = 15\phi'_{ij}/2$. The mechanical anisotropy in terms of normal force anisotropy and tangential force anisotropy are defined as follows:

$$\chi^n_{ij} = \frac{1}{4\pi}\int_{\Theta} \bar{f}^n(\Theta) n_i n_j d\Theta, \quad \bar{f}^n(\Theta) = \bar{f}^0[1 + a^n_{ij} n_i n_j], \tag{2}$$

$$\chi^t_{ij} = \frac{1}{4\pi}\int_{\Theta} \bar{f}^t(\Theta) t_i n_j d\Theta, \quad \bar{f}^t_i(\Theta) = \bar{f}^0[a^t_{ik} n_k - (a^t_{kl} n_k n_l) n_i], \tag{3}$$

where $\bar{f}^0 = \chi^n_{ii}$ is the average normal force. The second invariants of the three anisotropy tensors defined above are used to quantify the degree of anisotropy:

$$a_c = \sqrt{\frac{3}{2} a^c_{ij} a^c_{ij}}, \quad a_n = \sqrt{\frac{3}{2} a^n_{ij} a^n_{ij}}, \quad a_t = \sqrt{\frac{3}{2} a^t_{ij} a^t_{ij}} \tag{4}$$

A negative sign is attributed to a_c if the major principal direction of a^c_{ij} is more close to the perpendicular direction of the major principal stress direction, i.e., a_c has the same sign as $a^c_{ij} \sigma'_{ij}$.

3 Results and Discussion

Undrained triaxial compression tests are conducted with *free rolling* particles ($\mu_r = 0$) for Series I samples. The material responses are shown in Fig. 1. Note that the pore water pressure Δu is calculated by assuming a constant total confining pressure σ_3. As is shown, the three cases demonstrate perfectly the typical characteristic states a sand may experience, namely, peak strength (UM and UL), phase transformation state (UM and UD), liquefaction state (UL) and the critical state (UL and UD). The observations are in qualitative agreement with laboratory results (Mitchell and Soga 2005). The evolution of a_c and a_n in the strong/weak force subnetworks (Γ_{strong} and Γ_{weak}), and the entire force network (Γ_{total}) for UM and UL is shown in Fig. 2. The response of UD is similar to UM and will not be presented here. The anisotropic behavior of medium and dense samples (UM and UD) agrees favorably with the bimodal theory as found in (Radjaï et al. 1998). As is shown in Fig. 2a, c, the weak force network remains largely isotropic during the loading course (a_c is approximately zero and a_n is constantly small in the weak network during the loading course), whereas the strong force network is more anisotropically comprised of column-like chains which sustain nearly all deviatoric stress applied to the system. Note that the magnitude of a_t is small compared to

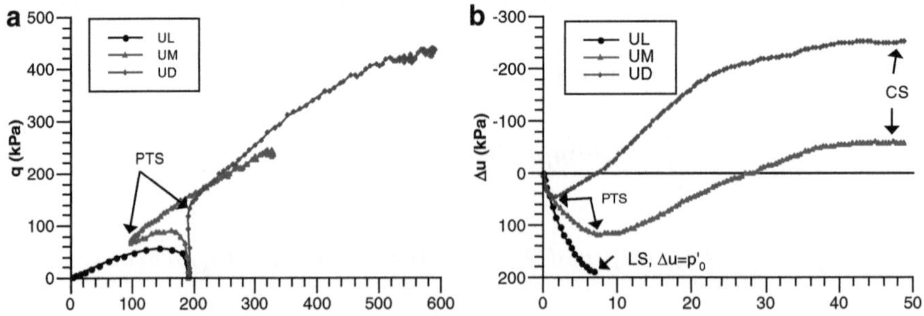

Fig. 1 Sand responses under undrained shear (**a**) $p' - q$; (**b**) $\varepsilon_1 - \Delta u$

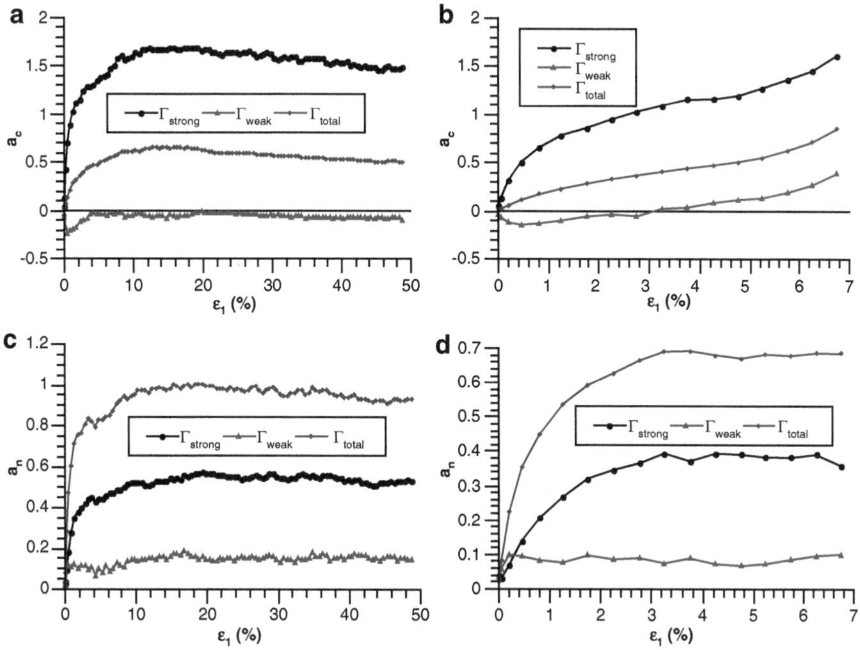

Fig. 2 Evolution of a_c and a_n for samples UM (**a, c**) and UL (**b, d**)

a_c and a_n, and are not presented here. As for the liquefiable loose sample UL, a_c in Γ_{weak} evolves gradually from a negative value to positive and finally reaches $a_c = 0.39$, which is very high compared to the cases of UM and UD. This indicates that in the loose case, the weak contacts have to sustain significant deviatoric loads due to the lack of efficient amount of strong force chain formed. This somehow may sacrifice its function of acting as the lateral support for the strong contact force columns. The overall structure may be more susceptible to abrupt buckling which leads to liquefaction. Though not presented here, it is meanwhile found that the (total) geometrical anisotropy contributes a much higher proportion to the overall strength in the UL case than in UM and UD, whereas in the latter two cases the mechanical anisotropy dominates the soil strength. Specifically, in the cases of UM and UD, a_n contributes more than 50% to the total strength and a_c contributes about 30%, whilst the tangential force anisotropy a_t contributes less than 20%. In the case of UL, the contributions from a_c and a_n are comparable. In some very loose cases a_c may even outweighs a_n in the shear strength. a_t remains at a very low proportion in all cases. For details see Guo and Zhao (2010).

The effect of rolling resistance on the undrained behavior of sand has also been examined by testing on samples in Series II in Table 1. Figure 3 shows the responses with different μ_r. It appears that the increase in μ_r generally leads to improved shear strength. Note that Samples UR0, UR1 and UR3 that are all otherwise regarded as loose and liquefiable in the free-rolling case manage to pull back their strength

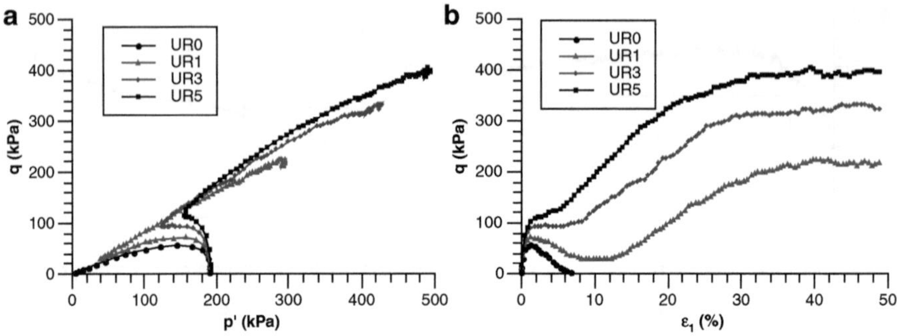

Fig. 3 Undrained behavior with rolling resistance: (**a**) p'-q; (**b**) ε_1-q

after reaching the phase transformation point. Liquefaction is observed only for the extremely loose case of UR5. It is also found that the critical state stress ratio in Series II is higher than that in Series I. This is reasonable as the rolling resistance generally leads to higher interlocking and higher internal friction angle for the samples. In the presence of rolling resistance, the distribution of anisotropy in the weak and strong networks is reminiscent to the case of free rolling case. Nevertheless, the magnitudes of a_c, a_n and a_t are greater than in the free rolling case, which explains the improved shear strength observed.

4 Conclusions

The behavior of anisotropy in a liquefiable sand under undrained shear has been studied using a three dimensional Discrete Element Method. Significant anisotropy is found in the weak force network when a loose sample is sheared towards liquefaction. Rolling resistance may generally help to increase the shear strength of a sand and to render a loose sand sample less vulnerable to be liquefied.

Acknowledgements This work was supported by RGCHK (Grant No. 622910, DAG08/09.EG04).

References

E. Azéma, F. Radjaï, R. Peyroux, Force transmission in a packing of pentagonal particles. Phys. Rev. E. **76**, 011–301 (2007)

J. Christoffersen, M.M. Mehrabadi, S. Nemat-Nasser, A micromechanical description of granular material behavior. J. Appl. Mech. **48**, 339–344 (1981)

N. Estrada, A. Taboada, F. Radjaï, Shear strength and force transmission in granular media with rolling resistance. Phys. Rev. E. **78**, 021–301 (2008)

N. Guo, J. Zhao, Characterization of granular anisotropy in sand under shear. J. Mech. Phys. Solids. Under review (2010)

K. Ishihara, M. Oda, Rolling resistance at contacts in simulation of shear band development by dem. J. Eng. Mech. **124**(3), 285–292 (1998)

M. Jefferies, K. Been, *Soil Liquefaction: A Critical State Approach* (Taylor & Francis, New York, 2006)

J. Mitchell, K. Soga, *Fundamentals of Soil Behavior*, 3rd edn. (John Wiley & Sons, New Jersey, 2005)

M. Oda, Fabric tensor for discontinuous geological materials. Soil. Found. **22**(4), 96–108 (1982)

M. Oda, K. Iwashita, *Mechanics of Granular Materials: An Introduction* (Taylor & Francis, Balkema/Rotterdam, 1999)

H. Ouadfel, L. Rothenburg, Stress-force-fabric relationship for assemblies of ellipsoids. Mech. Mater. **33**, 201–221 (2001)

F. Radjaï, D.E. Wolf, M. Jean, J.J. Moreau, Bimodal character of stress transmission in granular packings. Phys. Rev. Lett. **80**, 61–64 (1998)

M. Satake, Fabric tensor in granular materials, in *Deformation and Failure of Granular Materials*, ed. by P.A. Vermeer, H.J. Luger, A.A. Balkema, pp. 63–68 (1982)

T.G. Sitharam, J.S. Vinod, B.V. Ravishankar, Post-liquefaction undrained monotonic behaviour of sands: experiments and dem simulations. Géotechnique **59**(9), 739–749 (2009)

Friction and Localization Associated with Non-spherical Particles

A.V. Dyskin and E. Pasternak

Abstract Effect of rotations of particles of non-spherical shape is considered. Under combined action of shear and compressive forces, the moment balance produces apparent negative stiffness. In the case when friction is controlled by rotating particles non-sphericity leads to oscillating friction force. In the case of particulate materials in compression this may lead, under certain magnitudes of compression, to the initiation of global negative stiffness and, subsequently, strain localization.

Keywords Particle rotation • Negative stiffness • Effective characteristics • Self-consistent method • Instability

1 Introduction

Sliding over the shear cracks or faults is controlled by friction which is known to be rate and path-dependent. In many cases friction is associated with rotating particles between the sliding surfaces, either pre-existing (as in fault gouge or in granular materials) or developed due to wear. This phenomenon can micromechanically be modelled based on the conceptual (e.g., the Cosserat-type models) or direct computer representation (e.g., discrete element methods) of the gouge particles as spheres (or disks in 2D). When modelling complex phenomena involving real-shape grains or fragments one needs however to take into account the fact that the grains usually have shapes which are different from spherical as the influence of shape

A.V. Dyskin (✉)
School of Civil and Resource Engineering, University of Western Australia, Perth, WA, Australia
e-mail: arcady@civil.uwa.edu.au

E. Pasternak
School of Mechanical Engineering, University of Western Australia, Perth, WA, Australia
e-mail: elena@mech.uwa.edu.au

S. Bonelli et al. (eds.), *Advances in Bifurcation and Degradation in Geomaterials*, Springer Series in Geomechanics and Geoengineering 11,
DOI 10.1007/978-94-007-1421-2_7, © Springer Science+Business Media B.V. 2011

can be profound ranging from a considerable increase in the stresses and reduction in the rates of granular flow (Cleary and Sawley 2002) to post-peak softening in interlocking structures without involvement of failure or damage as experimentally demonstrated by Estrin et al. (2003).

Another phenomenon related to particle non-sphericity can be inferred from the mechanism of frictionless shear in rock mass at great depths (Tarasov and Randolph 2008) whereby columns produced in process zone of shear crack at high stresses can rotate and thus decrease the friction resistance (effect of negative friction). We realized that one does not need the concept of rotating columns to mimic the effect of negative friction: any non-spherical particle can produce the same effect. Furthermore, the action of non-sphericity can be interpreted in terms of local negative stiffness, which under some circumstances can effect a global transition to (incremental) negative stiffness of the whole volume and the initiation of strain localization.

Here we discuss the effect of non-sphericity of rolling particles on both friction and strain localization.

2 Effect of Non-sphericity of Rolling Particles on Friction

We introduce the concept using an example of particles with a square section, Fig. 1. The effect of particle shape can be seen from the moment equilibrium with respect to point 0, which gives the following relationship between the tangential force T, pressure P and displacement u of the left corner of the particle expressed in terms of length variables normalized the length parameters with respect to a:

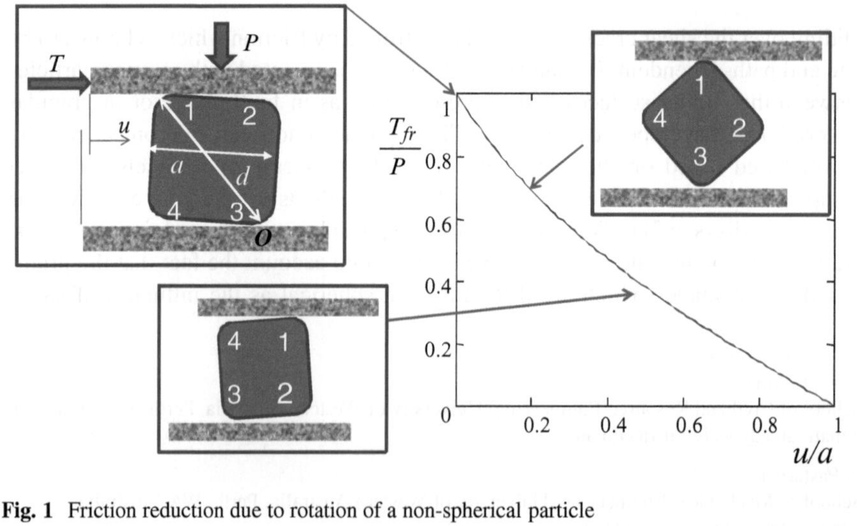

Fig. 1 Friction reduction due to rotation of a non-spherical particle

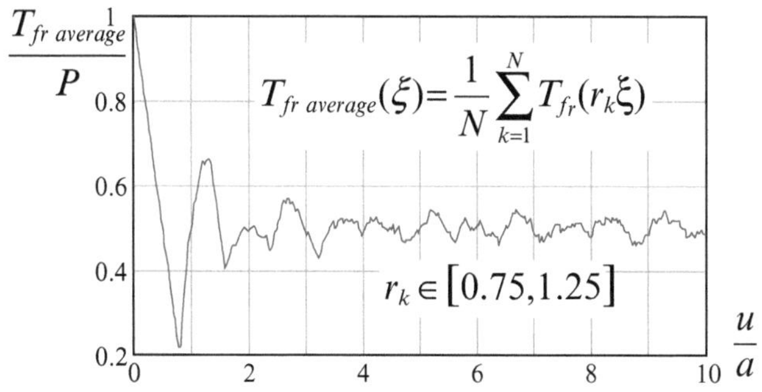

Fig. 2 Simulation of collective effects of rolling particles with uniform and independent random variations of rolling periods. Here the number of particles $N = 500$, r_k are uniformly distributed independent random variables

$$T\sqrt{\delta^2 - (1-\xi)^2} = P(1-\xi), \quad \xi = u/a, \quad \delta = d/a \tag{1}$$

Here δ is a shape parameter which controls the effect. (For particles with a square section, $\delta = 2^{1/2}$; for general shape particles, δ can be different for different parts of the whole period of rotation.)

As seen from Fig. 1, the force needed to effect rotation (friction force, T_{fr}) decreases with displacement u, within a quarter period of the particle rotation where it resembles negative friction. After that, the friction force surges to its initial level resulting in a saw-like friction law.

When large numbers of particles are involved in the process of rolling with randomly varying sizes and subsequently randomly varying rolling periods, the friction law is a result of a collective effect of their rolling. It leads to oscillating friction with high initial values whose magnitude then decreases inversely proportional to the displacement, Fig. 2. Thus in the cases when sliding involves rolling of particles, their non-spherical shapes can lead to complex patterns of friction vs. displacement dependence observable in laboratory experiments (e.g., Di Toro et al. 2004).

3 Negative Stiffness Caused by Rotating Non-spherical Particles

Consider a particulate material under compression and assume that, at an advanced stage of loading, some grains get detached from the matrix or break and acquire the ability to rotate. We also assume that the compression is high enough for the rotating grain to keep its contacts with the matrix (or the rest of the particulate material) in one direction, such that the diagram in Fig. 1 holds.

From (1) one can find the tangential shear stiffness, which is negative:

$$k = dT/du = -Pa^{-1}\delta^2\left[\delta^2 - (1-\varsigma)^2\right]^{-3/2} \tag{2}$$

Furthermore, as far as the stability analysis is concerned, only the initial value (when $u = 0$) of the negative stiffness is important, which gives

$$k = -Pa^{-1}\delta^2\left[\delta^2 - 1\right]^{-3/2} \tag{3}$$

It is seen that the value of negative stiffness is proportional to pressure force P applied to the particle. Thus, by increasing pressure, the value of negative stiffness can be made fairly large, even is the concentration of rotating particles can be low. We now investigate the influence of this fact on the possibility of localization.

4 Instability in Particulate Materials Under Compression Caused by Rotating Non-spherical Particles

We illustrate the interrelation between the concentration of rotating particles and the pressure in their influence of the effective modulus of the particulate material using an anti-plane strain simple model of isotropic matrix with cylindrical inclusions of negative shear modulus proportional to stiffness k determined in the previous section. So, suppose the material has the (positive) shear modulus μ_0, and the inclusions have the negative shear modulus μ_1. We assume that the negative modulus is proportional to the applied compression $\mu_1 = -\kappa p$, where p is the first principal compressive stress and κ is a factor representing the particle geometry.

In the approximation of dilute concentration of inclusions, we take the available solution (e.g., Germanovich and Dyskin 1994) for effective shear modulus and replace the positive shear modulus with the negative one (see reasoning in Dyskin and Pasternak 2010). The effective modulus reads

$$\mu_* = \mu_0\left[1 + 2\frac{\kappa p + \mu_0}{\kappa p - \mu_0}c\right] \tag{4}$$

Here c is the concentration (volumetric fraction) of the inclusions. We note that the solution has a singularity at $\kappa p = \mu_0$, indicating that the interaction between inclusions becomes long range such that it cannot be neglected no matter how small the concentration is. Therefore, non-interacting solution (4) is not applicable and a method which takes into account interaction between inclusions is required.

We use the differential self-consistent method based on successive embedding of groups of randomly distributed inclusions of infinitesimal (and equal) concentrations into effective medium determined by previously embedded inclusions (e.g. Bruggeman 1935). This procedure leads, in our case, to the following differential equation

$$\frac{d\mu}{dc} = 2\mu\frac{m+\mu}{m-\mu}\frac{1}{1-c}, \quad \mu|_{c=0} = 1, \quad \mu = \frac{\mu_*}{\mu_0}, \quad m = \frac{\kappa p}{\mu_0} \tag{5}$$

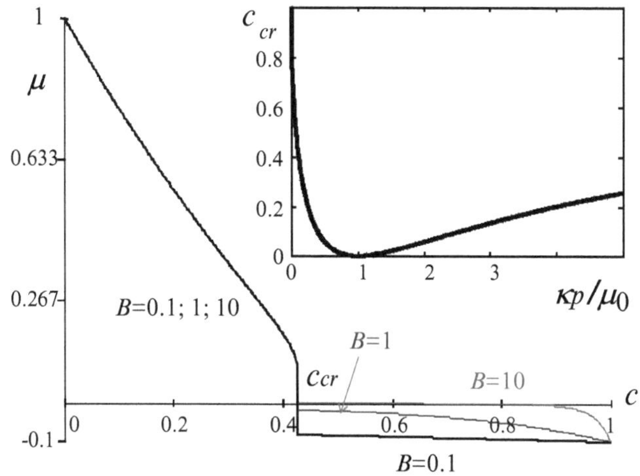

Fig. 3 Normalised effective shear modulus vs. concentration of negative stiffness inclusions for $\kappa p = 0.1$. The inset shows the dependence of the critical concentration upon the principal compressive stress

whose solution read (see Dyskin and Pasternak 2010, for details)

$$\mu = \begin{cases} \alpha(c)^2 - m - \alpha(c)\sqrt{\alpha(c)^2 - 2m}, & \alpha(c) = 2^{-1/2}(1+m)(1-c), \quad \text{if } c \le c_{cr} \\ -\alpha(c)^2 - m + \alpha(c)\sqrt{\alpha(c)^2 + 2m}, & \alpha(c) = 2^{-1/2}B(1+m)(1-c), \quad \text{if } c > c_{cr} \end{cases} \tag{6}$$

where $c_{cr} = 1 - 2\sqrt{m}\,(m+1)^{-1}$. Here B is a constant which cannot be determined since, due to instability of the material with negative stiffness its value is dictated by the properties of the loading device.

Figure 3 shows this dependence for different values of B. It is seen that as the concentration of negative stiffness inclusions increases, the effective modulus reminds positive decreasing from its initial value which is the shear modulus of the matrix. After reaching the critical concentration, the effective stiffness abruptly drops to a negative value. (The particular value depends upon the loading device.) The negative effective modulus causes strain localization in the sense of Rudnicki and Rice (1975). In this respect, c_{cr} is the concentration of negative stiffness inclusions that causes the localization. The inset in Fig. 3 shows the dependence of critical concentration upon the first principal compressive stress, p. It is seen that initially the increase in p reduces the critical concentration down to zero (when $m = 1$, i.e. $\kappa \text{p} = \mu_0$). Reaching zero critical concentration of negative stiffness inclusions physically means that a single rotating particle is sufficient to topple the whole material into the negative stiffness state and produce strain localization. Under larger magnitudes of applied load p, the critical concentration increases. Therefore under very high compressive loads, which happens when the confining pressure is very large, no strain localization could be produced by the mechanism considered.

5 Conclusions

Under combined action of shear and compressive forces, the moment balance of rotating of particles of non-spherical shape produces the effect of apparent negative stiffness whose value depends upon the magnitude of the compressive force. In the case of particles rotating between sliding surfaces this leads to oscillating friction force and the overall friction reduction. In the case of particulate materials in compression this may lead to the initiation of global negative stiffness and, subsequently, strain localization. This happens in a certain range of magnitudes of the first principal compressive stress which depend upon the concentration of rotating particles. When the stress is too low or too high, the critical concentration of local negative stiffness zones is not achievable. (The concentration of local negative zones is made up by particles that have enough degrees of freedom to rotate.) Hence the global negative stiffness cannot be produced. In other words, as the stress increases it initially lowers the critical concentration and increases the concentration of rotating particles (due to damage accumulation). If the rotating particles reach the critical concentration, the particulate material abruptly drops its stiffness to a negative value producing strain localization.

Acknowledgments This research is supported by ARC Discovery Grants DP0988449, DP0773839 and WA Geothermal Centre of Excellence.

References

D.A.G. Bruggeman, Berechnung verschidener Phisikalischer Konstanten von Heterogenen Substanzen. Ann. Phys. **24**, 636 (1935) (in German)

P.W. Cleary, M.L. Sawley, DEM modelling of industrial granular flows: 3D case studies and the effect of particle shape on hopper discharge. Appl. Math. Model. **26**, 89–111 (2002)

G. Di Toro, D.L. Goldsby, T.E. Tullis, Friction falls towards zero in quartz rock as slip velocity approaches seismic rates. Nature. **427**, 436–439 (2004)

A.V. Dyskin, E. Pasternak, Effective anti-plane shear modulus of a material with negative stiffness inclusions, in *Proceedings of the 9th HSTAM International Congress on Mechanics Vardoulakis Mini-symposia*, Limassol, Cyprus, p. 129 (2010)

Y. Estrin, A.V. Dyskin, E. Pasternak, S. Schaare, S. Stanchits, A.J. Kanel-Belov, Negative stiffness of a layer with topologically interlocked elements. Scripta Mater. **50** 291–294 (2003)

L.N. Germanovich, A.V. Dyskin, Virial expansions in problems of effective characteristics. Part II. Anti-plane deformation of fibre composite. Analysis of self-consistent methods. J. Mech. Compos. Mater. **30**, 234–243 (1994)

J.W. Rudnicki, J.R. Rice, Conditions for the localization of deformation in pressure-sensitive dilatant materials. J. Mech. Phys. Solids **23**, 371–394 (1975)

B.G. Tarasov, M.F. Randolph, Frictionless shear at great depth and other paradoxes of hard rocks. Int. J. Rock Mech. Min. Sci. **45**, 316–328 (2008)

Length Scales for Nonaffine Deformation in Localized, Granular Shear

Amy L. Rechenmacher and Sara Abedi

Abstract We offer experimental observations of meso-scale deformation and kinematic activity within sheared granular layers to investigate the nature and spatial periodicity of nonaffine displacement fields within shear bands in a granular material. Prismatic specimens of sands and glass beads are subjected to plane strain deformation in which zero-strain conditions are enforced by translucent glass walls. We use the Digital Image Correlation (DIC) to track movements of small, overlapping particle clusters. By subtracting a superimposed first-order shear displacement field from the observed non-affine displacement fields, co-rotational vortices appear and coordinate with previously-observed kinematic patterns. We undertake a preliminary assessment of the spatial periodicity of such patterns to glimpse the nature of an underlying length scale for granular material deformation.

Keywords Granular material • Shear bands • Non-affine deformation • Vortices • Length scales

1 Introduction

The quest to determine an internal length scale governing granular material deformation has occupied much recent research (Arslan and Sture 2008). For example, for shear banding in sands, shear band thickness is commonly thought to be a function of median grain size (Muhlhaus and Vardoulakis 1987). Recently, attention has been drawn to the existence of coherent structures present within shear bands. For example, vortices have been detected within shear bands from experiments on photo-elastic disks (Utter and Behringer 2008) and in discrete

A.L. Rechenmacher (✉) · S. Abedi
Department of Civil and Environmental Engineering, University of Southern California, Los Angeles, CA 90089, USA
e-mail: arechenm@usc.edu; abedimas@usc.edu

S. Bonelli et al. (eds.), *Advances in Bifurcation and Degradation in Geomaterials*, Springer Series in Geomechanics and Geoengineering 11,
DOI 10.1007/978-94-007-1421-2_8, © Springer Science+Business Media B.V. 2011

element modeling studies (Alonso-Marroquin et al. 2006; Tordesillas et al. 2008). Such vortex structures are thought to induce local force chain buckling events (Tordesillas et al. 2008). Yet, detection and characterization of such structures in real granular materials is lacking. Recently, Rechenmacher (2006) and Rechenmacher et al. (2010), using digital imaging methods, detected spatial variations in various kinematic quantities along shear bands in dilative sands undergoing plane strain compression. At the softening-critical state transition, a systematic pattern emerged: in the direction of shear, volumetric dilation was followed by high rotational strains, followed by high magnitudes of perpendicular displacement (to the shear band axis), followed by volumetric contraction. This pattern signified a collective and coordinated multi-force chain collapse event microstructurally underlying the macroscopic behavioral transition from softening to critical state. The pattern was spatially periodic, which suggests that the internal physical processes related to this multi-force chain collapse event may relate to an underlying material length scale.

We compare the kinematic patterns observed by Rechenmacher and co-workers with residual displacement fields to assess the role of vortex structures in the force chain collapse process. Then, we measure the spacing between vortices and ensuing kinematic fluctuations in a first-run attempt to characterize the spatial period, and compare to approximations of shear band thickness for sands of four different median grain sizes. The results offer preliminary insight regarding the manifestation of an internal material length scale governing dense granular shear.

2 Experimental and Analytical Methods

Dense, dilative sand specimens were tested in plane strain compression. The specimen base rests on a one-way linear bearing which allows for the offset required for unconstrained shear band growth, thus minimizing boundary interference. The zero strain conditions are enforced by glass walls that enable digital imaging of in-plane deformations. Tests were conducted on four different sands to enable behavioral characterizations as a function of grain size and sand gradation (Table 1). Specimens are placed in a large-diameter acrylic confining cell, saturated, consolidated anisotropically, and sheared under drained, strain-controlled compression. Digital images are collected at frequent, regular time intervals throughout shear

The Digital Image Correlation (DIC) technique is used to compute the in-plane displacements of the sand specimens as they deform along thelubricated, glass,

Table 1 Properties of sands tested

Sand type	D_{50}(mm)	C_u	C_c
Delaware Beach (DB)	0.40	1.3	1.0
Silica-Coarse (SC)	0.42	1.2	1.0
Concrete (C)	0.62	3.8	0.7
Masonry-Coarse (MC)	0.85	1.2	1.1

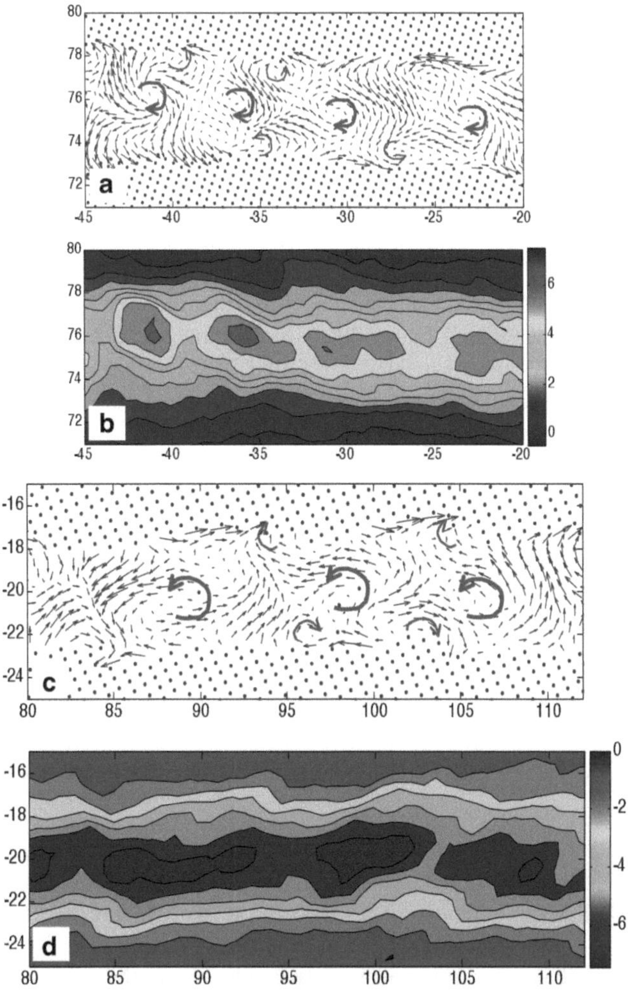

Fig. 1 Residual displacement vectors (**a** and **c**) and kinematic rotation (**b** and **d**) at the softening-critical state transition for SC sand (**a** and **b**; sense of shear is *top* moving *right* relative to *bottom*) and MC sand (**c** and **d**; sense of shear is *top* moving *left* relative to *bottom*)

plane-strain confining walls. DIC derives local displacements by mapping between subsequent digital images gray level values within small pixel subsets. Physically, the DIC data reflect average incremental displacements of approximately 6-particle-diameter clusters, overlapped, and spaced center-to-center about one grain diameter apart. The software VIC-2D, by Correlated Solutions, Inc. is used to perform the DIC analysis herein.

The DIC method limits accurate displacement calculation to small strain increments. In order to analyze the kinematics of shear band behavior over larger strains, as is necessary to track force chain buckling events, we have devised a methodology to accumulate sequential DIC analyses (Rechenmacher et al. 2010; Chupin et al. 2011). Typical analysis increments, designed to optimize capture of the collective force chain collapse event mentioned above, represent about 10–15% gross shear strain along the shear band. From the DIC accumulations, the deformation gradient, **F**, is computed. Local volume changes (assumed to be reflective of in-plane area changes) are quantified in terms of the Jacobian, J, of **F**. Macro-rotations, Ω, that correspond to rigid body rotations are obtained from the polar decomposition of **F**.

The accumulated DIC measurements are in the form of a matrix. For each column of data, the local displacement gradient is calculated. Assuming that only the material within the shear band is subject to a significant nonzero displacement gradient (Rechenmacher and Finno 2004; Rechenmacher et al. 2010), the lower and upper boundaries of shear band are found by applying a small threshold to the displacement gradient. The procedure is performed for all columns in the analysis area, and linear regression is used to obtain a best fit line demarcating each upper and lower shear band boundaries. Shear band thicknesses were roughly constant in a global sense across the lengths of the shear bands; however, locally, fluctuations in thickness were often observed (e.g. Fig. 1). We then superimpose an affine, first-order shear displacement field inside the shear band. The mean, affine displacement field is then subtracted from the observed displacements to obtain a "residual" displacement field. Typical examples of residual displacements are given in Fig. 1a, c.

3 Nonaffine Deformation in Shear Bands in Sands

Figure 1a, b compare the residual displacements and kinematic rotation obtained at the softening-critical state transition during a test on SC sand (Table 1), and Fig. 1c, d compare similar quantities for MC sand. First, we observe rotational vortices centered at the peaks (absolute value) in kinematic rotation. The spatial peaks in rotational strain denote locations where a force chain is in the process of buckling. These force chains were initially formed at peak stress, and while they are in the process of buckling throughout softening, it is only at the softening-critical state transition where vortex structures first appear (Abedi et al. 2011). This suggests that the vortices are a consequence of the force chain collapse process.

Also seen in Fig. 1a, c are opposite trending rotational entities at the shear band boundaries. These derive from the inertia of exciting the localized shear deformation, producing a local, relative displacement in the opposite direction of shear motion, due. At the conflux of adjacent vortices, where directionally opposite displacement fields are converging, local jamming is seen (which produces local regions of volumetric contraction, see (Abedi et al. 2011)). Thus, we do not see in Fig. 1a, c the "microbands" that have been suggested to occur from numerical simulations on circular particles (Kuhn 1999; Tordesillas et al. 2008). This is likely due to the stronger interlocking between sand grains.

4 Length Scales for Nonaffine Deformation

The data of Fig. 1a, c suggest that shear band thickness may be related to vortex diameter. Further, comparing Fig. 1a, b with c and d, the spacing between adjacent vortices appears wider for the larger-grained sand. How vortex size, spacing and shear band thickness are related may shed light on underlying material length scales. We compare each the spacing of kinematic rotational peaks (which align with vortex centers), shear band thickness and median grain size. To measure shear band thickness, we utilize the displacement gradient in a similar manner to the process for determining the mean shear field described in Sect. 2. However, we note that DIC suffers a systematic problem of always over-estimating shear band thickness, as follows. Consider an arbitrary DIC subset, of which only a small portion is inside the shear band. While only a portion of the subset will undergo straining, DIC will approximate a first-order deformation estimate to the entire subset, and an ensuing subset center displacement will be obtained, even though the subset center is outside the shear band. By systematically using small subset sizes, this overestimation error is minimized. Here, we proportioned subsets so that they represent the same physical size for each test analyzed, such that the degree of overestimation should be on the order of at most only tenths of a millimeter, and consistent for each sand analyzed.

Figure 2a first compares shear band thickness with median grain size for five tests conducted on the four sands listed in Table 1 (two tests on the MC sand). The dotted line represents the commonly assumed relationship of thickness equals $10 \times D_{50}$. The data clearly do not follow this trend: for $D_{50} \approx 0.4$ mm, thickness is about $13 \times D_{50}$; and for $D_{50} \approx 0.8$ mm, thickness is about $8 \times D_{50}$. These results suggest, preliminarily, that shear band thickness *as a function of grain diameter* decreases as median grain size increases.

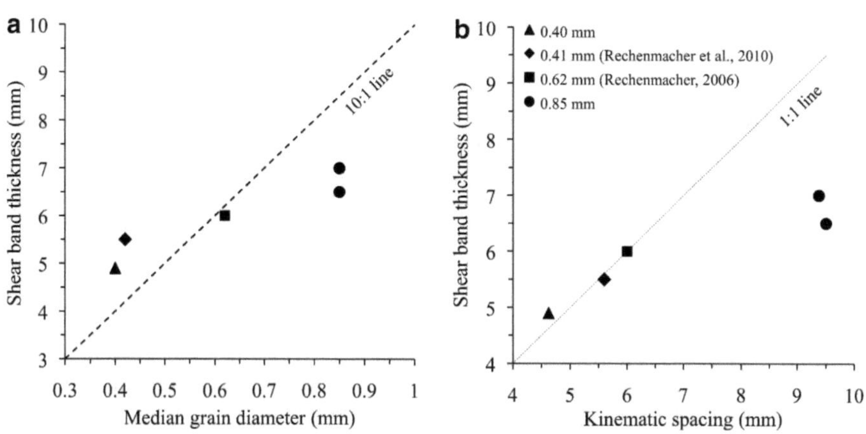

Fig. 2 (**a**) shear band thickness versus median grain size, and (**b**) shear band thickness versus spacing of kinematic fluctuations (or vortices)

From Fig. 1, it qualitatively seems that shear band thickness should relate to vortex size, and thus vortex spacing. Figure 2b compares the average spacing between vortices (or the spacing at which the kinematic variations cycle) for the same five tests. The kinematic spacing also increases with shear band thickness, and while the smaller grain sizes follow a 1:1 proportion between spacing and thickness, the larger grained sands do not: for the 0.85-mm sands, vortex spacing is about 1.3 times the band thickness. Part of this discrepancy may be due to grain shape: the MC sand was slightly more sub-angular than the other sands tested; and, thus jamming at the conflux between adjacent vortices as is seen in Fig. 1 may be particularly acute for more angular sands, causing increased vortex spacing. While more data are needed to substantiate these observations, what is clear is that neither shear band thickness nor vortex spacing are strict, one-to-one functions or a constant proportion of median grain size.

5 Conclusions

Meso-scale kinematics and residual displacements (actual displacements subtracted from mean-field displacements) have been analyzed within shear bands in dense sand specimens undergoing plane strain compression. Distinct, co-rotating, vortices are seen to spatially coordinate with previously-observed periodically varying peaks in kinematic rotation. Microbands have not been observed between adjacent vortices, likely because of significant grain interlocking and jamming, highlighting a key difference in behavior between real and idealized materials. A preliminary assessment of the spatial periodicity of such patterns has been performed to glimpse the nature of an underlying length scale for granular material deformation. The preliminary analysis suggests that while shear band thickness and spacing between vortices increase as median grain size increases, neither spacing nor thickness exhibit a consistent relationship to grain diameter for the grain sizes tested here.

Acknowledgements The authors would like to thank the National Science Foundation (grant CMMI-0748284), USC and the USC WiSE program for their financial support of this research.

References

S. Abedi, A.L. Rechenmacher, I. Faoro, Vortex structures inside shear bands in sands, in *Proceedings Geo-Frontiers 2011*, Dallas, TX, 13–16 Mar 2011

F. Alonso-Marroquin, I. Vardoulakis, H.J. Herrmann, D. Weatherley, P. Mora, Effect of rolling on dissipation in fault gouges. Phys. Rev. E 74, 031306 (2006)

H. Arslan, S. Sture, Evaluation of a physical length scale for granular materials. Comput. Mater. Sci. 42(3), 525–530 (2008)

O. Chupin, A. Rechenmacher, S. Abedi, Finite strain analysis of non-uniform deformations inside shear bands in sands. Int. J. Numer. Anal. Methods Geomech. (2011) [in press]

M.R. Kuhn, Structures deformation in granular materials. Mech. Mater. 31(6), 407–429 (1999)
H.B. Muhlhaus, I. Vardoulakis, The thickness of shear bands in granular materials. Géotechnique 37(3), 271–283 (1987)
A.L. Rechenmacher, Grain-scale processes governing shear band initiation and evolution in sands. J. Mech. Phys. Solids 54, 22–45 (2006)
A.L. Rechenmacher, R.J. Finno, Digital image correlation to evaluate shear banding in dilative sands. Geotech. Test. J. 27(1), 13–22 (2004)
A. Rechenmacher, S. Abedi, O. Chupin, Evolution of force chains in shear bands in sand. Geotechnique 60(5), 343–351 (2010)
A. Tordesillas, M. Muthuswamy, S.D.C. Walsh, Mesoscale measures of nonaffine deformation in dense granular assemblies. J. Eng. Mech. 134(12), 1095–1113 (2008)
B. Utter, R.P. Behringer, Experimental measures of affine and nonaffine deformation in granular shear. Phys. Rev. Lett. 100(20), 208302–1–4 (2008)

Discovering Community Structures and Dynamical Networks from Grain-Scale Kinematics of Shear Bands in Sand

Antoinette Tordesillas, David M. Walker, Amy L. Rechenmacher, and Sara Abedi

Abstract The quest to understand the connections between the triumvirate of structure, dynamics and function continues to drive the forefront of research in Complex Systems. Crucial to these explorations is the development of graph-theoretic techniques that: (i) can detect communities and associated boundaries in the underlying network or graph, which represents the interactions of constituent units, and (ii) quantify shortest paths and related network measures within this graph. We report on a new study using data from high resolution digital image correlation (DIC) measurements of grain-scale kinematics in sand under shear. Preliminary results show that the nodes of the network in the shear band region exhibit high *closeness centrality* – a network measure of how efficient a given node is in spreading information to all the other nodes in the graph. It is thus reasonable to expect that the most efficient routes for spread of kinematical information within this network are those from nodes that correspond to the grid points that lie along the shear band. We believe these studies will ultimately lead to an improved understanding of self-organization, the nature of energy flow and dynamics in the critical state regime in the presence of persistent shear bands.

Keywords Digital image correlation • Localization band • Grain-scale kinematics • Networks • Closeness centrality

A. Tordesillas (✉) · D.M. Walker
Department of Mathematics & Statistics, University of Melbourne, Parkville, VIC 3010, Australia
e-mail: atordesi@ms.unimelb.edu.au; dmwalker@unimelb.edu.au

A.L. Rechenmacher · S. Abedi
Department of Civil & Environmental Engineering, University of Southern California, Los Angeles, CA 90089-2531, USA
e-mail: arechenm@usc.edu; abedimas@usc.edu

S. Bonelli et al. (eds.), *Advances in Bifurcation and Degradation in Geomaterials*, Springer Series in Geomechanics and Geoengineering 11,
DOI 10.1007/978-94-007-1421-2_9,

1 Introduction

Data on particle kinematics from discrete element (DEM) simulations and photoelastic disk experiments of idealized granular assemblies (Tordesillas 2007; Tordesillas et al. 2009; Zhang et al. 2010) along with high resolution experiments on sand (Rechenmacher 2006) have uncovered considerable detail on structural evolution inside shear bands. While these have shed light on possible kinematical routes to and during shear banding – these findings have also raised more questions than answers. A key aspect that is common to all the data on shear bands is the emergence of coherent mesoscopic structures, i.e. sequence of vortex-like patterns interlaced with zones of high shear intermittently emerge along the shear band. The kinematics of these structures have been characterized for idealized assemblies, using data from both DEM analysis and experiments on photoelastic disks, particularly from the standpoint of micropolar nonaffine deformation (Tordesillas et al. 2009, 2008, 2010a). The pattern uncovered is consistent with earlier observations from combined simulations and experiments undertaken by Oda and Kazama (1998). In particular, the continual formation of new and collapse of old load-bearing force chains do indeed govern shear band dynamics and that the collapse by buckling of force chains leads to the development of relatively large voids in the band. More recently, two strands of developments have emerged from our research groups, and we review these briefly below to render transparent how these new developments can be usefully fused to deliver yet another layer of new information on the rheology and failure mechanism of sand under shear.

In the first of these developments, from the theoretical front, the processes leading up to and during strain localization are reexamined through the prism of Complex Systems Theory. The shear band is envisioned to be a process of self-organization and quantitatively characterized using an amalgam of techniques from Complex Networks and Dynamical Systems (Walker and Tordesillas 2010; Tordesillas et al. 2010b,c, 2011). A core element of these studies is the representation of the quasistatically deforming granular medium as a complex network (Newman 2003; Costa et al. 2007). The deforming medium is analysed as a sequence of equilibrium states. In each state, the contact network is mapped to an abstract graph wherein the nodes or vertices represent the particles, while links or edges represent the physical contacts (Walker and Tordesillas 2010; Tordesillas et al. 2010c). This abstraction affords not only the quantification of many aspects of connectivity between the particles, but also their influence on bulk behaviour, as the assembly deforms. Both an unweighted and weighted graph can be constructed. In the former, the connections (i.e. topology of contacts) are all that matters in the analysis; in the latter, the connections carry an additional information, for example, the contact force magnitude. The evolution of the sample in response to the applied load may be subsequently quantified by tracking the changes in the properties of the complex network from one equilibrium state to the next. This strategy has proved to be very effective at extracting new information, in particular, on the possible building blocks for self-organization and the coevolutionary nature of functional structures of force chains and their associated minimal cycle membership (Tordesillas et al. 2010b,c, 2011).

The second of these developments is from the experimental front. Rechenmacher (2006) and Rechenmacher and coworkers (2010) have developed a means to non-destructively and continuously track grain-scale displacements in real sands using the Digital Image Correlation (DIC) technique. Dense, dilative sand specimens are subjected to plane strain compression. Specimen-apparatus interfaces are systematically configured to allow shear bands to grow free of boundary interference. Persistent shear bands form at peak stress, and DIC analysis of displacements in and around the shear band are conducted throughout post-peak softening and into the global critical state (a state of continual shearing at constant stress and volume). The DIC data have helped reveal the presence of a systematic kinematical pattern along the shear band at the softening-critical state transition, which is thought to be indicative of a coordinated, multi-force chain buckling event (Rechenmacher 2006; Rechenmacher et al. 2010). The findings shown here were derived from data obtained by Rechenmacher and co-workers, similar to that shown by (2010). The data represent the beginning of the macroscopic critical state, and just follows the coordinated force chain collapse event identified in Rechenmacher et al. (2010).

The paper is arranged as follows. We briefly discuss the methods used for constructing a network based on data from experiment in Sect. 2, followed by the detection of communities in this network in Sect. 3. We next present the results on closeness centrality of nodes in the various communities in Sect. 4, before concluding in Sect. 5.

2 Constructing Dynamical Networks Without Seeing the Particles

The particular experiments of Rechenmacher (2006) and Rechenmacher and co-workers (2010) described above provides an evolving displacement vector field on a fixed grid. No information on contacts is provided. In contrast, the graph-theoretic approaches of Walker and Tordesillas (2010) and Tordesillas et al. (2010b,c, 2011) rely on explicit knowledge of the evolving connections between particles. The challenge therefore is to construct a relevant network that still retains information on the important dynamics but without need for information on particle contacts.

One such network representation consists of assigning to each network node a distinct observation grid point. Nodes are connected by a link if the observed properties (displacement vectors) are similar, i.e., the norm of the difference between their displacement vectors is small. In this way grid points exhibiting morphologically similar behaviour are closely connected within the network. This method of network construction is inspired by dynamical systems methods of obtaining a complex network from time series based on closeness in a reconstructed phase space (Xu et al. 2008). An issue is how many similar nodes to connect, i.e., what threshold for connecting nodes should be used? Here, we follow Xu et al. (2008) and connect a node to its k closest nodes. We set the threshold to be the minimum k where the resulting network forms one giant component (a single connected graph). At each observed strain stage this resulted in a k ranging from 4 to 8. We then examine the network properties of the constructed network, in particular, its community structure and a centrality index known as closeness centrality to reveal relevant structure within the deforming material.

3 Detection of Communities

Many real world networks are highly heterogeneous. Some nodes have more connections than others. This is especially true in networks underlying deforming dense granular materials which are highly anisotropic and particularly prone to self-organized pattern formation. In such networks, groups of nodes, i.e. so-called *communities*, can be found within which there are high densities of connections between constituent nodes, while far fewer connections exist between members of distinct communities. Algorithms that have been developed to identify communities include spectral graph partitioning and partitioning using a modularity metric (Newman 2006). Here we determine local communities by optimization, i.e. maximizing a local modularity measure (Clauset 2005). The local modularity gauges the extent to which the number of intracommunity links is greater than intercommunity links. The resulting communities consist of nodes densely connected to each other relative to the rest of the network. Thus, in the case of Rechenmacher's displacement data, grid points in the same community exhibit morphologically similar properties compared to others.

4 Closeness Centrality

Shortest paths play a crucial role in many phenomena notably in communication, transport and organizational systems (Newman 2003; Costa et al. 2007). To compute closeness centrality, one first computes the shortest path between a node i and all other nodes reachable from it. (We note here that by our choice of threshold k all nodes in the network are reachable by design.) The closeness centrality of a node i is then the reciprocal of the sum of geodesic distances to all other nodes j of the graph, i.e.,

$$C(i) = 1 \Big/ \sum_{j \neq i} d_{ij} \tag{1}$$

where d_{ij} is the path length between node i and node j. Nodes with high (relative) closeness centrality act as nexus or bridging points within the network. As such

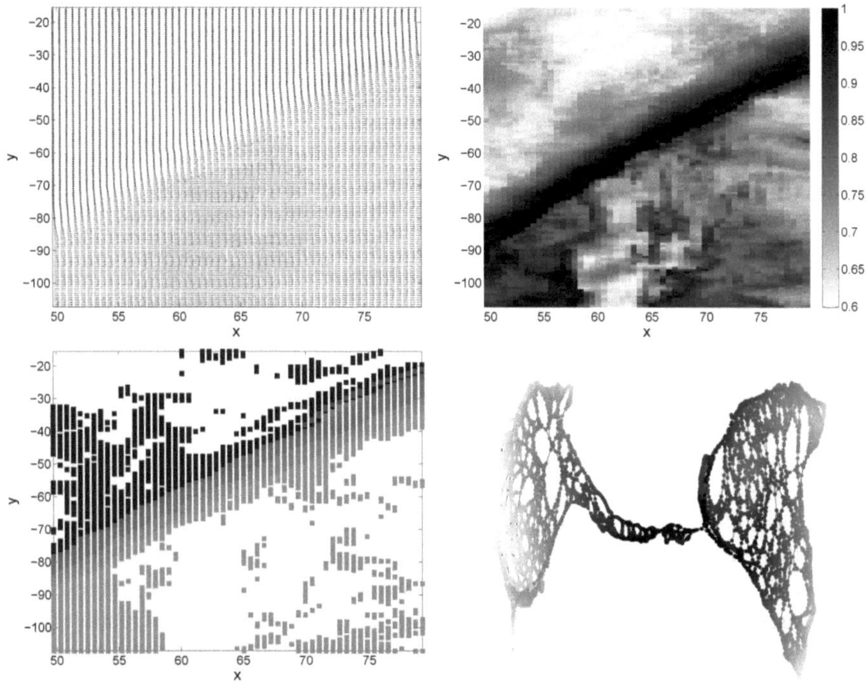

Fig. 1 Clockwise from *top left*: raw displacement field; grid points shaded by normalized closeness centrality; representation of the constructed network with nodes coloured by community membership; grid points shaded by community membership but showing only those communities encompassing the shear band to aid visualization

they typically lie in the centre of the network or on the boundaries of network communities. Figure 1 (top-left) shows the raw data from experiment wherein the shear band can be seen to form along a diagonal from the lower left to the top right boundary of the sample domain. The corresponding closeness centrality of each grid point is shown in Fig. 1 (top-right). The grid points with the highest closeness centrality mainly lie along the shear band indicating that the paths from their corresponding nodes to all others in the network (Fig. 1, bottom-right) are shorter compared to nodes that do not lie along the band. Note that only the connections in the constructed network matters; the shape of this network is not relevant to the present study although the shape give insight into the nodes of high closeness centrality as previously noted (central thin zone are the grid points where shear band is evident). There are around 10 communities that encompass the shear band out of a total 36 for the whole sample at this strain stage (Fig. 1, bottom-left, and centre of network in Fig. 1, bottom-right). Although difficult to visualize from the map and network without aid of color, these ten communities are heterogeneously spread along the band. Quantifying their spatial distributions may give insight into the kinematical events that govern shear banding in the critical state. Furthermore that these communities exhibit essentially the highest closeness centrality implies that the kinematics recorded at the grid points along the band provide a "central hub" for kinematical activity throughout the sample.

5 Conclusion

We have constructed a network from grain-scale measurements of kinematics in a deforming sand. This network is constructed by connecting nodes of similar displacement vector magnitudes and directions until a single graph is formed. We have shown preliminary results on community structures and network properties of this constructed network. We uncovered a rich community structure throughout the sample. We computed the closeness centrality of each node. The higher the closeness centrality, the shorter are the paths from this node to all others reachable from it. We found that the communities within the shear band are among those with the highest closeness centrality implying that information flow from the network nodes that correspond to the band grid points is relatively more efficient than from those nodes that correspond to grid points elsewhere in the sample. Current research is focussed on extracting more detailed information on shear band evolution from the constructed network and community structures over a wider range of strain states to determine robustness of these findings. The questions for the future are: to what extent is this distribution of shortest path length, summarized by closeness centrality, responsible for the high efficiency in the spread of "kinematical information" on the constructed network relevant to the evolution of kinematics in the critical state regime? As kinematics govern energy flow, what other properties of this constructed network as well as the various communities uncovered can give insights into the nature of self-organization and energy flow in the critical state regime in the presence of persistent shear bands? We believe the pursuance of such questions via such dynamical networks is an interesting new direction for granular mechanicians.

Acknowledgements AT and DMW are funded by: US Army Research Office (W911NF-07-1-0370), Australian Research Council (DP0986876), VPAC. AR and SA are funded by the National Science Foundation, Grant CMMI-0748284.

References

A. Clauset, Phys. Rev. E **72**, 026132 (2005)
L. da F. Costa, F.A. Rodrigues, G. Travieso, P.R.V. Boas, Adv. Phys. **56**, 167 (2007)
M.E.J. Newman, SIAM Rev. **45**, 167 (2003)
M.E.J. Newman, Phys. Rev. E **74**, 036104 (2006)
M. Oda, H. Kazama, Geotechnique **48**, 465 (1998)
A.L. Rechenmacher, J. Mech. Phys. Solids **54**, 22 (2006)
A. Rechenmacher, Geotechnique **60**(5), 343–351 (2010)

A. Tordesillas, Philos. Mag. **87**, 4987 (2007)
A. Tordesillas, M. Muthuswamy, S.D.C. Walsh, ASCE J. Eng. Mech. **134**, 1095 (2008)
A. Tordesillas, J. Zhang, R.P. Behringer, Geomech. Geoeng. **4**, 3 (2009)
A. Tordesillas, S.D.C. Walsh, M. Muthuswamy, Math. Mech. Solids **15**, 3 (2010a)
A. Tordesillas, P. O'Sullivan, D.M. Walker, Paramitha, C. R. Mecanique **338**, 556–569 (2010b)
A. Tordesillas, D.M. Walker, Q. Lin, Phys. Rev. E **81**, 011302 (2010c)
A. Tordesillas, Q. Lin, J. Zhang, R.P. Behringer, J. Shi, J. Mech. Phys. Solids **59**(2), 265–296 (2011)
D.M. Walker, A. Tordesillas, Int. J. Solids Struct. **47**, 624 (2010)
X. Xu, J. Zhang, M. Small, Proc. Natl. Acad. Sci. U.S.A. **105**, 19601 (2008)
J. Zhang, T.S. Majmudar, A. Tordesillas, R.P. Behringer, Granul. Matter **12**, 159 (2010)

Slip and Dislocation in Crystalline Solids as Precursors to Localized Deformation

Ronaldo I. Borja, Helia Rahmani, and Fushen Liu

Abstract We use crystal plasticity to describe the slip and dislocation in crystalline solids under mechanical loading. The constitutive formulation involves linearly dependent slip systems from which we extract a group of linearly independent slip systems using the ultimate algorithm advocated by Borja and Wren (Int J Numer Methods Eng 36:3815–3840, 1993). We implement the ultimate algorithm into a 3D nonlinear finite element code with infinitesimal deformation. We use the code to compare the deformation fields generated by crystal plasticity formulation and the classic J_2 plasticity model.

Keywords Crystal plasticity • Micromechanics • Rocks • Ultimate algorithm

1 Introduction

It is generally recognized that crystal deformation is responsible for some brittle micro-failure processes in crystalline solids such as rocks, including the loss of cohesion along grain boundaries and across individual crystals, a process called cataclasis (Price and Cosgrove 1990). Crystal deformation is responsible for other mechanical processes such as creep and pressure solution, albeit the actual mechanism for the latter process is not so well understood. In brittle crystals, such as quartz crystals, dislocation generation and movement are only possible under atmospheric pressure at temperatures above 820°K, but plastic flow is possible at lower

R.I. Borja (✉) · H. Rahmani
Department of Civil and Environmental Engineering, Stanford University, Stanford, CA 94305, USA
e-mail: borja@stanford.edu; rahmani@stanford.edu

F. Liu
ExxonMobil Research and Engineering, 1545 Route 22 East, Annandale, NJ 08801, USA
e-mail: fushen.liu@exxonmobil.com

S. Bonelli et al. (eds.), *Advances in Bifurcation and Degradation in Geomaterials*, Springer Series in Geomechanics and Geoengineering 11,
DOI 10.1007/978-94-007-1421-2_10,

temperatures if superimposed hydrostatic pressure prevents fracture (McLaren and Retchford 1969). Normal pressures near the grain-to-grain contacts are expected to be very high to allow dislocation, but these pressures disappear outside the contact zone thereby enhancing brittle fracturing. Such complex interplay between crystal deformation and micro-fracturing can only be investigated by simultaneously capturing both processes using a robust mathematical model.

We utilize crystal plasticity theory and a robust numerical algorithm to accommodate plastic slip and dislocation in crystalline solids. Crystal plasticity deals with multiple active constraints at the grain level representing yield conditions on glide planes. Some of these constraints are linearly independent, whereas others are redundant. The constraints are represented by multiple yield surfaces intersecting in a non-smooth fashion (Simo et al. 1988). We use the ultimate algorithm advocated by Borja and Wren (1993) to identify the independent active constraints in crystals slipping on glide planes. We cast the algorithm in a three-dimensional finite element framework appropriate for solving well-posed boundary-value problem with infinitesimal deformation. As an example, we compare the deformation fields in a 3D solid generated by the crystal plasticity formulation and the much simpler (but very approximate) J_2 plasticity model.

2 Crystal Micromechanics

Within the context of infinitesimal deformation, we denote by $\dot{\boldsymbol{\epsilon}}_{\mathrm{c}}$ the homogeneous strain rate in a crystal, which is composed of elastic and plastic parts,

$$\dot{\boldsymbol{\epsilon}}_{\mathrm{c}} = \dot{\boldsymbol{\epsilon}}_{\mathrm{c}}^{\mathrm{e}} + \dot{\boldsymbol{\epsilon}}_{\mathrm{c}}^{\mathrm{p}}. \tag{1}$$

The plastic component $\dot{\boldsymbol{\epsilon}}_{\mathrm{c}}^{\mathrm{p}}$ arises from slips on crystallographic planes. We denote by $\boldsymbol{n}^{(\beta)}$ the unit normal to a crystallographic plane containing the β-slip system, and by $\boldsymbol{m}^{(\beta)}$ the corresponding direction of plastic slip. If $\dot{\gamma}^{(\beta)}$ is the rate of crystallographic glide strain, then a point on the slip plane with position vector $\boldsymbol{x}$ will move at a velocity

$$\boldsymbol{v}^{(\beta)} = \dot{\gamma}^{(\beta)}(\boldsymbol{x} \cdot \boldsymbol{n}^{(\beta)})\boldsymbol{m}^{(\beta)}. \tag{2}$$

The velocity gradient contributed by glide strain β can be evaluated from the expression

$$\nabla \boldsymbol{v}^{(\beta)} = \dot{\gamma}^{(\beta)}\boldsymbol{m}^{(\beta)} \otimes \boldsymbol{n}^{(\beta)}, \tag{3}$$

where $\boldsymbol{m}^{(\beta)} \otimes \boldsymbol{n}^{(\beta)}$ is the slip tensor. Summing over all active crystallographic slips results in the following expression for the plastic strain rate

$$\dot{\boldsymbol{\epsilon}}_{\mathrm{c}}^{\mathrm{p}} = \sum_{\beta\ \text{active}} \dot{\gamma}^{(\beta)}\boldsymbol{\alpha}^{(\beta)}, \tag{4}$$

where

$$\boldsymbol{\alpha}^{(\beta)} = \frac{1}{2}\left(\boldsymbol{n}^{(\beta)} \otimes \boldsymbol{m}^{(\beta)} + \boldsymbol{m}^{(\beta)} \otimes \boldsymbol{n}^{(\beta)}\right). \tag{5}$$

Note that $\mathrm{tr}(\boldsymbol{\alpha}^{(\beta)}) = \boldsymbol{n}^{(\beta)} \cdot \boldsymbol{m}^{(\beta)} = 0$, producing a volume-preserving deformation.

We denote the overall crystal stress by $\boldsymbol{\sigma}_{\mathrm{c}}$. The elastic rate constitutive equation for the crystal takes the form

$$\dot{\boldsymbol{\sigma}}_{\mathrm{c}} = \boldsymbol{c}^{\mathrm{e}} : \left(\dot{\boldsymbol{\epsilon}}_{\mathrm{c}}^{\mathrm{e}} - \sum_{\beta \text{ active}} \dot{\gamma}^{(\beta)} \boldsymbol{\alpha}^{(\beta)}\right), \tag{6}$$

where $\boldsymbol{c}^{\mathrm{e}}$ is the elasticity tensor. The problem lies in identifying the active slip systems.

For a crystal with $2N$ potentially active slip systems, which include both the 'forward' and 'reverse' slips, there can be at most N actual active slip systems since if a slip direction is active then its conjugate direction must be latent. For example, f.c.c. crystals possess eight $\{111\}$ planes in the unit cell and three $\langle 110 \rangle$ possible slip directions on each plane, and so $N = 12$ for this crystal structure. Let τ_Y represent the yield stress for each slip system. Without loss of generality, we shall assume that the yield stress is constant (perfect plasticity); see Taylor (1938) for hardening-plasticity crystals. The system is potentially active if, at a given overall crystal stress $\boldsymbol{\sigma}_{\mathrm{c}}$, at least one of the following conditions is satisfied:

$$f^{(\beta)} = \begin{cases} \boldsymbol{\sigma}_{\mathrm{c}} : \boldsymbol{\alpha}^{(\beta)} - \tau_Y = 0, & \beta = 1, 2, \ldots, N \\ -\boldsymbol{\sigma}_{\mathrm{c}} : \boldsymbol{\alpha}^{(\beta)} - \tau_Y = 0, & \beta = N+1, N+2, \ldots, 2N \end{cases}. \tag{7}$$

The yield conditions described above represent $2N$ hyperplanes bounding the elastic region

$$\boldsymbol{E} = \{(\boldsymbol{\sigma}_{\mathrm{c}}, \tau_Y) \in R^6 \times R^1 \mid f^{(\beta)} < 0,\ \beta = 1, 2, \ldots, 2N\}. \tag{8}$$

Loading in a corner could activate two or more yield constraints and trigger multiple components of plastic flow. We use the associative flow rule and write the plastic strain rate in Koiter's form (1960) as

$$\dot{\boldsymbol{\epsilon}}_{\mathrm{c}}^{\mathrm{p}} = \sum_{\beta=1}^{2N} \dot{\gamma}^{(\beta)} \frac{\partial f^{(\beta)}}{\partial \boldsymbol{\sigma}_{\mathrm{c}}} = \sum_{\beta=1}^{2N} \dot{\gamma}^{(\beta)} \boldsymbol{\alpha}^{(\beta)}, \tag{9}$$

where $\boldsymbol{\alpha}^{(\beta)} = -\boldsymbol{\alpha}^{(\beta-N)}$ for $N < \beta \leq 2N$. The plastic multipliers $\dot{\gamma}^{(\beta)}$ satisfy the classical Kuhn-Tucker conditions (Budiansky and Wu 1962)

$$\dot{\gamma}^{(\beta)} \geq 0, \qquad f^{(\beta)} \leq 0, \qquad \dot{\gamma}^{(\beta)} f^{(\beta)} = 0 \tag{10}$$

for all β.

3 Ultimate Algorithm and 3D Simulations

We use the 'ultimate algorithm' described by Borja and Wren (1993) to identify all active constraints and eliminate the redundant ones. The algorithm can accommodate multislip processes up to a maximum of five linearly independent slip systems. In a nutshell, the algorithm assumes that a given strain increment $\Delta\boldsymbol{\epsilon}_{\mathrm{c}}$ has already activated $m \leq 4$ linearly independent slip systems $\beta_1, \ldots, \beta_m$, and we want to identify the $(m+1)$st active system. We apply a ray of deformation $\kappa\Delta\boldsymbol{\epsilon}_{\mathrm{c}}$ on the crystal and calculate the associated stress from the equation

$$\boldsymbol{\sigma}_{\mathrm{c}}(t) = \boldsymbol{\sigma}_{\mathrm{c},n} + \boldsymbol{c}^{\mathrm{e}} : \left(\kappa\Delta\boldsymbol{\epsilon}_{\mathrm{c}} - \sum_{i=1}^{m} \Delta\gamma^{(\beta_i)}\psi^{(\beta_i)}\boldsymbol{\alpha}^{(\beta_i)}\right). \tag{11}$$

The slips are determined from imposing a total of m independent consistency conditions,

$$\psi^{(\beta_i)}\boldsymbol{\alpha}^{(\beta_i)} : \boldsymbol{\sigma}_{\mathrm{c}}(t) - \tau_Y = 0, \qquad i = 1, \ldots, m, \tag{12}$$

which gives

$$\Delta\gamma^{(\beta_i)} = \kappa \sum_{j=1}^{m} g_{ij}^{-1}\psi^{(\beta_j)}\boldsymbol{\alpha}^{(\beta_j)} : \boldsymbol{c}^{\mathrm{e}} : \Delta\boldsymbol{\epsilon}_{\mathrm{c}}, \tag{13}$$

where

$$g_{ij} = \psi^{(\beta_i)}\psi^{(\beta_j)}\boldsymbol{\alpha}^{(\beta_i)} : \boldsymbol{c}^{\mathrm{e}} : \boldsymbol{\alpha}^{(\beta_j)} \tag{14}$$

and $\det(g_{ij}) > 0$ from the assumption of linear independence of the active slip systems.

We have implemented the ultimate algorithm as a stress-point integrator in a quasi-static nonlinear finite element program. Nonlinear equation solving is done by Newton iteration, which requires evaluation of the consistent tangent operator reflecting the contributions from all active constraints detected in the current iteration. The example presented in this section consists of a cubical sample $1 \times 1 \times 1\,\mathrm{m}^3$ in volume clamped at both its top and bottom surfaces and stretched vertically. The volume is defined by 216 nodes and 600 four-node tetrahedral elements integrated with a one-point rule. The elastic parameters are Young's modulus $E = 15\,\mathrm{GPa}$ and Poisson's ratio $\nu = 0.37$; the resolved yield stress is $\tau_Y = 20\,\mathrm{MPa}$.

We assume an f.c.c. crystal structure that is uniform throughout the mesh, with orientation #1 given by Euler angles $\theta = 9°$ (rotation about the y-axis) and $\phi = 27°$ (rotation about the z-axis), see Borja and Wren (1993). Figure 1 depicts the resulting deformed mesh and horizontal displacement contours after applying a nominal vertical strain of 3%. Note that the above crystal orientation does not produce 'symmetry' with respect to the x and y axes. Consequently, the displacement contour in the x-direction does not mirror the one generated in the y-direction.

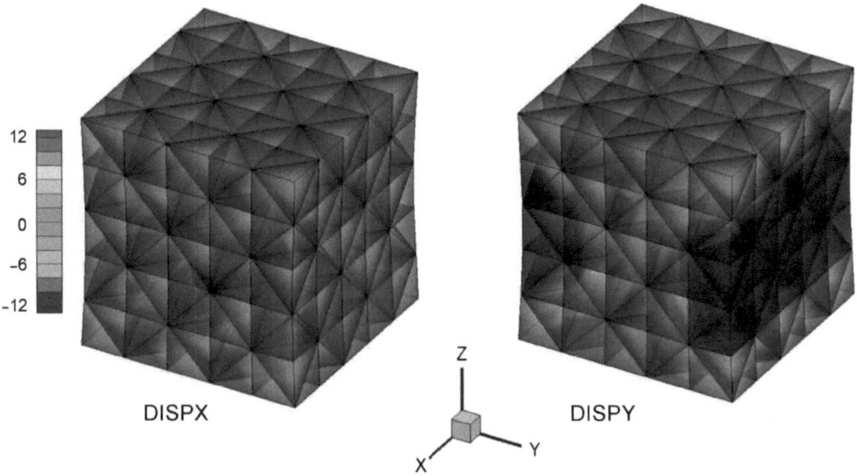

Fig. 1 Elastoplastic cube clamped on *top* and *bottom* surfaces and subjected to *vertical stretching*. The material is modeled as a perfectly plastic f.c.c. crystal with orientation #1. Note that the lateral contraction is greater in the y-direction than in the x-direction. *Color bar* is displacement in millimeters

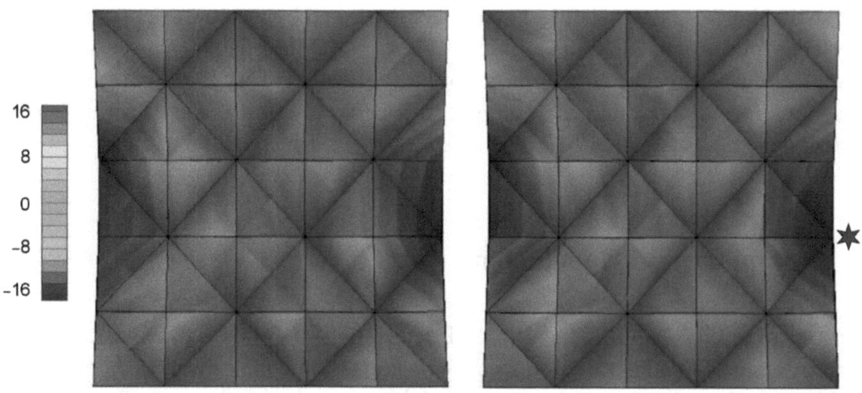

Fig. 2 Two dimensional view of f.c.c. crystal deformation on yz plane. *Left*: uniform crystal at orientation #1; *right*: crystal with 86.7% of the total volume at orientation #1 and 13.3% at orientation #2 (denoted by the *red star*). Note that perturbation in the crystal orientation alters the deformation field. *Color bar* is y-displacement in millimeters

In the second simulation we assume the same cubical volume but now rotate a small crystal volume (13% of the total volume) to orientation #2 defined by $\theta = 63°$ and $\phi = 45°$, while the rest of the crystal remains at orientation #1. Figure 2 compares the deformation profiles for the uniform and nonuniform crystal orientations. We see that the perturbation in the crystal orientation enhances a nonhomogeneous deformation field. By comparison, Fig. 3 shows the deformed

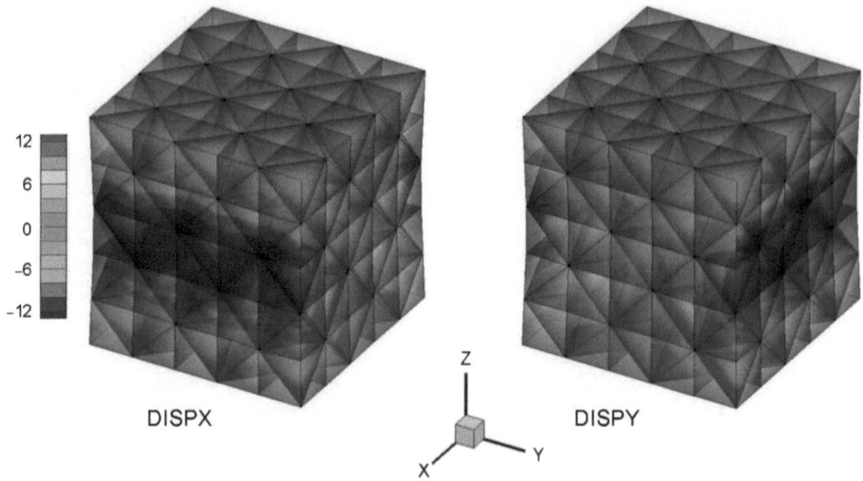

Fig. 3 Elastoplastic cube clamped on *top* and *bottom* surfaces and subjected to *vertical stretching*. The material is modeled with J_2 perfect plasticity resulting in symmetric displacement patterns in the x and y directions. *Color bar* is displacement in millimeters

mesh and horizontal displacement contours generated by the equivalent $J2$ elastic-perfectly plastic model (uniaxial yield stress $= 40\,\text{MPa}$). Note that the displacements in the x direction mirror those in the y direction. The $J2$ model smears the plastic deformation within the volume, so it cannot represent the effect of lattice orientation.

4 Summary and Conclusions

Crystal plasticity captures the slip and dislocation in solids with a well-defined microstructure. Spatial variation in crystal orientation could alter the deformation field within the solid even if the resolved yield stress remains the same. Perturbation in the crystal orientation could result in a nonhomogeneous deformation field and enhance the onset of localized deformation. This paper utilized a robust stress-point integration algorithm to track the evolution of slip systems in crystalline solids. Work is in progress to accommodate more complex slip systems.

Acknowledgments This work is supported by the US Department of Energy under Grant No. DE-FG02-03ER15454 to Stanford University.

References

R.I. Borja, J.R. Wren, Discrete micromechanics of elastoplastic crystals. Int. J. Numer. Methods Eng. **36**, 3815–3840 (1993)

B. Budiansky, T.T. Wu, Theoretical prediction of plastic strains of polycrystals, in: *Proceedings of the Fourth U.S. National Congress on Applied Mechanics*, ed. by R.M. Rosenberg (American Society of Mechanical Engineers, New York, 1962), pp. 1175–1185

W.T. Koiter, General theorems for elastic plastic solids. Prog. Solid Mech. **1**, 182–221 (1960)

A.C. McLaren, J.A. Retchford, Transmission electron microscope study of the dislocations in plastically deformed synthetic quartz. Physi. Status Solidi(a) **33**, 657–658 (1969)

N.J. Price, J.W. Cosgrove, *Analysis of Geological Structures* (Cambridge University Press, Cambridge, 1990)

J.C. Simo, J.G. Kennedy, S. Govindjee, Non-smooth plasticity and viscoplasticity. Loading/ unloading conditions and numerical algorithms. Int. J. Numer. Methods Eng. **26**, 2161–2185 (1988)

G.I. Taylor, Plastic strain in metals. J. Inst. Met. **62**, 307–324 (1938)

Part II
Geo-environmental Applications

A Chemo-Thermo-Mechanically Coupled Behavior During Gas Hydrate Dissociation and Its Numerical Analysis

S. Kimoto, F. Oka, Y. Miki, T. Fukuda, and H. Iwai

Abstract Gas hydrates, especially methane hydrates are viewed as a potential energy resource since a large amount of methane gas is trapped mainly within ocean sediments and regions of permafrost. In the present study, gas production process by heating-depressurizing method was simulated. The simulation was conducted for the model with inclined seabed ground with hydrate bearing layer in order to investigate the mechanical behavior during dissociation. The method has been developed based on the chemo-thermo-mechanically coupled analysis, taking into account of the phase changes from solids to fluids, that is, water and gas, the flow of fluids, heat transfer, and the ground deformation (Kimoto et al. 2010). As for the constitutive model for hydrate-bearing sediments, an extended elasto-viscoplastic model for unsaturated soils considering the effect of hydrate bonding is used.

Keywords Gas hydrate • Unsaturated soil • Numerical analysis

1 Introduction

Gas hydrates are ice-like materials composed of natural gas and water, i.e. gas is trapped inside cage-like crystal structures made up of water molecules. The gas hydrates are stable as a solid under high pressure and low temperature conditions. Recently, gas hydrates, especially methane hydrates are viewed as a potential energy resource since a large amount of methane gas is trapped mainly within ocean sediments and regions of permafrost. The dissociation process follows the phase changes from solids to fluids, i.e., from hydrates to water and gas, and the ground will be under unsaturated conditions. In addition, heat transfer becomes important

S. Kimoto (✉) · F. Oka · Y. Miki · T. Fukuda · H. Iwai
Department of Civil and Earth Resources Engineering, Kyoto University, Kyoto 615-8540, Japan
e-mail: kimoto.sayuri.6u@kyoto-u.ac.jp; oka.fusao.2s@kyoto-u.ac.jp

S. Bonelli et al. (eds.), *Advances in Bifurcation and Degradation in Geomaterials*, 85
Springer Series in Geomechanics and Geoengineering 11,
DOI 10.1007/978-94-007-1421-2_11,

during the dissociation process, since the phase equilibrium is controlled by temperature and pressure and the dissociation reaction is an endothermic reaction. The extreme change of environment during the gas production may induce the similar behavior. Some geologists have recently mentioned that some landslides in the seabed ground can be triggered by gas hydrates dissociation during and since the last deglaciation. Sultan et al. (2004) has indicated that dissociation of gas hydrates is a possible reason of the giant Storegga Slide on the Norwegian margin. Wong et al. (2003) showed that one of the possible trigger for mass wasting in the northwestern Sea of Okhtsk is gas hydrate instability. These phenomena had occurred in the geological period, and, the extreme change of environment during the gas production may induce the similar behavior. Hence, we need to study the simulation method for simulating the ground deformation associated with the dissociation of methane hydrate.

Recently, several numerical simulators have been developed to evaluate gas production values. Rutqvist et al. (2009) presented a numerical simulator for analyzing the geomechanical performance of hydrate-bearing permafrost. They combined numerical simulators of hydraulic behavior and mechanical behavior by the staggered technique. Although other numerical simulators have been developed, the solid phase was assumed to be rigid in most of them.

In the present study, we have analyzed gas hydrate dissociation in hydrate-bearing sediments by the chemo-thermo-mechanically coupled simulation method proposed by Kimoto et al. (2010). Simulations are conducted on a production process in the inclined seabed ground by heating-depressurizing method.

2 Governing Equations

In the proposed simulation method, the behavior of multiphase materials has been described within the framework of a macroscopic continuum approach through the use of the theory of porous media, that is, the ground is modeled to be superposition of four phases of hydrate, soil, water, and gas. The method has been developed based on the chemo-thermo-mechanically coupled analysis, taking into account of the phase changes from solids to fluids, that is, water and gas, the flow of fluids, heat transfer, and the ground deformation. As for the constitutive model for hydrate-bearing sediments, an extended elasto-viscoplastic model considering the effect of hydrate bonding is used. Also, the dependencies of hydrate saturation on the permeability are considered in the analysis. The governing equations are precisely described in Kimoto et al. (2010). Chemo-thermo-mechanically coupled reactions considered in the present study are summarized in Fig. 1. When the dissociation occurs, phase changes from solid to fluids, i.e., water and gas. This results in the change in volume fraction for each phase, and results in the change in stresses on each phase. Also the hydrate bonding between the soil particles is lost due to dissociation. These phenomena lead to deformation of soil and flow of water and gas. Conversely, the pressure change affects to the stability condition

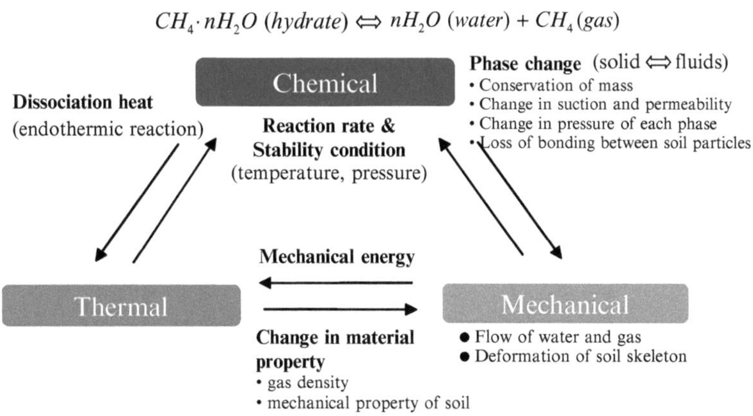

Fig. 1 Chemo-thermo-mechanical coupled reactions considered in the proposed method

and the reaction rate of the dissociation. As for mechanical- thermal reaction, the mechanical energy is considered in the energy balance, whereas, temperature change affects to the material properties, such as, gas density and viscosity of material. As for thermal-chemical reaction, the dissociation is an endothermic reaction, and the dissociation heat is considered in the energy balance. Conversely temperature controls the stability and the reaction rate of dissociation.

Weak forms of conservations of the mass for water and gas, conservation of momentum, conservation of energy are discretized in space and solved by the finite element method. As for the stress variable, the skeleton stress for the multiphase material is used. For the finite element method, an updated Lagrangian method with the objective Jaumann rate of Cauchy stress is used (Kimoto et al. 2010; Oka et al. 2006). The unknown variables are nodal velocity, pore water pressure, pore gas pressure, and temperature. The backward finite difference method is used for the time discretization. The stability condition for hydrate dissociation is evaluated in each element at each step, that is, the boundary of the dissociated area and undissociated area moves depending on pressure and temperature distribution.

3 Simulations

3.1 Initial and Boundary Conditions

We have simulated the production process by heating and depressurizing. Initial and final values of the pore pressure and the temperature are shown in Fig. 2 with the methane hydrate equilibrium curve. The finite element meshes and the boundary conditions for the simulation are shown in Fig. 3, in which plane strain condition is assumed. The seabed ground at around 200 m depth from the bottom of the

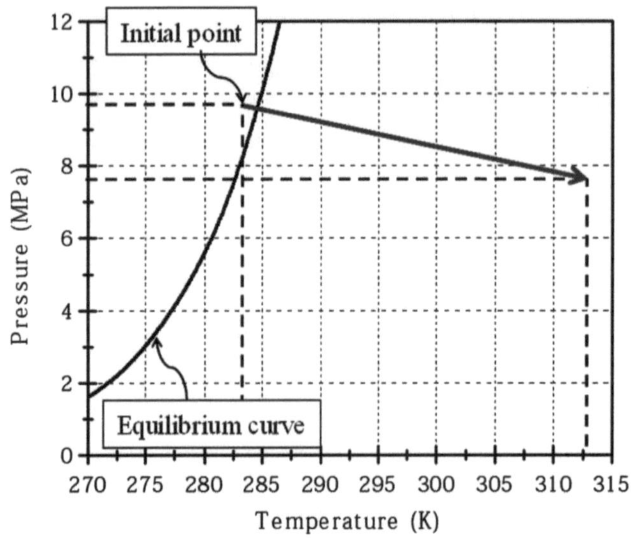

Fig. 2 Initial conditions with methane hydrate equilibrium curve

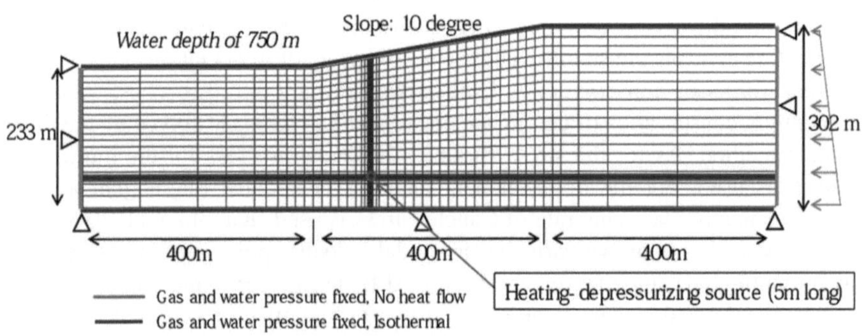

Fig. 3 Finite element meshes and the boundary conditions

sea at a water depth of 750 m is modeled. The ground is assumed to consist of silty clay with hydrate-bearing sediment of 10 m in thickness. We assume that a drilling rig exists under the seabed slope with gradient of 10° as shown in Fig. 3. Heating-depressurizing source exists in the center of the hydrate layer. The length of the source is 5 m. Initial mean effective stress in the center of the hydrate layer is about 1,400 kPa. Temperature increases from 283 to 313 K in 5 h, and at the same time the pressure decreases from 9.6 to 7.6 MPa at the source as illustrated in Fig. 2. The static pressures are given at the top, bottom and the right boundaries. Initial volume fraction of void, that is, the sum of water and hydrate phase, is 0.47, and initial hydrate saturation in the void is 0.51. Material parameters are mainly determined from the results of triaxial tests of samples obtained from the field

Table 1 Material parameters for the flow and the diffusion of soil, water, gas and hydrate

Permeability of water	k^W	$1.0 \times 10^{-5}(m/s)$
Permeability of gas	k^G	$1.0 \times 10^{-4}(m/s)$
Thermal conductivity of soil	λ^S	$1.9 \times 10^{-3}(kW/mK)$
Thermal conductivity of water	λ^W	$5.8 \times 10^{-4}(kW/mK)$
Thermal conductivity of gas	λ^G	$3.0 \times 10^{-5}(kW/mK)$
Thermal conductivity of hydrate	λ^H	$2.1 \times 10^{-3}(kW/mK)$
Specific heat of soil	c^s	800 (kJ/tK)
Specific heat of water	c^w	4,200 (kJ/tK)
Specific heat of gas	c^G	2,100 (kJ/tK)
Specific heat of hydrate	c^H	2,700 (kJ/tK)

Table 2 Material parameters for the soil skeleton

Initial shear elastic modulus	G_0	53800 (kPa)
Swelling index	κ	0.017
Compression index	λ	0.0169
Viscoplastic parameter	C_0	$1.0 \times 10^{-12}(1/s)$
Viscoplastic parameter	m'	23
Stress ratio af failure	M_m^*	1.08
Overconsolidation ratio	(OCR)	1
Structural parameter	β	0
Parameter for suction effect	${P^C}_i$	100 (kPa)
Parameter for suction effect	S_I	0.2
Parameter for suction effect	s_d	0.25
Parameter for MH effect	${S^H}_{ri}$	0.51
Parameter for MH effect	n_m	0.6
Parameter for MH effect	n_d	0.75

research conducted along the Nankai Trough.Material parameters for the flow and the diffusion are shown in Table 1 and material parameters for soil skeleton are shown in Table 2.

3.2 Simulation Results

Figure 4a, b show the distributions of pore water and pore gas pressures around the source after 360 h, respectively. The excess value from the initial value of the static pressure is shown for the gas pressure. When depressurizing has been finished, the pore water pressure decreases around the source. Produced gas exists in the dissociated elements, and the value is almost −2 MPa which equals to the value of the heating-depressurizing source. Figure 5a, b show stress paths in elements around the seabed surface and the depressurizing source, respectively. The mean

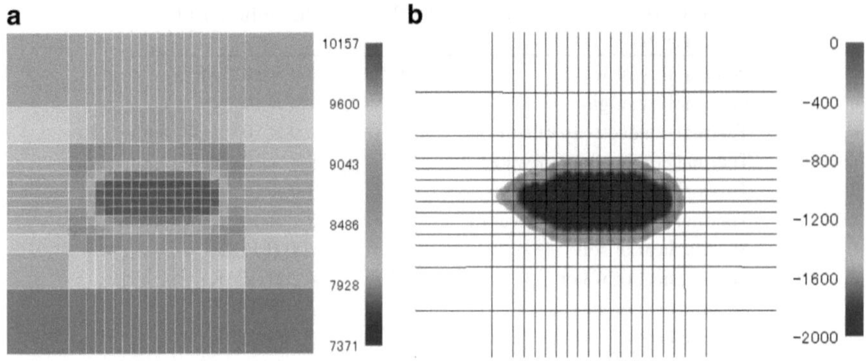

Fig. 4 Distributions of pore pressures around the source (after 360 h). (**a**) Pore water pressure (kPa), (**b**) Excess pore gas pressure (kPa)

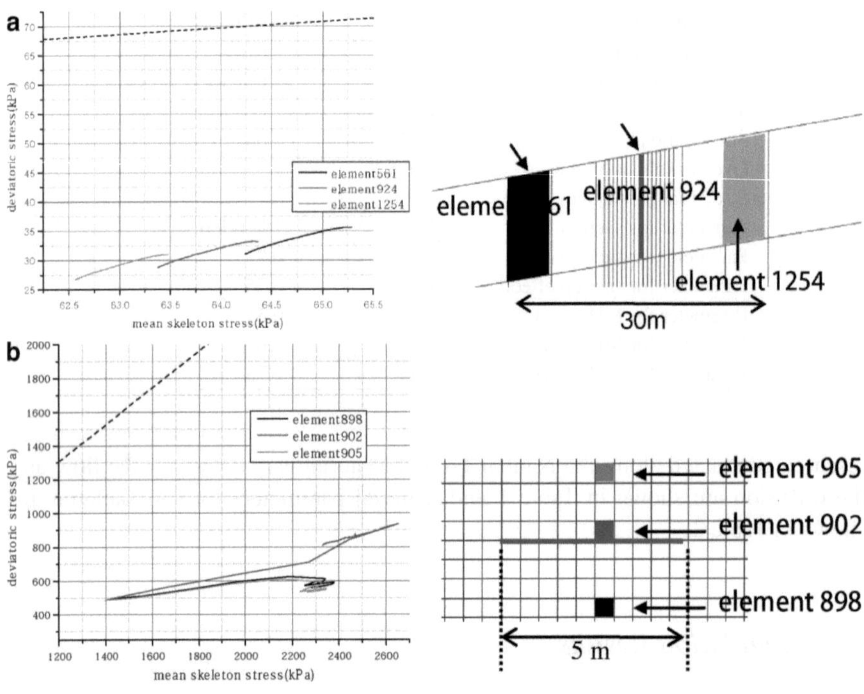

Fig. 5 (**a**) Stress paths from the elements around seabed surface, (**b**) Stress paths from the elements around the source

value of skeleton stress increases due to the depressurization both around surface and the depressurizing source, which results in volumetric compression around the depressurizing source. The final value of the settlement on the ground surface is about 0.017 m in this case.

4 Conclusions

We have conducted a two-dimensional dissociation analysis of seabed ground in order to predict the deformation behavior of hydrate-bearing sediments. The simulation has been conducted by using the chemo-thermo-mechanically coupled method that can take into account coupling process of the dissociation, deformation and the heat transfer. The results show that deformation occurs around seabed surface as well as the around the heating-depressurizing source due to dissociation.

References

S. Kimoto, F. Oka, T. Fushita, A chemo–thermo–mechanically coupled analysis of ground deformation induced by gas hydrate dissociation. Int. J. Mech. Sci. 52 365–376 (2010)

F. Oka, T. Kodaka, S. Kimoto, Y.-S. Kim, N. Yamasaki, An Elasto-viscoplastic model and multiphase coupled FE analysis for unsaturated soil, in Proceedings of the 4th International Conference on Unsaturated Soils, Carefree, AZ, 2–6 Apr 2006, Geotechnical Special Publication, No. 147, vol. 2 (ASCE, Reston, 2006), pp. 2039–2050.

J. Rutqvist, G.J. Moridis, T. Grover, T. Collett, Geomechanical response of permafrost-associated hydrate deposits to depressurization-induced gas production. J. Petrol. Sci. Eng. 67, 1–12 (2009)

N. Sultan, P. Cochonat, J.-P. Foucher, J. Mienert, Effect of gas hydrates melting on seafloor slope instability. Mar. Geol. 213, 379–401 (2004)

H.K. Wong, T. Ludmann, B.V. Baranov, B. Ya Karp, P. Konerding, G. Ion, Bottom current-controlled sedimentation and mass wasting in the northwestern Sea of Okhotsk. Mar. Geol. 201, 287–305 (2003)

Stability of Nail Reinforced Boreholes

Euripides Papamichos

Abstract Rock nailing of the borehole is considered as a means to reinforce the rock and increase its borehole failure strength. The technique is modeled using Cox's original shear-lag method. In the continuum sense, nail reinforcement is viewed as a body force that acts as a confinement. Borehole stability is analyzed using an analytical solution that couples the effects of the nails and the rock. Results for a reinforced with a given nail density and an un-reinforced borehole are presented. They show that the nail length and the nail-rock contact parameter play an important role in the stability. Rock dilation is also important since the action of the nails is mobilized from the difference in displacement between the rock and the nail. The results show the potential for a significant increase in borehole strength.

Keywords Rock nailing • Rock reinforcement • Borehole stability • Free surface • Contact force

1 Introduction

Nail reinforcement is used successfully in geotechnical engineering for the stabilization of slopes and embankments. The basic concept of soil nailing (Byrne et al. 1998) is to reinforce and strengthen the existing ground by installing closely spaced steel bars, called nails, into a slope or excavation as construction proceeds from the top down. This process creates a reinforced section that is itself stable and able to

E. Papamichos (✉)
Department of Civil Engineering, Aristotle University of Thessaloniki,
Thessaloniki GR-54124, Greece

SINTEF Petroleum Research, Trondheim N-7465, Norway
e-mail: epapamic@civil.auth.gr

S. Bonelli et al. (eds.), *Advances in Bifurcation and Degradation in Geomaterials*, Springer Series in Geomechanics and Geoengineering 11,
DOI 10.1007/978-94-007-1421-2_12,

retain the ground behind it. The reinforcement is passive and develops its reinforcing action through nail-ground interaction as the ground deforms during and following construction.

The technique of rock nailing in a smaller scale has been proposed by Papamichos and Vardoulakis (2011) for the reinforcement of a free surface. In the following the same technique is applied for improving the stability of wellbores for petroleum production. A vertical wellbore of radius r_i is considered under initial vertical and horizontal in situ effective stresses σ'_V and σ'_H, respectively. The pore pressure is constant and equal to the reservoir pressure p_{res}. The radial stress at the wellbore is reduced by $\Delta\sigma_{ri} = -\sigma'_H > 0$ (i.e. it is reduced to zero effective radial stress) to simulate the situation after the well has been drilled and the fluid pressure in the well is in balance with the pore pressure in the reservoir. In the vertical direction, plane strain is assumed since a long wellbore is considered. The wellbore is reinforced with radial nails at a given density and the objective is to calculate the effect of the reinforcement on the rock stresses and strains. The rock is assumed to be linearly elastic. The contact law between nails and rock is also assumed to be linear. In future analysis, borehole stability will be analyzed using finite elements enhanced with a Cosserat continuum for post-failure calculations. The results will then be compared with the unreinforced wellbore.

2 Formulation and Solution

A wellbore section of length H is reinforced by a number of radially placed rigid nails of length L_n in a regular pattern with nail density m, i.e. m is the number of nails per unit surface. Figure 1 shows a horizontal section of the considered wellbore. The total number of nails in the wellbore section is $m{\cdot}2\pi r_i \cdot H$. The section can be divided into $m{\cdot}2\pi r_i \cdot H$ equally spaced segments with a nail in each segment. For m_h segments per unit height and m_θ segments per unit circumference of the wellbore, then $m = m_h m_\theta$.

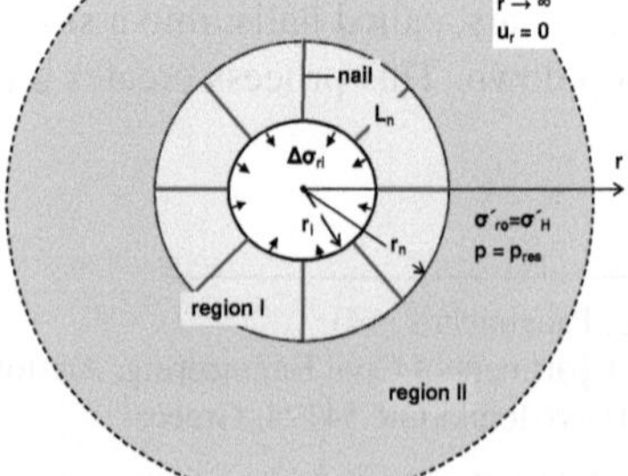

Fig. 1 Horizontal section of a vertical wellbore of radius r_i reinforced with radial nails of length L_n creating a reinforced zone of radius r_n

Inside the segment which is the volume of nail influence, an infinitesimal segment dr is considered with volume

$$dV = \frac{2r}{m_\theta r_i} dr \frac{1}{m_h} = \frac{2r}{m r_i} dr \tag{1}$$

Inside this annulus, the rock exerts into the nail a force $q\,(r)\,dr$ in the radial direction and the nail onto the rock an equal and opposite force. Force equilibrium on an infinitesimal length dr of the nail gives $-qdr = dN$. It is assumed that the force $-qdr$ is distributed uniformly inside the volume dV of the annulus and corresponds to a radial body force f_r (force per unit volume) that is introduced to model the action of the nails in a continuum sense. With Eq. 1 this gives

$$f_r = \frac{-q\,(r)\,dr}{dV} = \frac{m r_i}{2r} \frac{dN}{dr} = \frac{m r_i}{2r} \frac{dN}{dr} \tag{2}$$

The reinforcement creates a reinforced region I of radius $r_n = r_i + L_n$ and an unreinforced region II without nails. Stress equilibrium in the radial direction in the rock requires that

$$d\sigma_r/dr + (\sigma_r - \sigma_\theta)/r + f_r = 0 \tag{3}$$

where σ_r and σ_θ are the radial and tangential stresses, respectively, induced by the loading. This equation is solved for the stresses and strains in the rock with the boundary conditions

$$u_r\,(r \to \infty) = 0, \qquad \sigma_r\,(r = r_i) = \Delta\sigma_{ri} \tag{4}$$

where u_r is the radial deformation of the rock. Since f_r is applied only in the reinforced region I, Eq. 3 is solved separately in regions I and II under the requirement of continuity of the radial displacement u_r and radial stress σ_r at $r = r_n$

$$u_r^I = u_r^{II}, \quad \sigma_r^I = \sigma_r^{II} \qquad \text{at} \quad r = r_n \tag{5}$$

where superscripts *I* and *II* signify quantities in region I and II, respectively. For a linear, elastic rock, the stress equilibrium Eq. 3 with boundary conditions Eq. 4 can be solved analytically. Use of the elasticity relations under plane strain $\varepsilon_z = 0$ and $\varepsilon_r = u_{r,r}$, $\varepsilon_\theta = u_r/r$ for the radial and tangential strains, respectively, yields

$$u_{r,rr} + \frac{1}{r} u_{r,r} - \frac{1}{r^2} u_r = -\frac{1-2\nu}{2\,(1-\nu)\,G} f_r \tag{6}$$

where G and ν are the elastic shear modulus and Poisson's ratio of the rock, respectively. The boundary conditions Eq. 4 become

$$u_r\,(r \to \infty) = 0, \quad (1-\nu)\,u_{r,r}\big|_{r=r_i} + \frac{\nu}{r} u_r\Big|_{r=r_i} = \frac{1-2\nu}{2G} \Delta\sigma_{ri} \tag{7}$$

Substitution of f_r from Eq. 2 in Eq. 6 results in

$$u_{r,rr} + \frac{1}{r}u_{r,r} - \frac{1}{r^2}u_r = \begin{cases} -\dfrac{(1-2\nu)\,mr_i}{4(1-\nu)\,Gr}\dfrac{dN}{dr} & \text{for} \quad r_i \le r \le r_n \\ 0 & \text{for} \quad r \ge r_n \end{cases} \tag{8}$$

Following Cox (1952), it is assumed that

$$dN/dr = M\,(u_n - u_r) \qquad \text{for} \quad r_i \le r \le r_n \tag{9}$$

where N is the (internal) force in the nail (equal force in the opposite direction is applied from the nail to the rock) and M is a contact law parameter with dimension of stress. According to this law, the force per unit length transmitted from the nail to the rock and vice versa due to their stiffness difference is linearly related to their relative displacement. For rigid nails, there is no strain in the nails, i.e.

$$du_n/dr = 0 \qquad \text{for} \quad r_i \le r \le r_n \tag{10}$$

The problem to be solved comprises thus the differential equations (8)–(10). These equations are solved for the two unknown displacements u_n of the nail and u_r of the rock and the unknown force on the nail N, under the boundary conditions that $N = 0$ at the two ends of the nail and Eq. 7 for the deformation of the rock. Equation 10 means that $u_n = const$. Equation 8 can be written as

$$\begin{aligned} & r^2 u_{r,rr} + r u_{r,r} - (1 + Ar)\,u_r = -Au_n r && \text{for} \quad r_i \le r \le r_n \\ & r^2 u_{r,rr} + r u_{r,r} - u_r = 0 && \text{for} \quad r \ge r_n \end{aligned} \tag{11}$$

with $A = \frac{m(1-2\nu)M}{4(1-\nu)Gr_i}$ a positive constant with dimension inverse length. Equation 11 are solved first for the unknown displacement u_r as a function of u_n under the boundary conditions Eq. 7. Then with known u_r as a function of u_n, Eq. 9 is solved to obtain the force N and displacement u_n of the nail. The solution of Eq. 11 is

$$u_r = \begin{cases} c_1 I_2\left(2\sqrt{Ar}\right) + c_2 K_2\left(2\sqrt{Ar}\right) + u_n - \dfrac{u_n}{Ar} & \text{for} \quad r_i \le r \le r_n \\ c_3/r & \text{for} \quad r \ge r_n \end{cases} \tag{12}$$

where I_2 and K_2 are the second order modified Bessel functions of the first and second kind and c_1, c_2 and c_3 are integration constants. In the second of Eq. 12, the first boundary condition in Eq. 7 was used to eliminate the solution proportional to r. The second boundary condition in Eq. 7 together with the continuity conditions Eq. 5 are used to obtain the remaining three integration constants

$$c_1 = \frac{Q(r_i)}{P(r_i)}c_2 - \frac{R(r_i)}{P(r_i)} + (1-2\nu)\frac{\Delta\sigma_{ri}}{2GP(r_i)}, \quad c_3 = [Sc_2 + T]\,r_n$$

$$c_2 = \frac{\dfrac{1-2\nu}{r_n}T + \dfrac{1-2\nu}{2G}\Delta\sigma_{ri}\dfrac{P(r_n)}{P(r_i)} + R(r_n) - R(r_i)\dfrac{P(r_n)}{P(r_i)}}{Q(r_n) - \dfrac{1-2\nu}{r_n}S - Q(r_i)\dfrac{P(r_n)}{P(r_i)}} \tag{13}$$

Table 1 Parameters for a typical case of drilling of a vertical wellbore. Compressive stresses and strains are negative

Young's modulus [MPa]	$E = 2000$
Poisson's ratio [−]	$\nu = 0.3$
UCS [MPa]	$UCS = 5$
Friction angle [deg]	$\varphi = 35$
Contact parameter [MPa]	$M = 6000$
Borehole radius [m]	$r_i = 0.05$
Nail density [nails/m]	$m_h = 1000$
Nail length [m]	$L = 0.05$
In-situ eff. vertical stress [MPa]	$\sigma_V = -15$
In-situ eff. horizontal stress [MPa]	$\sigma_H = -10$

$$P(r) = (1-\nu)\sqrt{\frac{A}{r}} I_1\left(2\sqrt{Ar}\right) - \frac{1-2\nu}{r} I_2\left(2\sqrt{Ar}\right)$$

$$Q(r) = (1-\nu)\sqrt{\frac{A}{r}} K_1\left(2\sqrt{Ar}\right) + \frac{1-2\nu}{r} K_2\left(2\sqrt{Ar}\right)$$

$$R(r) = \frac{u_n}{r}\left(\frac{1-2\nu}{Ar} + \nu\right), \quad S = \frac{Q(r_i)}{P(r_i)} I_2\left(2\sqrt{Ar_n}\right) + K_2\left(2\sqrt{Ar_n}\right)$$

$$T = \left(\frac{1-2\nu}{P(r_i)}\frac{\Delta\sigma_{ri}}{2G} - \frac{R(r_i)}{P(r_i)}\right) I_2\left(2\sqrt{Ar_n}\right) + u_n - \frac{u_n}{Ar_n} \tag{14}$$

Equation 9 is now solved with the boundary condition $N = 0$ at $r = r_i$ and r_n to obtain the force N and displacement u_n of the nail as

$$\begin{aligned} N = {} & c_1\left[I_0\left(2\sqrt{Ar}\right) - I_0\left(2\sqrt{Ar_i}\right) - \sqrt{Ar}\, I_1\left(2\sqrt{Ar}\right) + \sqrt{Ar_i}\, I_1\left(2\sqrt{Ar_i}\right)\right] M/A \\ & + c_2\left[K_0\left(2\sqrt{Ar}\right) - K_0\left(2\sqrt{Ar_i}\right) + \sqrt{Ar}\, K_1\left(2\sqrt{Ar}\right) - \sqrt{Ar_i}\, K_1\left(2\sqrt{Ar_i}\right)\right] M/A \\ & + u_n M \ln(r/r_i)/A \end{aligned} \tag{15}$$

Finally, the boundary condition $N = 0$ at $r = r_n$ gives

$$\begin{aligned} & u_n \ln r_n/r_i + c_1\left[I_0\left(2\sqrt{Ar_n}\right) - I_0\left(2\sqrt{Ar_i}\right) - \sqrt{Ar_n}\, I_1\left(2\sqrt{Ar_n}\right) + \sqrt{Ar_i}\, I_1\left(2\sqrt{Ar_i}\right)\right] \\ & + c_2\left[K_0\left(2\sqrt{Ar_n}\right) - K_0\left(2\sqrt{Ar_i}\right) + \sqrt{Ar_n}\, K_1\left(2\sqrt{Ar_n}\right) - \sqrt{Ar_i}\, K_1\left(2\sqrt{Ar_i}\right)\right] = 0 \end{aligned} \tag{16}$$

The nail displacement u_n is then found as the value that satisfies Eq. 16. A typical field case of drilling of a vertical wellbore is considered with the problem parameters listed in Table 1. Figure 2 compares radial profiles for the (a) radial nail body force, (b) wellbore closure, and (c) the radial, tangential and axial stress profiles during

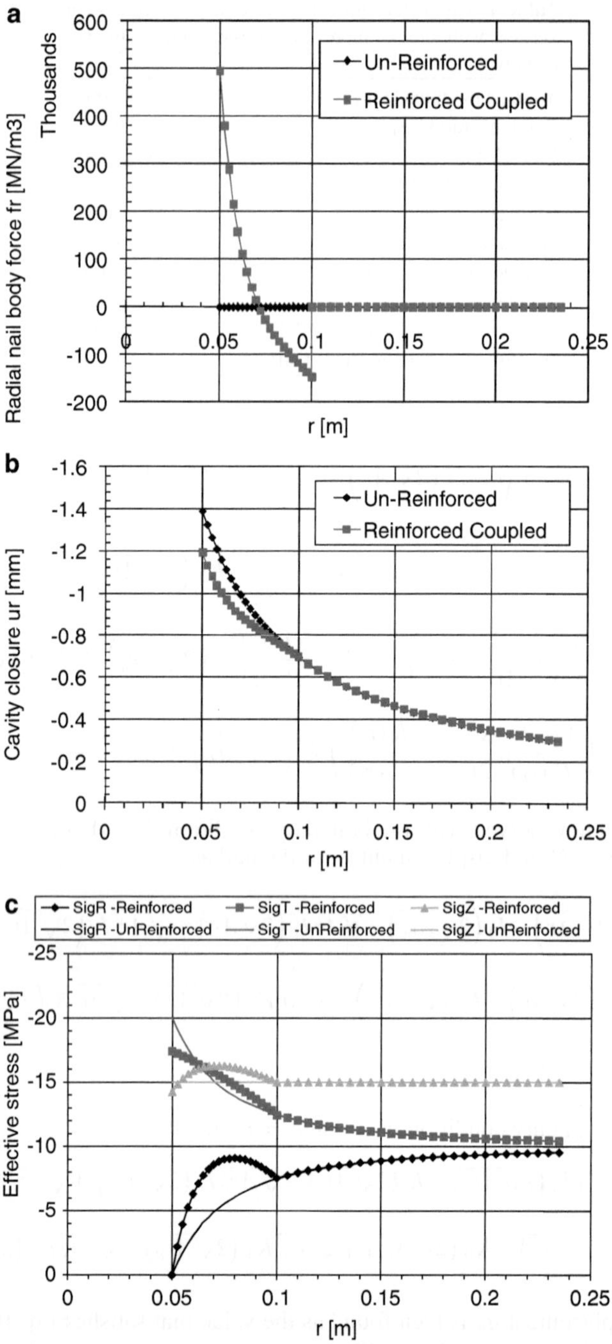

Fig. 2 (**a**) Radial nail body force, (**b**) Wellbore closure, and (**c**) Radial, tangential and axial stress profiles after drilling in reinforced and un-reinforced wellbores. Compression is negative

drilling of a reinforced and an un-reinforced wellbore. The reinforcement affects mainly the nailed region of the wellbore so it is important to have as long nails as practically possible. The results show the positive effect of reinforcement by both increasing the radial stress σ_r (acts as confinement) and by reducing the magnitude of the tangential stress σ_θ which causes failure. The nail-rock contact parameter play also an important role in the stability. Rock dilation is also important since the action of the nails is mobilized from the difference in displacement between the rock and the nail.

Acknowledgements The author wishes to thank Statoil for supporting this research and in particular JH Jøranson and JS Andrews for many discussions and their continuing support.

References

R.J. Byrne, D. Cotton, J. Porterfield, C. Wolschlag, G. Ueblacker, Manual for design and construction monitoring of soil nail walls. U.S. Department of Transportation, Federal Highway Administration, Report No. FHWA-SA-96–069R, Nov. 1996 (Revised October 1998) (1998)

H.L. Cox, The elasticity and strength of paper and other fibrous materials. Br. J. Appl. Phys. 3, 72–79 (1952)

E. Papamichos, I. Vardoulakis, Rock nail reinforcement of a free surface. Int. J. Numer. Anal. Methods Geomech. 35: n/a. doi:10.1002/nag.1003 (2011)

Stability Analysis of Creeping Faults: The Role of Chemical Decomposition of Minerals

Jean Sulem and Nicolas Brantut

Abstract The stability of creeping faults is studied under the effect of shear heating, pore fluid pressurization and mineral decomposition. Such reactions enhance the pore fluid pressurization because they release fluid, but they limit the temperature rise because they are endothermic. The stability of stationary slip is investigated by performing a linear perturbation analysis. It is shown that chemical reactions change a stable behaviour into an unstable one when the pore pressure effect is larger than the endothermic effect. It is shown that the opposite effect can also be observed when the dehydration reactions can trigger an arrest of the fault. Mineral decomposition can thus strongly modify the nucleation of seismic slip.

Keywords Fault mechanics • Shear heating • Dehydration of minerals • Chemo-mechanical couplings • Seismic slip

1 Introduction

Earthquake nucleation and slip instability occur because the frictional resistance to slip on the fault walls decreases, causing an acceleration of sliding. The presence of fluids greatly affects the frictional resistance of a fault. Pore fluid pressurization under shear heating is a mechanism of thermal softening. The principle of slip weakening by thermal pressurization is based on the fact that pore fluids trapped inside the fault zone are put under pressure by shear heating, thus inducing a reduction of the effective mean stress, and of the shearing resistance of the fault

J. Sulem (✉)
CERMES UR Navier, CNRS UMR 8205, Ecole des Ponts ParisTech, Marne-la-Vallée, France
e-mail: jean.sulem@enpc.fr

N. Brantut
Laboratoire de Géologie, CNRS UMR 8538, Ecole Normale Supérieure, Paris, France

S. Bonelli et al. (eds.), *Advances in Bifurcation and Degradation in Geomaterials*, Springer Series in Geomechanics and Geoengineering 11,
DOI 10.1007/978-94-007-1421-2_13,

(Lachenbruch 1980; Mase and Smith 1985; Rice 2006; Sulem et al. 2005, 2007). This mechanism has also been suggested for weakening in catastrophic landslides (Voight and Faust 1982; Vardoulakis 2002; Veveakis et al. 2007).

Recent data on fault rocks mineralogy (Sulem et al. 2004; Solum et al. 2006) have revealed the presence of a significant amount of clays and hydrous phyllosilicates along major subsurface fault zones. These minerals are thermally unstable and can release their adsorbed and/or structural water. Consequently, a dehydration reaction can increase the pore pressure as well as the porosity of the rock. Part of the frictional heat is consumed due to the endothermic character of such reactions. This process has been shown to influence faults mechanical behaviour during seismic slip (Sulem and Famin 2009; Brantut et al. 2010). In these papers, it was shown that the co-seismic temperature rise in the slipping zone is limited while the mineral decomposition progresses. This can provide an explanation to the notorious absence of strong positive heat flow anomaly on active crustal faults such as San Andreas (Lachenbruch and Sass 1980), and to the relative scarcity of frictional melts in exhumed faults: a large part of the heat produced by friction is consumed by endothermic reactions.

2 Problem Statement and Governing Equations

Figure 1 shows the geometry of the problem and the far field loading conditions. The configuration is similar to the one of Garagash and Rudnicki (2003). A crustal block of thickness $2L$ contains a shear band of thickness h with $h << L$. Considering that the length scales in the direction parallel to the fault over which the thermo-poro-mechanical fields vary are much larger than in the direction normal to it, we analyse here a 1D problem.

The governing equations are obtained from the conservation and transport equations (Vardoulakis and Sulem 1995). The reader can refer to Sulem and Famin (2009) and Brantut et al. (2011) for the details of the derivations. These equations are summarized below.

Because we neglect inertial effects, the shear stress τ and the total normal stress σ_n are uniform throughout the layer. The material inside the shear band ($|y| \leq h$) undergoes inelastic loading whereas the material inside the crustal block ($h < |y| \leq L + h$) behaves elastically: $\tau = G(v_L - v_h)t/L$ where G is the elastic modulus of the crustal block. Inside the shear band, it is assumed that the frictional resistance is proportional to the mean effective stress inside the band: $\tau = f(\sigma_n - p(t))$ where f is the friction coefficient of the material inside the band. From the requirement of uniform shear stress, we obtain the slip velocity v_h at the shear band boundary

$$v_h = v_L + \frac{L}{G} f \frac{d}{dt} p(t) \tag{1}$$

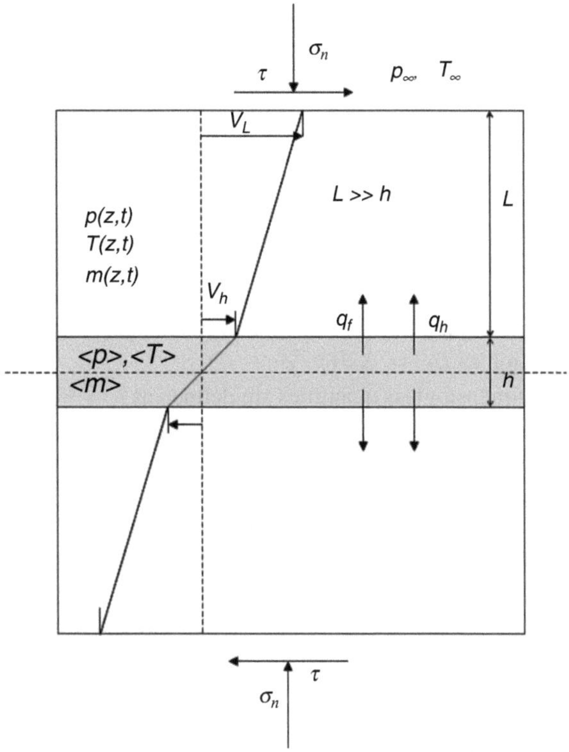

Fig. 1 Model of a deforming shear band with heat and fluid fluxes

Conservation of the fluid and solid masses leads to the pore pressure production/-diffusion equation

$$\frac{\partial p}{\partial t} = \alpha_{hy} \frac{\partial^2 p}{\partial z^2} + \Lambda \frac{\partial T}{\partial t} + \frac{1}{\rho_f \beta^*} \frac{\partial m_d}{\partial t} - \frac{1}{\beta^*} \frac{\partial n_d}{\partial t} \tag{2}$$

where ρ_f is the density of the saturating fluid, β^* is storage coefficient of the material, α_{hy} is the hydraulic diffusivity, Λ is the thermal pressurization coefficient, m_d is the mass per unit volume of the fluid released in the system by mineral decomposition. In Eq. 2 $\frac{\partial n_d}{\partial t}$ is the rate of porosity change due to the decomposition of the solid phase and is expressed as proportional to the reaction rate: $\frac{\partial n_d}{\partial t} = \xi \frac{\partial m_d}{\partial t}$ where ξ corresponds to the ratio of the volume of voids formed over the volume of fluid released by the reaction. The pore volume created during the reaction can be plastically compacted at high effective stress, which affects the evolution of n_d. Assuming a first order reaction mechanism and an Arrhenius law for the temperature dependency, the mass m_d of released water per unit volume is expressed as $\partial m_d/\partial t = (m_0 - m_d) A \exp(-E_a/RT)$ where A is a constant (pre-exponential term of the Arrhenius law), E_a is the activation energy of the reaction, R is the gas constant (8.314 J K^{-1} mol^{-1}).

Conservation of energy is expressed as

$$\rho C \frac{\partial T}{\partial t} = \alpha_{th} \frac{\partial^2 T}{\partial z^2} + E_F - E_C \tag{3}$$

where ρC is the specific heat per unit volume of the fault material and α_{th} is the thermal diffusivity. It is assumed that all the plastic work is converted into heat and thus $E_F = \tau \frac{\partial v}{\partial z}$. On the other hand, the rate of heat used in the chemical reaction can be expressed from the enthalpy change of the reaction and the rate of reacted fraction as $E_C = \Delta_r H_T^0 \frac{\partial m_d}{\partial t}$.

Dimensionless quantities are defined as $\tilde{t} = \frac{\alpha_{th}}{L} t$; $\pi = \frac{p}{\sigma_n}$; $\theta = \Lambda \frac{T - T_c}{\sigma_n}$; $u = \frac{v_h}{v_L}$; $\mu = \frac{m_d}{m_0}$ where T_c is a threshold temperature below which no reaction occurs. As shown by Sulem and Famin (2009) and Brantut et al. (2010) the temperature is efficiently buffered around the equilibrium temperature T_{eq} of the reaction during a slip event. Thus, the temperature is thus not expected to increase far from T_{eq} when the reaction occurs. Therefore the Arrhenius law can be linearized around T_{eq}:

$$(\partial_t \mu = 0 \text{ if } \theta \leq 0) \text{ and } \left(\partial_t \mu = c^* \theta \text{ if } \theta > 0\right)$$

Considering the extreme thinness of the shear band as observed in faults zones (from few hundreds of microns to few centimeters) (Rice 2006), it is relevant to consider mean values for the pore pressure and temperature fields inside the slip zone. Let us denote by $\bar{\pi}$ and $\bar{\theta}$ the averaged dimensionless mean pore pressure and temperature inside the shear band. From Eqs. 1–3 and by approximating the pore pressure and temperature gradients by finite difference we obtain:

$$\begin{aligned} u &= 1 + \mathcal{A} \partial_t \bar{\pi} \\ \partial_t \bar{\pi} &= \frac{\sqrt{\varepsilon}}{\delta} (\pi_\infty - \bar{\pi}) + \partial_t \bar{\theta} + \mathcal{P} c^* \bar{\theta} \\ \partial_t \bar{\theta} &= \frac{1}{\delta} \left(\theta_\infty - \bar{\theta} + \mathcal{T} c^* \bar{\theta} + \mathcal{B}(1 - \bar{\pi}) u\right) \end{aligned} \tag{4}$$

where $\mathcal{A}$, $\mathcal{B}$, $\mathcal{P}$, $\mathcal{T}$, ε, δ are dimensionless parameters derived from Eqs. 1–3. The stationary solution of the above system can be derived analytically;

$$\begin{aligned} u_0 &= 1;\ \pi_0 = \frac{\pi_\infty \sqrt{\varepsilon} (1 - \mathcal{T} c^* \delta) + \theta_\infty \mathcal{P} c^* \delta + \mathcal{P} c^* \mathcal{B} \delta}{\sqrt{\varepsilon} (1 - \mathcal{T} c^* \delta) + \mathcal{P} c^* \mathcal{B} \delta}; \\ \theta_0 &= \frac{\sqrt{\varepsilon} (\theta_\infty + \mathcal{B} (1 - \pi_\infty))}{\sqrt{\varepsilon} (1 - \mathcal{T} c^* \delta) + \mathcal{P} c^* \mathcal{B} \delta} \end{aligned} \tag{5}$$

3 Stability Analysis

The stability of the system above linear system of equations is controlled by the sign of the eigenvalues. If at least one eigenvalue has a positive real part, the system is unstable. One can show that the stability condition is

$$(\delta - \mathcal{AB}(1-\bar{\pi})) > 0 \text{ and } -\left(1+\mathcal{B}+\sqrt{\varepsilon}\right)+\mathcal{AB}(1-\bar{\pi})\mathcal{P}c^{*}+\delta\mathcal{T}c^{*} < 0 \quad (6)$$

If we neglect the contributions of the chemical kinetic parameters $\mathcal{P}$ and $\mathcal{T}$, we retrieve the solution derived by Garagash and Rudnicki (2003). We observe that the sign of parameter $(\delta - \mathcal{AB}(1-\bar{\pi}))$ fully determines the stability of the system. For $(1 - \mathcal{AB}(1-\bar{\pi})/\delta) < 0$ (Fig. 2 in black), the stationary solution is unstable and is a saddle point: the pore pressure and temperature (and thus the slip velocity) will evolve exponentially along the unstable eigendirection. The position of the perturbation with respect to the separatrix controls on the sense of the instability: a negative perturbation in pressure or temperature stops the fault motion whereas a positive perturbation exponentially increases the temperature and pore pressure. Interestingly, the slope of the unstable eigendirection is rather high, which means that the instability is mainly a thermal one for the set of parameters used. Conversely, if $(1 - \mathcal{AB}(1-\bar{\pi})/\delta) > 0$ (Fig. 2 in gray) the stationary solution is stable and all the small perturbations will tend to be dissipated. However, this case might not be realistic as it would need an extremely low $\mathcal{A}$ value, i.e., an anomalously high stiffness for the crustal block.

The effect of the reaction is investigated by considering nonzero values of parameters $\mathcal{P}$ and $\mathcal{T}$. On Fig. 2 (right), the stability diagram is plotted for the particular case $\mathcal{P} = -\mathcal{T}$. In that case the stability condition is unchanged as compared to the case without reaction, as seen in Eq. 6 but the relative evolution of pore pressure and temperature is significantly different from the case without reaction. The eigendirections tend to be perpendicular to each other, the unstable direction being almost parallel to the pore pressure axis. The instability evolves at

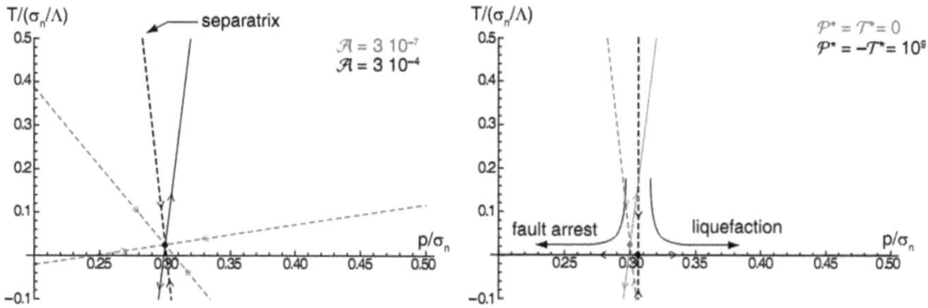

Fig. 2 (*left*) Eigenvectors and growth rates in a stable case (in *gray*, for $\mathcal{A} = 3\ 10^{-7}$) and unstable case (in black, for $\mathcal{A} = 3\ 10^{-4}$); (*right*) Linear stability diagram for the no reaction case (*gray*), and when dehydration occurs (*black*) ($\mathcal{A} = 3\ 10^{-4}$)

almost constant temperature, very close the dehydration temperature threshold T_c. Depending on the sign of the perturbation, the shear zone can evolve towards a quasi isothermal liquefaction (for a positive perturbation in pore pressure) or towards the arrest of the fault (for a negative perturbation in pore pressure). Again, this thermal buffering effect is shown to be a fundamental characteristic of a dehydrating system. If the pore fluid pressurization high enough as compared to the endothermic effect ($\mathcal{P} >> -\mathcal{T}$), the reaction will destabilize the system (Brantut et al. 2011).

Considering that the stationary solution is different whether the reaction is activated or not, an other interesting case arises when the perturbation in pore pressure is positive with respect to the stationary solution without reaction but negative with respect to the stationary solution with reaction. The slip is accelerated in absence of chemical reactions but is arrested when the reaction starts. Mineral decomposition can thus strongly modify the nucleation of seismic slip.

References

N. Brantut, A. Schubnel, J. Corvisier, J. Sarout, Thermo-chemical pressurization of faults during coseismic slip. J. Geophys. Res. **115**, B05314 (2010). doi:10.1029/2009JB006533

N. Brantut, J. Sulem, A. Schubnel, Effect of dehydration reactions on earthquake nucleation: stable sliding, slow transcients and unstable slip. J. Geophys. Res. **116**, B05304 (2011). doi:10.1029/2010JB007876

D.I. Garagash, J.W. Rudnicki, Shear heating of a fluid-saturated slip-weakening dilatant fault zone 1. limiting regimes. J. Geophys. Res. **108**(B2), 2121 (2003)

A.H. Lachenbruch, Frictional heating, fluid pressure and the resistance to fault motion. J. Geophys. Res. **85**, 6097–6112 (1980)

A.H. Lachenbruch, J.H. Sass, Heat flow and energetics of the San Andreas fault zone. J. Geophys. Res. **85**, 6185–6223 (1980)

J.R. Rice, Heating and weakening of faults during earthquake slip. J. Geophys. Res. **111**, B05311 (2006)

J.G. Solum, S.H. Hickman, D.A. Lockner, D.E. Moore, B.A. van der Pluijm, A.M. Schleicher, J.P. Evans, Mineralogical characterization of protolith and fault rocks from the Safod main hole. Geophys. Res. Lett. **33**, L21314 (2006)

J. Sulem, V. Famin, Thermal decomposition of carbonates in fault zones: slip-weakening and temperature-limiting effects. J. Geophys. Res. **114**, B03309 (2009)

J. Sulem, I. Vardoulakis, H. Ouffroukh, M. Boulon, J. Hans, Experimental characterization of the thermo-poro-mechanical properties of the Aegion fault gouge. C. R. Geo-Sci. **336**(4–5), 455–466 (2004)

J. Sulem, I. Vardoulakis, H. Ouffroukh, V. Perdikatsis, Thermo-poro-mechanical properties of the Aigion fault clayey gouge – application to the analysis of shear heating and fluid pressurization. Soils Found. **45**(2) 97–108 (2005)

J. Sulem, P. Lazar, I. Vardoulakis, Thermo-Poro-Mechanical Properties of Clayey Gouge and Application to Rapid Fault Shearing. Int. J. Num. Anal. Meth. Geomech. **31**(3), 523–540 (2007)

I. Vardoulakis, Dynamic thermo-poro-mechanical analysis of catastrophic landslides. Géotechnique **52**(3), 157–171 (2002)

I. Vardoulakis, J. Sulem, *Bifurcation Analysis in Geomechanics* (Blackie Academic and Professional, Glasgow, 1995)

E. Veveakis, I. Vardoulakis, G. Di Toro, Ther-moporomechanics of creeping landslides: The 1963 Vaiont slide, northern Italy. J. Geophys. Res. **112**, F03026 (2007)
B. Voight, C. Faust, Frictional heat and strength loss in some rapid landslides. Géotechnique **32**, 43–54 (1982)

[illegible]

Onset of Immersed Granular Avalanches by DEM-LBM Approach

Jean-Yves Delenne, M. Mansouri, F. Radjaï, M.S. El Youssoufi, and A. Seridi

Abstract We present 3D grain-fluid simulations based on the discrete element method interfaced with the lattice Boltzmann method and applied to investigate the initiation of underwater granular flows. We prepare granular beds of 800 spherical grains with different values of the initial solid fraction in a biperiodic rectangular box. In order to trigger an avalanche, the bed is instantaneously tilted to a finite slope angle above the maximum angle of stability. We simulate the dynamics of the transient flow for different solid fractions. In agreement with the experimental work of Iverson (Water Resour Res 36(7):1897–1910, 2000) and Pailha et al. (Phys Fluids 20(11):111701, 2008), we find that the flow onset is controlled by the initial solid fraction.

Keywords Fluid-grain interaction • Lattice Boltzmann • Discrete element method • Granular avalanches • Granular material

1 Introduction

Many geological hazards involve the presence of liquid flows through a granular material. Well-known examples are quick sands, liquefaction, landslides and submarine avalanches. Classical models that involve one or two fluids with a complex

J.-Y. Delenne (✉) · F. Radjaï · M.S. El Youssoufi
LMGC UMR 5508 CNRS-Université Montpellier 2 / MIST, IRSN DPAM-CNRS, cc. 048, Pl. E. Bataillon, 34095 Montpellier cedex 5, France
e-mail: jean-yves.delenne@univ-montp2.fr; radjai@lmgc.univ-montp2.fr

M. Mansouri
Département de génie civil, université Ferhat-bbas, 19000, Setif, Algeria

A. Seridi
Laboratoire de mécanique des solides et systèmes (LM2S), université M'hamed-Bougara, Boumerdes, Algeria

S. Bonelli et al. (eds.), *Advances in Bifurcation and Degradation in Geomaterials*, Springer Series in Geomechanics and Geoengineering 11,
DOI 10.1007/978-94-007-1421-2_14, © Springer Science+Business Media B.V. 2011

non Newtonian behavior deal mostly with steady flows and thus unable to account for such transients. In parallel, discrete element modeling of granular materials containing a liquid phase has received a great interest in recent years. In the case of non-saturated materials, capillary forces have been successfully modeled through their effect at the liquid bridges between grains (Richefeu et al. 2006; Soulié et al. 2006). For saturated materials, the straining leads to fluid flow in the pores spaces, therefore a convenient model of fluid-grain interactions is required in addition to discret element model for the grains. In the last decade, the Lattice Boltzmann Method (LBM) has emerged as a powerful tool to model fluid flows in complex geometries. This method has been recently coupled with the Discrete Element Method (DEM) to study saturated granular materials (Feng et al. 2007; Strack and Cook 2007; Mansouri et al. 2009).

In this paper, we introduce a 3D grain-fluid algorithm based on a Discrete Element Method interfaced with the Lattice Boltzmann Method (DEM-LBM). The relevance of this approach is illustrated by applying this approach to simulate the initiation and dynamics of underwater granular flows. We study the dynamics of the transient flow for different values of the solid fraction. Our numerical data are consistent with recent experimental results that evidence the role of the initial solid fraction and the subsequent dilation of the granular bed in connection with pore pressures (Iverson 2000; Pailha et al. 2008).

2 Sample Construction and DEM Computation

The DEM is used to build a densely-packed sample of polydisperse spherical grains. We use the statistical model proposed by Voivret et al. to obtain a set of grain diameters according to the cumulative beta distribution (Voivret et al. 2007). This distribution has the advantage to be bounded on both sides and capable of representing double-curved distributions similar to the soil grain-size distributions encountered in practice. We use the reduced diameter $d_r = \frac{d-d_{min}}{d_{max}-d_{min}}$ where $d \in [d_{min}, d_{max}]$ and $d_r \in [0,1]$. The size distribution is given by $\beta(d_r) = \frac{1}{B(a,b)}\int_0^{d_r} t^{a-1}(1-t)^{b-1}dt$, where $a > 0$, $b > 0$, Γ is the Gamma function and $B(a,b) = \frac{\Gamma(a)\Gamma(b)}{\Gamma(a+b)}$.

Once the diameters are generated, the grains are placed on a regular grid in a rectangular column as a dilute sample. In this study, we use biperiodic boundaries in both directions x and y (Fig. 1a). The grains are deposed under gravity into a square based periodic cell. The following procedure is used: The force $\mathbf{F}$ between two grains has a normal component $\mathbf{N}$ and a tangential component $\mathbf{T}$ due to friction. For the force law between a pair of grains i and j in contact, we use the linear-elastic approximation: $\mathbf{N} = \left(-k_n\,\delta + 2\alpha\sqrt{mk_n}\,\dot{\delta}\right) H(-\delta)\,\frac{\ell}{\|\ell\|}$, where $\delta = \|\ell\| - \frac{1}{2}(d_i + d_j)$ is the gap or the overlap between the two grains, ℓ is the branch vector,

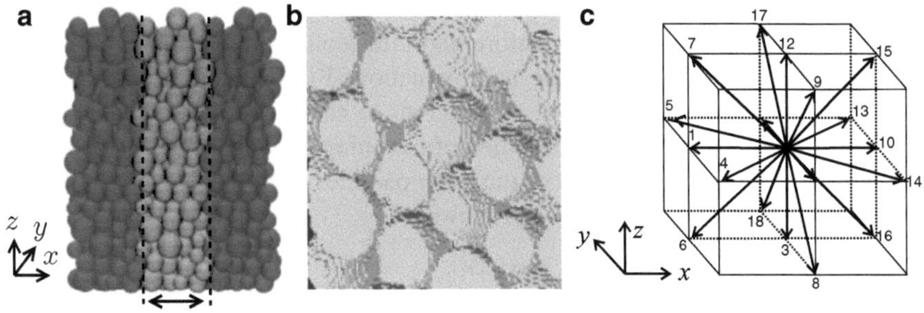

Fig. 1 (**a**) Sample generated by DEM; (**b**) Magnification of the sample meshing; (**c**) D3Q19 Lattice Boltzmann model

H is the Heaviside function, k_n is the normal stiffness, $m = \frac{m_i m_j}{m_i + m_j}$ is the reduced mass and $\alpha \in [0, 1]$ is a damping parameter which controls energy dissipation due to inelastic collision. For the friction, we use the simple Coulomb law expressed as a nonlinear relation between the friction force $\mathbf{T}$ and the sliding velocity $\dot{\delta}_t$ with a viscous regularization around the zero velocity: $\mathbf{T} = -\min\{\eta \|\dot{\delta}_t\|, \mu_f \|\mathbf{N}\|\} \frac{\dot{\delta}_t}{\|\dot{\delta}_t\|}$, where η is the tangential viscosity and μ_f is the coefficient of friction.

3 Fluid-Grains Interaction

We use the Lattice Boltzmann Method (LBM) in order to compute the fluid flow and the fluid-grain interactions (Mansouri et al. 2011). The LBM can be described as a time-stepping Eulerian micro-particle based procedure. Hence, the fluid is modeled as particle moving on a fixed regular grid.

We consider a 3D fluid flow within a periodic domain, with fluid density $\rho = 1{,}000\,\text{kg/m}^3$ and kinematic viscosity $\nu = 10^{-6}\,\text{m}^2/\text{s}$. The sample is discretized into a three-dimensional array of identical small cubes whose centers are the lattice nodes, hence a solid grain is represented by grouped cubes (Fig. 1b).

The fluid particles located at each node can naturally move in all directions. However, it is assumed that the number of degrees of freedom is finite, so that fluid motion is possible only through prescribed paths from one site to another. Hence, the fluid mass density at each node is understood as a sum of partial densities corresponding to the discrete selected velocities. In the present work, the D3Q19 scheme is used (Fig. 1). We note $Q = \{0, \ldots, 19\}$ and $f_i(\mathbf{x}, t)$ with $i \in Q$ the mass density distribution of the particles moving in direction $\mathbf{e}_i$, at time t and position $\mathbf{x}$.

The time is discrete with increments Δt and at each time step the particles located at all nodes move to the neighboring nodes along the corresponding directions. A "lattice speed" is defined by $c = \frac{\Delta x}{\Delta t}$. During particle motions to the neighboring sites, collisions may occur and lead to new particle distributions. The variation of density distributions before and after collisions is usually computed through the single relaxation time approximation (BGK kinetic model). In this way, computationally, there are two operations at each time step: streaming and collision. Let $f_i^{in}(\mathbf{x},t)$ be the distribution of incoming particles to the site $\mathbf{x}$ at time t^- and $f_i^{out}(\mathbf{x},t)$ the distribution of outgoing particles from the site after collision (at time t^+). The algorithm works as follows:

- collision operation (after BGK model):

$$f_i^{out}(\mathbf{x},t) = f_i^{in}(\mathbf{x},t) - \frac{1}{\tau}\big(f_i^{in}(\mathbf{x},t) - f_i^{eq}(\mathbf{u},\rho)\big).$$

  The parameters τ and f_i^{eq} are defined below.
- streaming operation:

$$f_i^{in}(\mathbf{x}+\mathbf{e}_i\Delta t, t+\Delta t) = f_i^{out}(\mathbf{x},t)$$

The parameter τ is the dimensionless relaxation time and the f_i^{eq} are the equilibrium distribution functions that depend on the macroscopic variables ρ and velocity $\mathbf{u}$ at position $\mathbf{x}$ and time t. The macroscopic fluid variables can then be obtained from the moments of the distribution functions:

$$\rho = \sum_{i\in Q} f_i \ ; \ \rho\mathbf{u} = \sum_{i\in Q} f_i\mathbf{e}_i \tag{1}$$

In the simulations, three types of boundary conditions are considered: periodic, non-slip and velocity imposed boundaries. In both horizontal directions, periodic conditions are imposed so that nodes of opposite boundaries are treated as neighboring. The fluid velocity on the horizontal bottom and upper boundaries is set to zero (Zou and He 1997). At the grain boundaries the non-slip condition of the fluid should be imposed. We use the interpolated bounce back boundaries described in Bouzidi et al. (2001). For moving solid grains the procedure proposed by Lallemand and Luo (2003) is employed.

The hydrodynamic forces acting on the grains are computed using the momentum exchange method proposed by Ladd (1994). The fluid particles that are bounced back on grain boundaries transmit forces to the grain proportionally to their momentum change. The implementation of this method is straightforward since the grain momenta are known at each time step. The total hydrodynamic force exerted on a solid grain is obtained by summing up the forces at all boundary nodes of that grain. Finally, these forces are used to compute the motion of grains according to the DEM algorithm (Sect. 2).

4 Onset of a Granular Avalanche

We prepare granular beds of 800 spherical grains with different values of solid fraction in a biperiodic rectangular box (see Sect. 2). Seven samples were built with solid fraction ϕ ranging from 0.57 to 0.63. For the fluid-grain simulation, the box is meshed using D3Q19 elements (Sect. 3) so that the liquid level exceeds the grain (Fig. 3). As in the experiments of Pailha et al. (2008), we consider the flow in a closed saturated box (i.e. with no liquid-gaz free surface at the top boundary). In our simulations, we set the fluid velocity to zero in the z direction at the bottom and top boundary. For the lateral wall of the box we use periodic conditions along x and y directions.

In order to trigger an avalanche, the beds are instantaneously tilted along x direction to a finite slope angle of 27° above the maximum angle of stability. We consider the evolution of the x component of the barycentric velocity V of the grains (Fig. 2a). In agreement with the experimental work of Iverson (2000) and Pailha et al. (2008), we find that the onset of the granular flow strongly depends on the initial solid fraction. For high solid fractions the avalanche is delayed (Pailha et al. 2008) while for high solid fractions the avalanche is triggered instantly.

Let us consider a dense sample ($\phi = 0.63$) and a loose sample ($\phi = 0.57$). At solid fractions above 0.60, the bed tends to dilate (Fig. 2b) and hence the avalanche is delayed due to development of negative pressure in the pores (Fig. 3a). Conversely, the flow is triggered instantly at lower solid fractions as a result of compaction (Fig. 2b) and a positive pressure in the pore (Fig. 3b).

The spatiotemporal evolution of this process is plotted regarding the evolution of pore pressure in the case of the dense sample (Fig. 4). The vertical pressure is obtained by averaging the local pore pressures in the whole sample. The horizontal dotted line shows the approximative bed height. We see that the flow is initiated at the bottom of the bed and spreads subsequently to the top.

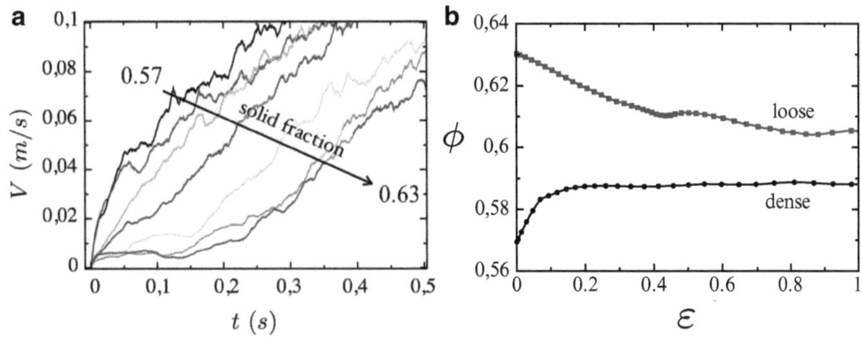

Fig. 2 (**a**) Velocity V of the granular beds as a function of time. (**b**) Solid fraction ϕ as a function of deformation for a loose ($\phi = 0.57$) and a dense sample ($\phi = 0.63$)

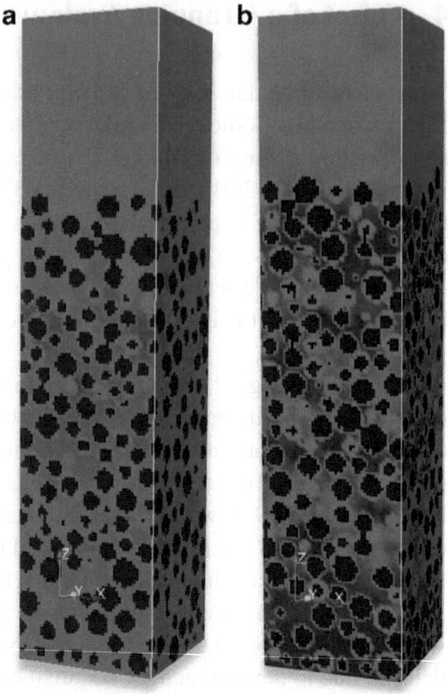

Fig. 3 Pore pressure at the onset of granular flow (color online www.cgp-gateway.org/ref008); (**a**) positive pressure for low solid fraction, (**b**) negative pressure for hight solid fraction

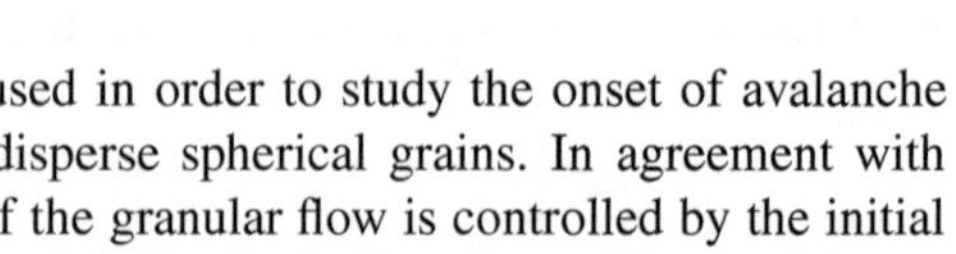

Fig. 4 Spatiotemporal evolutions of intergranular pressure; for the dense sample ($\phi = 0.63$)

5 Conclusions

A joint 3D DEM-LBM model was used in order to study the onset of avalanche of granular beds composed of polydisperse spherical grains. In agreement with experiments, we find that the onset of the granular flow is controlled by the initial solid fraction. At solid fractions above 0.60, the bed tends to dilate and hence the avalanche is delayed due to development of negative pressure in the pores. Conversely, the flow is triggered instantly at lower solid fractions as a result of compaction and a positive pressure in the pore.

References

M. Bouzidi, M. Firdaouss, P. Lallemand, Momentum transfer of a lattice Boltzmann fluid with boundaries. Phys. Fluids **13**, 3452 (2001)

Y.T. Feng, K. Han, D.R.J. Owen, Coupled Lattice Boltzmann and discrete element modelling of fluid-particle interaction problems. Int. J. Numer. Methods Eng. **72**, 1111–1134 (2007)

R.M. Iverson, Landslide triggering by rain infiltration. Water Resour. Res. **36**(7), 1897–1910 (2000)

A.J.C. Ladd, Numerical simulation of particular suspensions via a discretized Boltzmann equation, Part 2, Numerical results. J. Fluid Mech. **271**, 311–339 (1994)

P. Lallemand, L.-S. Luo, Lattice Boltzmann method for moving boundaries. J. Comput. Phys. **184**, 406–421 (2003)

M. Mansouri, J.-Y. Delenne, M.S. El Youssoufi, A. Seridi, A 3D DEM-LBM approach for the assessment of the quick condition for sands. C. R. Mécanique **337**, 675–681 (2009)

M. Mansouri, J.-Y. Delenne, A. Seridi, M.S. El Youssoufi, A numerical model for the computation of permeability of cemented granular material. Powder Technol. **208**(2), 532–536 (2011)

M. Pailha, M. Nicolas, O. Pouliquen, Initiation of underwater granular avalanches: influence of the initial volume fraction. Phys. Fluids **20**(11), 111701 (2008)

V. Richefeu, F. Radjai, M.S. El Youssoufi, Stress transmission in wet granular materials. Eur. Phys. J. E **21**, 359–369 (2006)

F. Soulié, F. Cherblanc, M.S. El Youssoufi, C. Saix, Influence of liquid bridges on the mechanical behaviour of polydisperse granular materials. Int. J. Numer. Anal. Meth. Geom. **30**(3), 213–228 (2006)

O.E. Strack, B.K. Cook, Three-dimensional immersed boundary conditions for moving solids in the Lattice Boltzmann method. Int. J. Numer. Meth. Fluids **55**, 103–125 (2007)

C. Voivret, F. Radjai, J.-Y. Delenne, M.S. El Youssoufi, Space-filling properties of polydisperse granular media. Phys. Rev. E Stat. Nonlin. Soft Matter Phys. **76**, 021301 (2007)

Q. Zou, X. He, On pressure and velocity boundary conditions for the Lattice Boltzmann BGK model. Phys. Fluids **9**, 1591–1598 (1997)

Multi-scale Periodic Homogenization of Ionic Transfer in Cementitious Materials

K. Bourbatache, O. Millet, and A. Aït-Mokhtar

Abstract In this communication, we perform a multi-scale periodic homogenization of Nernst-Planck and Poisson-Boltzmann equations describing the ionic transfer in saturated cementitious materials. Two models of ionic transfer are established successively. The first one by periodic homogenization from the Debye length scale to the capillary porosity scale, taking into account the electrical double layer phenomenon. The second model is obtained by upscaling the same ionic transfer equations from the capillary porosity scale to the material's scale. Numerical simulations on porous media with more or less complex periodic microstructures are then carried out. Comparisons with existing experimental data are also presented and discussed.

Keywords Ionic transfer • Multi-scale periodic homogenization • Electrical double layer • Cementitious materials

1 Introduction

In this work, we present two homogenized models of ionic transfer in saturated cementitious material, obtained by periodic homogenization (Auriault et al. 1996; Bensoussan et al. 1978; Sanchez Palencia 1980) of the Nernst-Planck and Poisson-Boltzmann equations.

K. Bourbatache
LTI, 48 rue Ostende 02100 Saint-Quentin

O. Millet (✉) · A. Aït-Mokhtar
University of La Rochelle, avenue Michel Crépeau, 17042 La Rochelle cedex, France
e-mail: olivier.millet@univ-lr.fr

S. Bonelli et al. (eds.), *Advances in Bifurcation and Degradation in Geomaterials*, Springer Series in Geomechanics and Geoengineering 11,
DOI 10.1007/978-94-007-1421-2_15,

The first multi-scale homogenization is carried out from the Debye length scale to the scale of the capillary porosity, taking into account the electrical double layer (EDL) phenomenon (Moyne and Murad 2006). The second homogenization procedure is conducted from the scale of the capillary porosity (micrometer scale) to the material's scale (centimeter scale). In the second homogenization procedure, the EDL effects are negligible at the scale of the capillary porosity.

Finally, comparisons are carried out between multi-scale homogenized diffusion coefficients calculated by numerical simulations on more or less complex periodic three dimensional microstructures, and macroscopic diffusion coefficients of chloride determined by electro-diffusion tests.

2 Ionic Transfer at the Scale of the Debye Length

The material studied occupies the domain S^* of the three-dimensional space $\mathbb{R}^3$, whose characteristic length is noted L (Fig. 1a). A point of the macroscopic domain S^* will be noted $x^* = (x_1^*, x_2^*, x_3^*)$. We suppose that the microstructure of the considered material is periodic and constituted of the repetition of the elementary cell $\Omega^* = \Omega_s^* \cup \Omega_f^*$ composed of the solid phase Ω_s^* and of the fluid phase Ω_f^* (Fig. 1b). The "nanostructure" (at the Debye length scale) is composed of the repetition of another periodic elementary cell $\Omega'^* = \Omega_s'^* \cup \Omega_f'^*$, where the solid and fluid phases are noted respectively $\Omega_s'^*$ and $\Omega_f'^*$ (Fig. 1c). The boundary of the domain Ω^* (respectively Ω'^*) is noted Γ^* (respectively Γ'^*). It is composed of the solid–fluid interface Γ_{sf}^* (respectively $\Gamma_{sf}'^*$) between the solid and the fluid phases, and of the fluid–fluid interface Γ_{ff}^* (respectively $\Gamma_{ff}'^*$) separating two neighbouring elementary cells. A point of the domain Ω^* at the microscale (respectively Ω'^* at the nanoscale) will be noted $y^* = (y_1^*, y_2^*, y_3^*)$ (respectively $z^* = (z_1^*, z_2^*, z_3^*)$). δ, l and L are respectively the characteristic lengths at the nanoscale, at the microscale and at the macroscale.

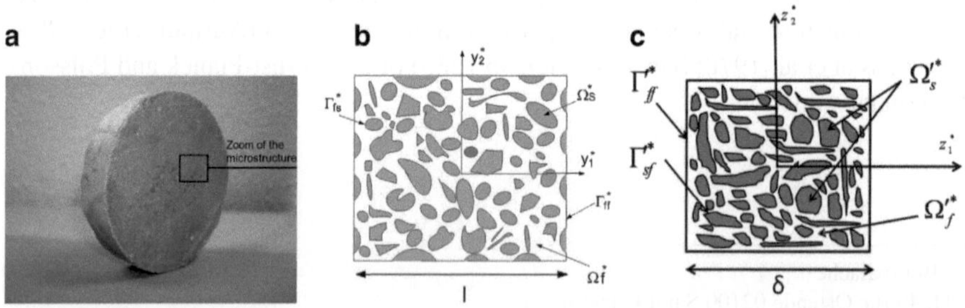

Fig. 1 Modelling the different scales of the material with periodic elementary cells. (**a**) Sample of cement paste (macroscale), (**b**) Elementary cell at microscale, (**c**) Elementary cell at nanoscale

The ionic transfer with EDL phenomenon is classically governed by the Nernst-Planck and Poisson-Boltzmann equations involving the EDL potential φ^* and the bulk potential ψ_b^* where the electroneutrality is satisfied. They write:

$$\frac{\partial c_\pm^*}{\partial t^*} + div^*\left(-D_\pm^*\left(grad^* c_\pm^* \pm \frac{F}{RT} z_\pm c_\pm^* grad^*\left(\psi_b^* + \varphi^*\right)\right)\right) = 0 \ in \ \Omega_f^{'*} \quad (1)$$

$$-D_\pm^*\left(grad^* c_\pm^* \pm \frac{F}{RT} z_\pm c_\pm^* grad^*(\psi_b^* + \varphi^*)\right).n = 0 \qquad on \quad \Gamma_{sf}^{'*} \quad (2)$$

$$\varepsilon_v \Delta^*(\psi_b^* + \varphi^*) = -(\rho_+^* - \rho_-^*) \quad in \quad \Omega_f^{'*} \quad (3)$$

$$\varepsilon_v grad^*(\psi_b^* + \varphi^*).n = \sigma^* \qquad on \quad \Gamma_{sf}^{'*} \quad (4)$$

where $c_\pm^*$, $D_\pm^*$ and $z_\pm$ denote respectively the concentration, diffusion coefficient and valence of cations and anions in the fluid phase $\Omega_f^{'*}$; F, R, T and ε_v respectively the Faraday constant, the ideal gas constant, the temperature assumed to be constant and the dielectric constant of the fluid phase. ρ_+^* and ρ_-^* represent respectively the volume electrical charge density of cations and anions in the fluid phase and σ^* the electrical surface charge density assumed to be constant.

3 Homogenized Model at the Capillary Porosity Scale

The periodic homogenization procedure performed is similar to that developed in Auriault et al. (1996), Bourbatache (2009), and Millet et al. (2008). We define the dimensionless variables as the ratio between the dimensional ones (with a star) and the reference data (with index r) as follows:

$$y = \frac{y^*}{l},\ z = \frac{z^*}{\delta},\ t = \frac{t^*}{t_r},\ D_\pm = \frac{D_\pm^*}{D_r},\ c_\pm = \frac{c_\pm^*}{c_r},\ \psi_b = \frac{\psi_b^*}{\psi_r},\ \varphi = \frac{\varphi^*}{\varphi_r},\ \sigma = \frac{\sigma^*}{\sigma_r} \quad (5)$$

Replacing (5) in the transfer equations (1)–(4), we obtain:

$$\tau \frac{\partial c_\pm}{\partial t} + div\left(-D_\pm(gradc_\pm \pm z_\pm c_\pm grad(\mathcal{R}^\psi \psi_b + \mathcal{R}^\varphi \varphi))\right) = 0 \quad in \quad \Omega_f^{'} \quad (6)$$

$$-D_\pm\left(gradc_\pm \pm z_\pm c_\pm grad(\mathcal{R}^{\psi_b} \psi_b + \mathcal{R}^\varphi \varphi)\right).n = 0 \qquad on \quad \Gamma_{sf}^{'} \quad (7)$$

$$\mathcal{A}^\psi \Delta\psi_b + \mathcal{A}^\varphi \Delta\varphi = 2c_b|z_\pm| \sinh(R^\varphi |z_\pm|\varphi) \qquad in \quad \Omega_f^{'} \quad (8)$$

$$\left(\mathcal{B}^\psi grad\psi_b + \mathcal{B}^\varphi grad\varphi\right).n = \sigma \qquad on \quad \Gamma_{sf}^{'} \quad (9)$$

This dimensional analysis of equations naturally lets appear the following dimensionless numbers characterizing the ionic transfer in a saturated porous medium:

$$\tau = \frac{\delta^2}{t_r D_r},\ \mathcal{R}^{\psi} = \frac{F\psi_r}{RT},\ \mathcal{R}^{\varphi} = \frac{F\varphi_r}{RT},\ \mathcal{A}^{\psi} = \frac{\varepsilon_v \psi_r}{\delta^2 F c_r},\ \mathcal{A}^{\varphi} = \frac{\varepsilon_v \varphi_r}{\delta^2 F c_r},$$
$$\mathcal{B}^{\psi} = \frac{\varepsilon_v \psi_r}{\delta \sigma_r},\ \mathcal{B}^{\varphi} = \frac{\varepsilon_v \varphi_r}{\delta \sigma_r} \tag{10}$$

To reduce the considered problem to a one scale problem, we must link the dimensionless numbers (10) to the perturbation parameter $\varepsilon = \delta/l$ using experimental data (Amiri et al. 2001). We obtain for a large applied external electrical field:

$$\tau = \mathcal{O}(\varepsilon),\ \mathcal{A}^{\psi} = \mathcal{O}\left(\frac{1}{\varepsilon}\right),\ \mathcal{A}^{\varphi} = \mathcal{O}(1),\ \mathcal{R}^{\psi} = \mathcal{O}\left(\frac{1}{\varepsilon}\right),\ \mathcal{R}^{\varphi} = \mathcal{O}(1),$$
$$\mathcal{B}^{\varphi} = \mathcal{O}(1),\ \mathcal{B}^{\psi} = \mathcal{O}\left(\frac{1}{\varepsilon}\right)$$

Then, the unknowns (c_b, ψ_b, φ) of the problem are assumed to admit a formal expansion with respect to ε:

$$(c_b, \psi_b, \varphi)(y, z, t) = (c_b^0, \psi_b^0, \varphi^0)(y, z, t) + \varepsilon(c_b^1, \psi_b^1, \varphi^1)(y, z, t)$$
$$+ \varepsilon^2(c_b^2, \psi_b^2, \varphi^2)(y, z, t) + \ldots \tag{11}$$

The classical asymptotic expansion of Eqs. 6–9, using (11), leads to a migration model at the scale of the capillary porosity taking into account the EDL effects:

$$\varepsilon_p \frac{\partial}{\partial t}\left\langle c_b^0 \exp(\mp|z_\pm|\varphi^0)\right\rangle \mp div_y\left(\mathbf{D}_{\pm}^{'\mathbf{hom}}|z_\pm|c_b^0 \frac{\partial \psi_b^0}{\partial y}\right) = 0 \tag{12}$$

where $\varepsilon_p = \frac{|\Omega_f'|}{|\Omega'|}$ denotes the nanoporosity of the porous medium and $\langle . \rangle$ the average in the fluid domain Ω_f'. The homogenized diffusion tensor $\mathbf{D}_{\pm}^{'\mathbf{hom}}$ is given by:

$$\mathbf{D}_{\pm}^{'\mathbf{hom}} = \frac{1}{|\Omega'|}\int_{\Omega_f'}\left(D_\pm \exp(\mp|z_\pm|\varphi^0)\left(I + \overline{\frac{\partial \chi}{\partial z}}\right)\right) d\Omega_f' \tag{13}$$

where the vector $\chi(z)$ is solution of the following boundary problem:

$$div_z\left(\mp D_\pm|z_\pm|\exp(\mp|z_\pm|\varphi^0)\left(I + \overline{\frac{\partial \chi}{\partial z}}\right)\right) = 0 \quad in \quad \Omega_f' \tag{14}$$

$$\mp D_\pm|z_\pm|\exp(\mp|z_\pm|\varphi^0)\left(I + \frac{\partial \chi}{\partial z}\right).n = 0 \quad on \quad \Gamma_{sf}' \tag{15}$$

The potential φ^0 induced by the EDL is solution of a non linear coupled problem:

$$\Delta_z \varphi^0 + \frac{\partial \psi_b^0}{\partial y} div_z \left(I + \overline{\frac{\partial \chi}{\partial z}} \right) = 2|z_\pm| c_b^0 \sinh(|z_\pm| \varphi^0) \quad in \quad \Omega_f^{'} \tag{16}$$

$$\left(\frac{\partial \varphi^0}{\partial z} + \frac{\partial \psi_b^0}{\partial y} \left(I + \frac{\partial \chi}{\partial z} \right) \right) .n = \sigma \quad on \quad \Gamma_{sf}^{'} \tag{17}$$

4 Homogenized Model at the Scale of the Material

In a second time, we perform a periodic homogenization procedure from the scale of capillary porosity to the scale L of the material. At the capillary scale, it can be proved that the EDL effects are negligible. So that, we rewrite the Nernst-Planck Poisson-Boltzmann equations without accounting for the EDL effects (with $\varphi = 0$ and $\sigma = 0$). The periodic homogenization of these equations in the case of an important electrical field leads to a homogenized pure migration model (see Bourbatache 2009; Millet et al. 2008, for more details). We prove in particular that the leading terms c_k^0 of the expansion of the concentrations c_k satisfy the electroneutrality assumption $\rho^0 = \sum_{k=1}^{N} z_k c_k^0 = 0$. Moreover, the macroscopic concentrations c_k^0 and potential Ψ^0 at the scale of the material are solution of the homogenized migration model:

$$\varepsilon_p \frac{\partial c_k^0}{\partial t} - div_x \left(\mathbf{D_k^{hom}} \left(\frac{\partial \Psi^0}{\partial x} z_k c_k^0 \right) \right) = 0 \tag{18}$$

where $\varepsilon_p = \frac{|\Omega_f|}{|\Omega|}$ denotes the porosity of the material. The homogenized diffusion tensor $\mathbf{D_k^{hom}}$ is given by:

$$\mathbf{D_k^{hom}} = \frac{1}{|\Omega|} \int_{\Omega_f} D_k \left(I + \overline{\frac{\partial \chi}{\partial y}} \right) d\Omega \tag{19}$$

The vector $\chi(y)$ is periodic, of zero average on Ω_f, and solution of:

$$div_y \left(D_k \left(I + \overline{\frac{\partial \chi}{\partial y}} \right) \right) = 0 \quad in \quad \Omega_f \tag{20}$$

$$D_k \left(I + \frac{\partial \chi}{\partial y} \right) .n = 0 \quad on \quad \Gamma_{sf} \tag{21}$$

5 Multi-scale Simulations and Comparison with Experimental Measurements

In order to validate the multi-scale periodic homogenization procedure performed, numerical simulations are carried out to determine the homogenized diffusion coefficient of chlorides in a cementitious material. To do this, we chose periodic microstructures at nanoscale and capillary porosity scale as representative as possible of the real microstructure. The mutli-scale homogenization allows to take into account the EDL effects and the geometrical properties of the microstructure at the nanoscale (first homogenization at the Debye length scale (Fig. 2a), and the geometrical properties of the microstructure at the microscale (second homogenization at the capillary porosity scale (Fig. 2b). Such a multi-scale homogenization allows the comparison between the experimental and the numerical results, integrating the transfer properties of the two main scales of the cementitious material.

We denote ε_{p1},ε_{p2} and $\varepsilon_p = \varepsilon_{p1} + \varepsilon_{p2}$ respectively the porosity at the Debye length scale, the porosity at the capillary scale and the total porosity of the cementitious material. The results of the double homogenization procedure are presented in Table 1. $D_1^{'*hom}$ is the homogenized diffusion coefficient resulting from the first homogenization procedure, and D_2^{*hom} the one obtained after the second homogenization procedure. The second homogenized diffusion coefficient D_2^{*hom} will be compared to the experimental diffusion coefficient of chlorides $D_{Cl^-}^{*exp}$ obtained by a migration test (Amiri et al. 2001).

The comparison between the experimental and the homogenized diffusion coefficients is very satisfactory. The double homogenization performed leads to an important decrease of the homogenized diffusion coefficient whose values are very close to the experimental ones.

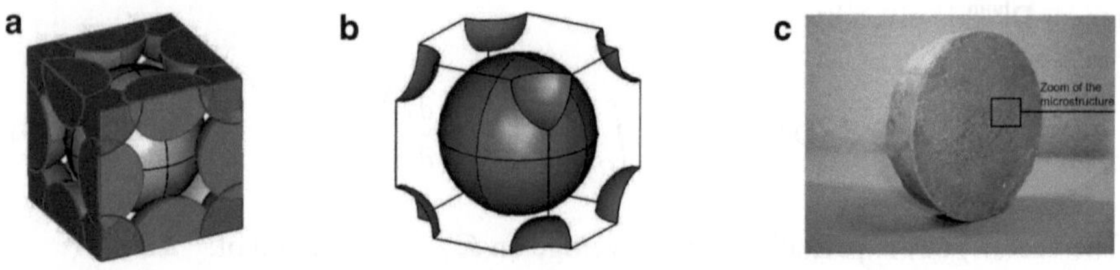

Fig. 2 Modelling the principal scales of a cement paste sample. (**a**) Debye length scale $\delta = 10$ nm, (**b**) Capillary porosity scale $l = 10\mu$m, (**c**) Scale of the material $L = 1$ cm

Table 1 Comparison between homogenized and experimental diffusion coefficients

ε_{p1}	$D_{1Cl^-}^{'*hom}$ (m^2/s)	ε_{p2}	$D_{2Cl^-}^{*hom}$ (m^2/s)	ε_p	$D_{Cl^-}^{*exp}$ (m^2/s)	$D_{2Cl^-}^{*hom}/D_{Cl^-}^{*exp}$
8%	3.810^{-11}	10%	6.8410^{-12}	18%	5.9510^{-12}	1.15
12%	8.710^{-11}	17%	15.6610^{-12}	29%	13.410^{-12}	1.17

6 Conclusion

We developed a multi-scale periodic homogenization procedure to model the ionic transfer accelerated with an important electrical field in cementitious materials. The first homogenization is carried out from the Debye length scale to the capillary porosity scale and takes into account the EDL phenomenon. The second periodic homogenization procedure is performed from the scale of the capillary porosity to the macroscopic scale of the material. The EDL effects are involved in the homogenized model at the capillary porosity scale through the EDL potential φ which contributes to slow down the chlorides transfer. Comparisons between the homogenized diffusion coefficient obtained by successive double homogenization procedure, and the experimental diffusion coefficient of chlorides obtained by a migration test, revealed very close results.

References

O. Amiri, A. Aït-Mokhtar, P. Dumargue, G. Touchard, Electrochim. Acta. **46**, 1267–1275 (2001)
J.L. Auriault, J. Lewandowska, Eur. J. Mech. A/Solids **15**(4), 681–704 (1996)
A. Bensoussan, J.L. Lions, G. Papanicolaou, Rockafellar Editors, North Holland, Amsterdam (1978)
K. Bourbatache, PhD thesis, La Rochelle University (2009)
O. Millet, A. Aït-Mokhtar, O. Amiri, Int. J. Multiphys. **2**(17), 129–145 (2008)
C. Moyne, M. Murad, Transp. Porous Media **62**, 333–380 (2006)
E. Sanchez Palencia, *Lecture Notes in Physics* (Springer, New York, 1980)

Chimney of Fluidization and "Sandboil" in a Granular Soil

P. Philippe and M. Badiane

Abstract In this paper, we report experimental results on the development in an immersed granular bed of a fluidized zone under the effect of a confined liquid flow upward. For this, two optical techniques (index-matching and laser induced fluorescence) have been combined to visualize the interior of a model granular medium made of glass beads. For small flow rates, the bed remains static while it gets fluidized in a vertical chimney at larger flow rates. For intermediate flow rates, a fluidized cavity can be observed in the vicinity of the injection hole. This cavity can either reach a steady-state or progressively expand up to the top at a small speed. The stability of such a cavity reveals a strong hysteretic behavior depending on whether the flow rate increases until fluidization or decreases after fluidization. We present here some characteristics of the phenomenon: phase diagram, regime of stable fluidized cavity, kinematics of fluidization front.

Keywords Fluidization • Instability • Sandboil • Granular medium • Refractive index matching

1 Introduction

Localized fluidization of a granular soil is observed in several industrial processes as for instance the spouted beds process for the mixing of a granular bed by an upward-fluid jet (Peng and Fan 1997), or the maintenance of navigable waterways by fluidizer systems (Weisman and Lennon 1994). In nature, some geological structures observed in layered sediments and called "fluid-escape structures" can be explained by an unstable fluidization giving rise to a growing water-filled crack (Nichols et al. 1994; Mörz et al. 2007). At a smaller scale, presumably similar phenomena referred

P. Philippe (✉) · M. Badiane
CEMAGREF d'Aix en Provence, 3275, Route de Cézanne CS 40061,
13182 Aix en Provence Cedex 5, France
e-mail: pierre.philippe@cemagref.fr; mdbadiane@yahoo.fr

S. Bonelli et al. (eds.), *Advances in Bifurcation and Degradation in Geomaterials*, Springer Series in Geomechanics and Geoengineering 11,
DOI 10.1007/978-94-007-1421-2_16,

to as "sandboils" are commonly observed on the downstream side of a dike: such a localized fluidization is induced by seepage water flow in the foundation and can potentially initiate piping, i.e. regressive pipe formation along the base of the levee (Ohja et al. 2003).

From a more fundamental view, all these examples refer more or less to the general situation of a confined ascending water flow inside an immersed granular layer. Only few studies were carried out on this problem, either with a punctual (Zoueshtiagh and Merlen 2007) or a homogeneous water injection (Rigord 2005; Wilhelm 2002). They all revealed the appearance of a localized instability that develops along a vertical fluidized pipe. The water flow intensifies inside this chimney where the porosity is significantly smaller than in the static bed. Such porosity waves have already been reported and modeled in fluidized sand column (Vardoulakis 2004; Vardoulakis et al. 1998).

As described in Sect. 2, we present in this paper an experimental device that relies on laser-induced fluorescence to probe locally within an index-matched granular medium. Section 3 is devoted to the description of the localized fluidization induced by an upward flow. A fluidized cavity is first observed in the vicinity of the injection hole while a chimney of fluidization over the entire height of the packing is obtained for larger flow rates. The difference between increasing the flow rate to fluidization and decreasing it back reveals a strong hysteresis. A phase-diagram is proposed to sum up these findings. Some preliminary results on the kinematics of the upward cavity expansion are presented in Sect. 4.

2 Optical Techniques and Experimental Device

Index-matching is generally used to make translucent a two-phase medium where both phases have the same refraction index. Here, we use a granular medium of spherical borosilicate glass beads immersed in a mixture of mineral oils which has the same refractive index as borosilicate glass, i.e. $n \approx 1.474$, and about 20 times the viscosity of water. Laser-induced fluorescence is implemented as follows: a fluorescent dye is added to the oil which, when illuminated by a 532 nm laser sheet, reemits light at a larger wavelength. Then, by interposing a high pass optical filter, it is possible to image the grains within the bed as can be seen in Figs. 1 and 2.

A given mass of borosilicate beads with diameter $d = 5\,mm$ is poured in a rectangular box with cross-section 200 × 80 mm and slowly saturated by the oil. A reproducible configuration of the packing is obtained by gently stirring the beads with a stick before each experiment. The corresponding solid volume fraction is about 0.62 and the height of the packing is denoted H. In the bottom of the box, an injection hole of inner diameter 14 mm is connected to a gear pump. The pump induces an ascending flow within the granular medium at constant rate Q. The liquid is discharged by overflowing at the top of the cell to maintain a constant level of oil.

More than one hundred tests have been carried with various heights H and recorded by a video camera for post-processing.

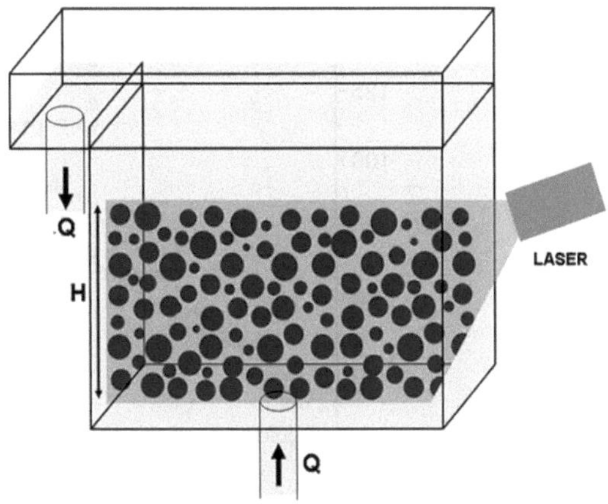

Fig. 1 Schematic of the experimental device

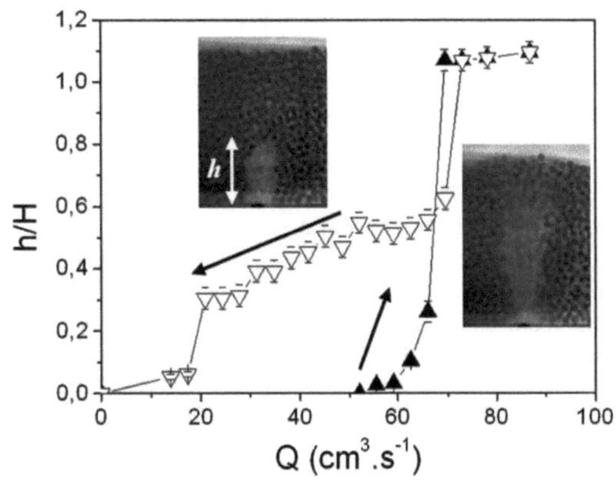

Fig. 2 Relative height of the fluidized zone *h/H* as a function of the flow rate *Q* during a sequence of increase and decrease of the flow ($H = 120\,\text{mm}$)

3 Fluidized Cavity and Chimney: Hysteresis and Phase-Diagram

As illustrated in Fig. 2, the granular medium remains static while the flow rate does not exceed a threshold Q_c^{up}. Beyond this threshold, a fluidized cavity appears in the vicinity of the injection hole. The size of this cavity increases slowly with the flow rate until a second threshold Q_f^{up} is reached when the fluidized zone joins the upper surface of the granular bed to form a chimney of fluidization over the entire height of the packing. Starting from this situation where the medium is fluidized throughout its height, if the flow is now reduced, the transition between chimney and fluidized cavity is observed for a flow rate Q_f^{dw}only slightly different from Q_f^{up}. Conversely, the return to an homogeneous static packing without cavity is only achieved for a low flow rate Q_c^{dw}. So the usual slight hysteretic behavior encountered between fluidization and de-fluidization is recovered for the chimney state but the very wide stability range for the fluidized cavity state is a far more surprising result.

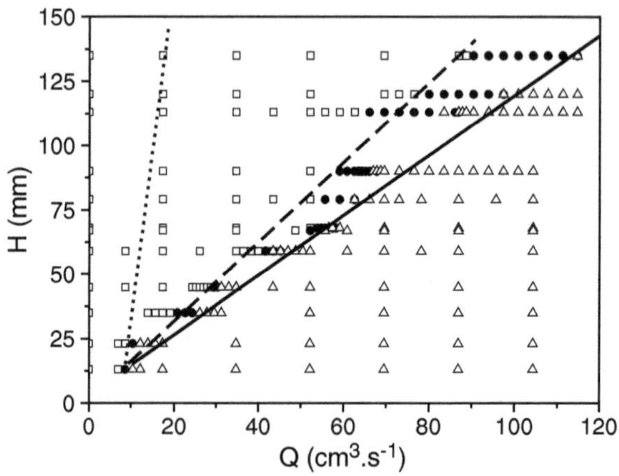

Fig. 3 Phase-diagram of the static (*squares*), fluidized cavity (*circles*) and chimney (*triangles*) states in the *Q-H* plane. Every mark depicts one test. The *plain line* denotes roughly chimney threshold while the *dashed* and *dotted lines* are approximately the borders between static and cavity domains observed respectively for an increasing and for a decreasing flow rate sequence

Fig. 4 Mean fluidization front velocity $V = H/T$ as a function of the imposed constant flow rate Q. T is the time required for the establishment of a fluidized chimney

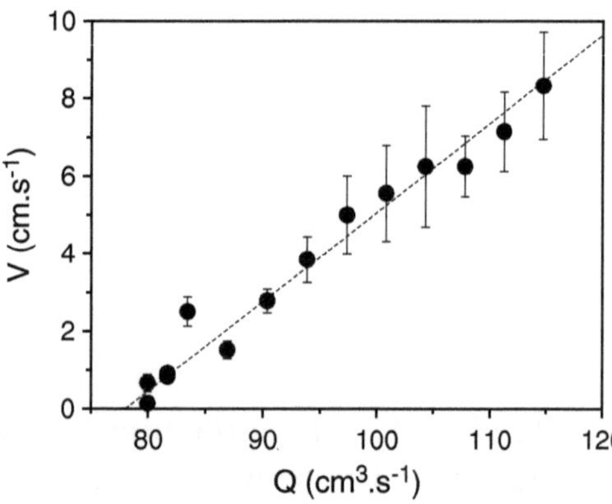

A phase-diagram has been built for the static, fluidized cavity and chimney states in a $Q - H$ plane and is shown in Fig. 3. Note that the relation between Q_f^{up} and H is roughly linear as previously observed and discussed (Zoueshtiagh and Merlen 2007).

4 Kinematics of Fluidization Front at Constant Flow Rate

Some preliminary results presented in Fig. 4 reveal that, when subjected to a constant flow rate, the velocity of the fluidization front, deduced from the time required for the establishment of the chimney, is almost linear with Q far enough beyond Q_f^{up} while the fluidization becomes even longer and poorly reproducible as the threshold is approached. In this case, the behavior is highly sensitive to the unavoidable variability of the initial configuration of the packing.

5 Conclusion

Fluidization of a granular medium through a localized ascending flow induces two distinct fluidized states: chimney at high flow rate and cavity at intermediate flow rate. A surprisingly difficult healing of the granular medium after a prior scar by chimney fluidization is revealed by the large stability domain of the cavity regime post-fluidization. This will be investigated more in depth in a future work with a systematic study of this small-scale sand-boil experiment.

References

T. Mörz, E.A. Karlik, S. Kreiter, A. Kopf, An experimental setup for fluid venting in unconsolidated sediments: new insights to fluid mechanics and structures. Sediment. Geol. **196**, 251–267 (2007)

R.J. Nichols, R.S.J. Sparks, C.J.N. Wilson, Experimental studies of the fluidization of layered sediments and the formation of fluid escape structures. Sedimentology **41**(2), 233–253 (1994)

C.S.P. Ohja, V.P. Singh, D.D. Adrian, Determination of critical head in soil piping. J. Hydraul. Eng. **129**(7), 511–518 (2003)

Y. Peng, L.T. Fan, Hydrodynamic characteristics of fluidization in liquid-solid tapered beds. Chem. Eng. Sci. **52**(4), 2277–2290 (1997)

P. Rigord, A. Guarino, V. Vidal, J.-C. Géminard, Localized instability of a granular layer submitted to an ascending liquid flow. Granul. Matter **7**(4), 191–197 (2005)

I. Vardoulakis, Fluidisation in artesian flow conditions: hydromechanically stable granular media. Geotechnique **54**(2), 117–130 (2004)

I. Vardoulakis, M. Stavropoulou, A. Skjaerstein, Porosity waves in a fluidized sand-column test. Philos. Trans. R Soc. Lond. A-Mathematical Phys. Eng. Sci. **356**(1747), 2591–2608 (1998)

R.N. Weisman, G.P. Lennon, Design of fluidizer systems for coastal environment. J. Waterway Port Coast. Ocean Tech. **120**(5), 468–487 (1994)

T. Wilhelm, K. Wilmanski, On the onset of flow instabilities in granular media due to porosity inhomogeneities. Int. J. Multiph. Flow **28**, 1929–1944 (2002)

F. Zoueshtiagh, A. Merlen, Effect of a vertically flowing water jet underneath a granular bed, Phys. Rev. E **75**(5), 053613 (2007)

Experimental Study of Contact Erosion at a Granular Interface

R. Beguin, P. Philippe, and Y.-H. Faure

Abstract Contact erosion appears at the interface of two soil layers subject to a groundwater flow. Particles of the finer soil are eroded by the flow and transported through the pores of the coarser layer. Fluvial dykes are often exposed to this phenomenon. Small-scale experiments combining Refractive Index Matching medium, and Particle Image Velocimetry were carried out to measure the flow characteristics close to the interface between a porous medium and a sandy layer. Velocity and shear-stress distributions were obtained. They underline the spatial variability of stress exerted by the flow on the fine soil which is directly related to the variability of the pore geometry. These distributions can be helpful in better modelling contact erosion by going beyond the simple use of global mean values, such as Darcy velocity, as is usually proposed in the literature.

Keywords Internal erosion • Granular interface • Particle image velocimetry • Refractive index matching • Porous medium

---

R. Beguin (✉)
Laboratoire d'étude des Transferts en Hydrologie et Environnement, Université de Grenoble, BP 53, 38041 Grenoble Cedex 09, France

EDF-Centre d'Ingénierie Hydraulique (CIH), 73373 Le Bourget du Lac Cedex, France
e-mail: remi.beguin@ujf-grenoble.fr

P. Philippe · R. Beguin
CEMAGREF d'Aix en Provence, 3275, Route de Cézanne CS, 40061 13182 Aix en Provence Cedex 5, France
e-mail: pierre.philippe@cemagref.fr

Y.-H. Faure
Laboratoire d'étude des Transferts en Hydrologie et Environnement, Université de Grenoble, BP 53, 38041 Grenoble Cedex 09, France
e-mail: yhfaure@ujf-grenoble.fr

S. Bonelli et al. (eds.), *Advances in Bifurcation and Degradation in Geomaterials*, Springer Series in Geomechanics and Geoengineering 11,
DOI 10.1007/978-94-007-1421-2_17,

1 Introduction

Dykes in fluvial valleys are structures sensitive to a particular type of internal erosion: contact erosion. Indeed, due to the structure of the dyke, or to the deposition of successive sediment layers in the subsoil, fine layers (silt or clay) can be in contact with coarser layers (gravel). These interfaces with a high permeability contrast are preferential locations for erosion. Seepages in the foundation may be important and particles of the fine soil can be removed by the flow and transported through the pores of the coarse soil. As soil is removed, cavities appear that can seriously jeopardise the safety of the dyke. Contact erosion can occur for a coarse layer either above or below a fine soil, and gravity will not have the same influence on erosion in both cases. In the following, we will only consider the case of a coarse layer above a fine layer. In this case, the main distinguishing feature of contact erosion compared to classical surface erosion is the presence of a coarse layer in contact with the eroded material. The local geometry of the coarse particle layer modifies the flow and the development of the erosion. If the constrictions connecting the pores of the coarse soil are too small, particles from the fine soil cannot pass the constrictions and cannot be transported. The coarse soil is said to be "geometrically closed". When this is not the case, a hydraulic criterion must be reached by the flow to generate erosion and transport. This paper deals with this hydraulic criterion for erosion. Several authors (Brauns 1985; Bezuijen et al. 1987; Wörman and Olafsdottir 1992; Schmitz 2007) have studied this criterion and proposed to adapt the classical surface erosion laws to this particular case, by empirical or semi-empirical adjustment. Nevertheless, except in the work of Den Adel (Den Adel et al. 1994), who took into account temporal variability, they all use mean characteristics of the flow to model the process, neglecting the variability of the porous medium. Experimental observations underline the importance of the heterogeneity of the contact erosion process linked to the spatial variability of the porous flow. In the present work, to identify the hydraulic stress exerted on the fine soil in a contact erosion situation, the Particle Image Velocimetry method has been used in combination with Refractive Index Matching. Statistical distributions of the hydraulic stress were obtained and used as a valuable input to improve contact erosion models.

2 State of the Art: Contact Erosion Experiments

Contact erosion experiments were conducted in a similar manner to previous authors (Guidoux et al. 2010). An interface between a fine soil layer and a coarse soil is reconstituted in an experimental device. A controlled flow is generated into the sample and the quantity of soil eroded is monitored. During these experiments, as done by previous authors, a critical velocity is identified, defined as the minimum velocity necessary for contact erosion development. For clay, this velocity can be superior to 10 cm/s, for non cohesive silt it is around 1 cm/s, and for sand this velocity seems to be proportional to the square root of the grain diameter. The knowledge of this threshold value can be very useful for engineers as it can be the criterion to know if contact erosion is likely to develop or not. To predict this threshold, previous authors focus on sand erosion, and they adapt Shields' criterion (Shields 1936) to fit with their critical velocity measurements. In this method the main difficulty is estimating the hydraulic shear-stress exerted on the particles of the fine soil, which is used as the loading variable in Shields' criterion. For example, Bezuijen make the hypothesis that we can assume a relation between the shear velocity $u^*(\mathrm{m/s})$, defined as $u_* = \sqrt{\rho/\tau}$ with $\rho(\mathrm{kg/m^3})$ being the specific mass of the fluid and τ (Pa) the hydraulic shear-stress, and the pore velocity (Bezuijen et al. 1987). Guidoux proposed the use of a capillary tube model to estimate the shear-stress in the coarse layer (Guidoux et al. 2010). Actually, the shear-stress at the interface is spatially variable because of the variability of the pore geometry. The erosion first begins and develops only in areas exposed to the highest hydraulic loading. In consequence, the use of mean values of the flow, as proposed previously, cannot faithfully represent contact erosion.

3 Small-Scale Flow Characterization

As can be concluded from previous work, it is necessary to link the continuous global characteristics of the flow, such as the Darcy velocity, to the local hydraulic stress on the fine soil between the coarse grains. An experimental device has been set up to measure the velocity field of the flow into the pores of the coarse layer close to the interface. A Refractive Index Matching medium was used, consisting of borosilicate beads immersed in mineral oil. As refractive index of the solid and the liquid are equal, visual observations can be made within the medium. The cell is filled with a layer of fine soil, consisting of sand with $d_{50} = 0.3$ mm, topped by a layer of borosilicate beads, diameter 9.7 mm or 7.6 mm, representing the coarse layer (Fig. 1). Both layers are about 40 mm thick.

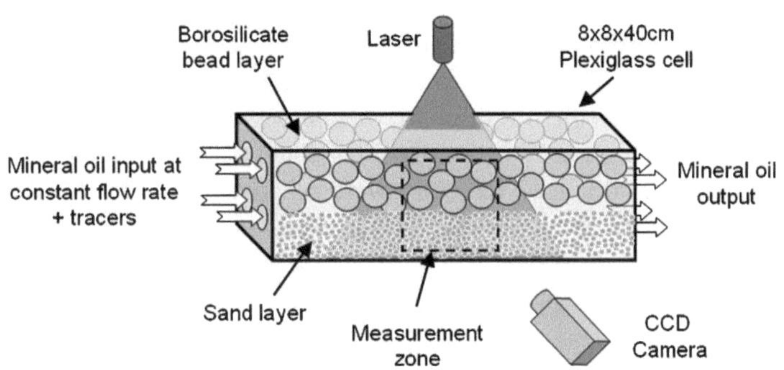

Fig. 1 Small scale measurement experimental device

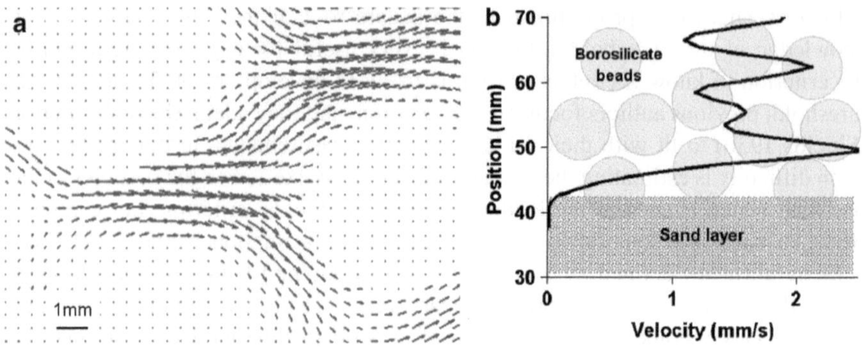

Fig. 2 (**a**) Velocity field in the coarse layer made of borosilicate beads (**b**) Mean horizontal velocity profile at the interface between the two layers

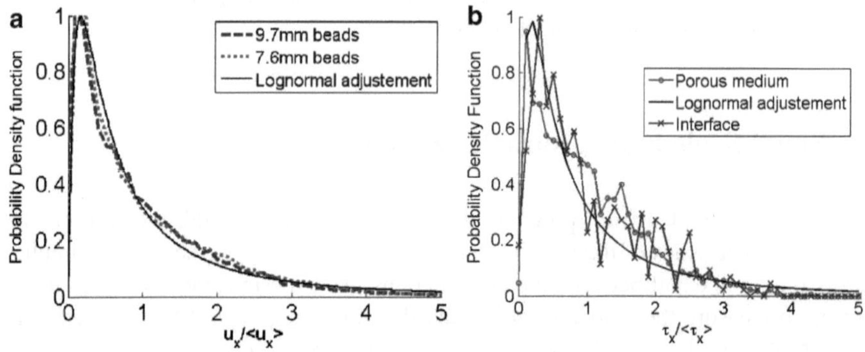

Fig. 3 (**a**) Longitudinal velocity distributions in the porous medium (**b**) Shear stress distribution at the interface and in the porous medium

Fluorescent tracers were incorporated in the oil. A section of the porous medium is illuminated by a laser sheet and filmed by a CCD camera. Recorded pictures are analyzed by Particle Image Velocimetry (PIV) and the velocity field of the liquid through the pores of the coarse layer is obtained (Fig. 2a). Three different flow rates were tested, corresponding to Darcy velocities between 0.5 and 5 mm/s and yielded similar results. The velocity profile across the interface was then computed (Fig. 2b), and highlights the transition between the two layers. Fluctuations in the velocity linked to the porosity heterogeneities could also be observed. Statistical distributions of the normalized velocities were calculated and reveal a log-normal distribution, in good agreement with several previous works (Fig. 3a) (Magnico 2003). Next, the shear-stress field was computed from the relation:

$$\tau_x = \mu \frac{du_x}{dy} \tag{1}$$

Only the shear-stress values at the interface can initiate the erosion process. Unfortunately, the PIV method is not very effective near a fixed interface, and the calculation of shear-stresses with these distorted velocities is unreliable. Consequently, a semi-empirical relation was found between the local maximum shear-stress value deduced from PIV measurement and the shear-stress obtained at the interface by a Poiseuille law adjustment. The values obtained with this method at the interface between the two layers and in the overlying coarse layer alone were then analyzed and reveal a similar variability to that of the velocity measurement (Fig 3b).

No significant differences were obtained between values at the interface and values in the porous medium. The normalized shear-stresses obtained are pretty well represented by a log-normal distribution. This means that high shear-stress values exist in the medium even if the mean shear stress is small, due to the large tail of the distribution. Erosion is initiated in these exposed areas and, indeed, a visual observation shows that erosion starts only in a few pores. With this kind of distribution as hydraulic input, contact erosion can be modelled, assuming, for example, a threshold erosion law: $\varepsilon = k_{er}(\tau - \tau_c)$ when $\tau > \tau_c$ with $\varepsilon(\mathrm{kg/s/m^2})$ being the erosion rate, k_{er}(s/m) a coefficient of erosion and τ_C (Pa) the critical shear stress.

4 Conclusion

Pore-scale measurement of velocity fields have been carried out in a model porous medium. Shear-stresses were deduced and seem to follow a log-normal statistical distribution. This result underlines the variability of the hydraulic loading exerted on fine particles located in contact with a porous medium, and that existing contact erosion models could be improved by using these shear-stress distributions instead of mean values. As the process seems influenced by heterogeneities, it would be more faithfully represented.

Acknowledgement Authors acknowledge the financial support provided by EDF-CIH.

References

A. Bezuijen, M. Klein-Breteller, K.J. Bakker, Design criteria for placed block revetments and granular filters, in *Proceedings of the 2nd International Conference on Coastal & Port Engineering in Developing Countries*, Beijing, China, 1987

J. Brauns, Erosionsverhalten geschichteten Bodens bei horizontaler Durchstromung. Wasserwirtschaft **75**, 448–453 (1985)

H. Den Adel, M.A. Koenders, K.J. Bakker, The analysis of relaxed criteria for erosion-control filters. Can. Geotech. J. **31**(6), 829–840 (1994)

C. Guidoux, Y.-H. Faure, R. Beguin, C.-C. Ho, Contact erosion at the interface between granular filter and various base-soils with tangential flow. J. Geotech. Geoenviron. Eng. **136**(5), 741–750 (2010)

P. Magnico, Hydrodynamic and transport properties of packed beds in small tube-to-sphere diameter ratio: pore scale simulation using an Eulerian and a Lagrangian approach. Chem. Eng. Sci. **58**, 5005–5024 (2003)

S. Schmitz, Zur hydraulischen Kontakterosion bei bindigen Basiserdstoffen. Ph.D. thesis, Universität der Bundeswehr, Munich, 2007

A. Shields, Anwendung der Ähnlichkeitsmechanik und Turbulenzforschung auf die Geschiebebewegung. Ph.D. thesis Versuchsanstalt für Wasserbau und Schiffbau 26, Berlin, 1936

A. Wörman, R. Olafsdottir, Erosion in a granular medium interface. J. Hydraul. Res. **30**(5), 639–655 (1992)

Numerical Modeling of Hydrofracturing Using the Damage Theory

Alice Guest and Antonin Settari

Abstract Hydraulic fracturing is a field technology widely used in the petroleum industry in order to increase the effective permeability of the reservoir and thus the production of gas by fracturing the rock by an injection of fluid. Hydrofracturing is often being monitored by detecting and analyzing microseismic events. We present a numerical technique that simulates the occurrence of microseismic events and their deformation modes during hydrofracturing and thus allows us to improve our understanding of hydrofracturing-related microseismicity. We can explain the time and spatial spread in the location of seismic events by the heterogeneity of the reservoir and the variability in the deformation modes as a natural process reflecting the reorganization of stresses in an elastic medium. We show that microseismic activity reflects the macroscopic description of hydrofracturing as a tensile crack even in highly heterogeneous reservoirs.

Keywords Hydraulic fracturing • Microseismicity • Seismic moment tensors • Numerical modeling • Isotropic damage theory

A. Guest (✉)
Department of Geoscience, University of Calgary, Calgary, AB, Canada

2500 University Dr., Calgary, AB, Canada, T2N 1N4
e-mail: alice.m.guest@gmail.com

A. Settari
Department of Chemical and Petroleum Engineering, The Schoolich School of Engineering, University of Calgary, 2500 University Dr., Calgary, AB, Canada, T2N 1N4
e-mail: asettari@ucalgary.ca

S. Bonelli et al. (eds.), *Advances in Bifurcation and Degradation in Geomaterials*, Springer Series in Geomechanics and Geoengineering 11,
DOI 10.1007/978-94-007-1421-2_18, © Springer Science+Business Media B.V. 2011

1 Introduction

Hydraulic fracturing is an in-situ tool widely used in the petroleum industry in order to increase the permeability of rock and thus the production of oil and gas. During hydrofracturing, a fluid is pumped in a reservoir through a wellbore with the goal to form a single fracture or fracture system of a maximum conductivity. Having control over hydrofracturing is crucial for the success of the whole process: it is important to estimate the final extent as well as the permeability of the fracture system, to determine the most important factors that influence the fracturing and to learn how to control them. Recently, the hydrofracturing has been monitored by detecting and analyzing the microseismic activity that arises as a result of changed stress conditions in the reservoir. So far, it is not clear what is the relationship between the microseismic events and the fracture system, and to which extent microseismicity represents and contributes to the forming fracture system. In this paper we present a numerical technique that allows us to study this relationship in detail.

2 Geomechanical Equations

The proper numerical simulation of hydrofracturing involves the interactive solution of mechanical equations and equations for fluid motion in porous medium because the fluid flow is coupled to the deformation through permeability. In this paper, for simplicity, the mechanical and fluid-flow solutions are not fully coupled, and instead a test pressure function representing an injection in an unfractured wellbore is considered. Then, our local 2-dimensional solution is based on the constitutive and geometrical equation (1), and mechanical equilibrium equation (2) under plain-strain conditions:

$$\sigma_{ij} = \frac{E_D \nu}{(1+\nu)(1-2\nu)} \varepsilon_{kk}\delta_{ij} + \frac{E_D}{1+\nu}\varepsilon_{ij} + \alpha p \delta_{ij}, \varepsilon_{ij} = \frac{1}{2}\left(\frac{\partial u_i}{\partial x_j} + \frac{\partial u_j}{\partial x_i}\right), \quad (1)$$

$$\frac{\partial \sigma_{ij}}{\partial x_j} = 0, \quad (2)$$

where σ_{ij} is the stress tensor, ε_{ij} is the strain tensor, x_i is the Cartesian coordinate, u_i is the displacement vector, E_D is the damaged Young's modulus, ν is Poisson's ratio, δ_{ij} is the Kronecker delta ($\delta_{ij} = 0$ except for $i = j$ when $\delta_{ij} = 1$), p is the pressure and α is the Biot's coefficient. The indexes i and j range from 1 to 2. The pressure distribution is input into the code in steps simulating time evolution. The initial stress conditions represent the regional stress state, and the boundary conditions are in zero displacements. The fracture formation is solved using damage theory (Tang and Kaiser 1998; Zhu and Tang 2004). The damage of material initiates when the Mohr-Coulomb or tensile failure criteria (3) are satisfied locally:

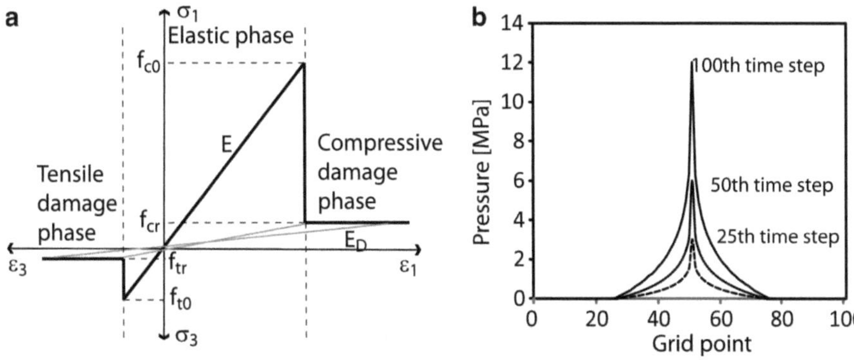

Fig. 1 (**a**) Evolution of Young's modulus during damage. (**b**) Pressure distribution on the numerical grid of 101 × 101 nodes is circularly symmetric. Full pressure profile is reached in 100 time steps

$$\sigma_1 - \frac{1+\sin\varphi}{1-\sin\varphi}\sigma_3 > f_{c0}, \quad \sigma_3 < f_{t0}, \tag{3}$$

where σ_1 and σ_3 are the maximum and minimum principal stress, respectively, f_{c0} is the compressive strength and f_{t0} is the tensile strength, and ϕ is the friction angle. Then the damage variable D and the damaged Young's modulus E_D are defined as:

$$D = 1 - \frac{f_{cr}}{E\,\varepsilon_1}, \quad E_D = (1-D)E, \tag{4}$$

where f_{cr} is the residual compressive strength and ε_1 is the maximum compressive strain and E is the initial undamaged Young's modulus. The damage parameter D ranges from 0 for the deformation before the failure criteria is reached, to 1 for very high compressive strains. For the failure under the tensional criterion, the damage is calculated the same way except that f_{cr} is replaced by f_{tr}, the residual tensional strength, and ε_1 by ε_3, the minimum tensile strain in (4) (Fig. 1a). So when failure occurs, the local Young's modulus value is determined using (4), stress equilibrium for the whole reservoir is recalculated using (1–2), and the failure criteria are newly tested (3). The non-failed elements are tested for original strength while failed elements are tested for the residual strength. Only after the stress equilibrium is reached everywhere in the reservoir, pressure representing the next time step is imported into the constitutive Eq. 1 and whole process is repeated.

Each point of failure is considered a source of seismic energy and is recorded in time and space. The released seismic energy is calculated using the change of the displacement during the failure as (Aki and Richards 2002):

$$M_{ij} = \left(\frac{E\nu}{(1+\nu)(1-2\nu)}\Delta u_{k,k}\delta_{ij} + \frac{E}{2(1+\nu)}(\Delta u_{i,j} + \Delta u_{j,i})\right)V, \tag{5}$$

where M_{ij} is seismic moment tensor for the individual grid element, Δu_k is the change of the k-component of the displacement during the failure and V is the volume of the element. The scalar seismic moment M_o and seismic magnitude m_w are (e.g., Hazzard and Young 2002):

$$M_0 = \left(\frac{1}{2} \sum_{i=1}^{3} \sum_{j=1}^{3} M_{ij}^2 \right)^{1/2}, \quad m_w = \frac{2}{3} \log_{10} M_0 - 6.1, \tag{6}$$

Because of the linearity of the Eq. 5 for the constant grid size, the individual seismic moment tensors can be summed along any surface or in time in order to represent seismic moment tensor of the chosen geometry or a time-span (see also Hazzard and Young 2002):

$$M_{fij} = \sum_{S,time} M_{ij}, \tag{7}$$

where M_{fij} is the moment tensor representing the chosen geometry and time-span.

3 Hydrofracturing

The hydraulic fracture length can reach several hundred meters and its opening (width) is on the order of mm to cm, while recorded microseismicity forms a belt of tens of meters around the fracture. Therefore, microseismicity distribution must be related to the broad stress changes around the propagating fracture (Settari et al. 2002) and the distribution of heterogeneities in the rock. As the stress around a propagating fracture can be estimated (Settari et al. 2002), it is not clear what the heterogeneities are related to. Based on the rock mechanics experiments (Wong et al. 2006), the failure in a rock starts as a propagation and coalescence of microscopic extensional cracks that progress to form a macroscopic fracture. The initial microscopic cracks are related to the joints (cracks) between the grains and matrix of the rock and the coalescence is related to the heterogeneity of strength defined through statistical description of crack densities, average sizes and maximum length (Wong et al. 2006). Using this analogy, microseismic events reflect the coalescence of the pre-existing fractures in the reservoir to form a macroscopic hydraulic fracture. Since the size of microseismic events is roughly on the order of ten meters, the heterogeneity should also reflect changes in distribution of pre-existing fractures in the order of ten meters. In case of a reservoir, ten-meter heterogeneity can be related to the depositional environment: the homogeneous reservoir could represent a massive shale while a heterogeneous reservoir could represent a turbidite depositional system.

Mathematically, the heterogeneity of the material can be introduced through a random function following the Weibull distribution (e. g., Zhu and Tang 2004):

$$f(\sigma) = \frac{m}{\sigma_0}\left(\frac{\sigma}{\sigma_0}\right)^{m-1} \exp\left[-\left(\frac{\sigma}{\sigma_0}\right)^{m}\right], \tag{8}$$

where σ_0 is the medium value of a variable σ and m is a parameter defining the heterogeneity: the lower m, the more heterogeneous medium.

4 Results and Summary

The parameters of our model are as follows: the size of the grid is 1,000 by 1,000 m, with 101 by 101 constant-size elements, 100 time steps, E = 24 GPa, $\nu = 0.2$, $\alpha = 1$, $\phi = 30$, regional stress field in the east-west direction 30 MPa, in the north-south direction 10 MPa, $f_{c0} = 10$ MPa, $f_{cr} = 1$ MPa, $f_{t0} = -1$ Mpa, $f_{tr} = -0.1$ MPa, $m = 3$. The pressure distribution is shown of Fig. 1b.

Figures 2 and 3 show the distribution of microseismicity for two different cases that differ by a degree of heterogeneity. The microseismicity in the first, homogeneous case (Fig. 2), forms a narrow belt parallel to the maximum horizontal stress, as predicted by macroscopic models, whereas the belt is much more spread not only spatially but also temporally in the second, heterogeneous case (Fig. 3). The results agree well with observed variability of microseismicity. It shows that if we are able to statistically characterize the reservoir distribution of fractures, we can predict the extent of microseismicity a priori.

Figure 4 shows the evolution of seismic moment tensors during each equilibrium recalculation and for every element of the grid. The moment tensors show significant variability of deformation modes even for the simplest homogeneous case, e.g., crack opening, crack closure, and simple shearing that are related to the stress redistribution during the fracture opening and propagation. It means that the observed variability of seismic moment tensors (Baig and Urbancic 2010) is natural to the process. In the heterogeneous case, the variability of deformation modes is the

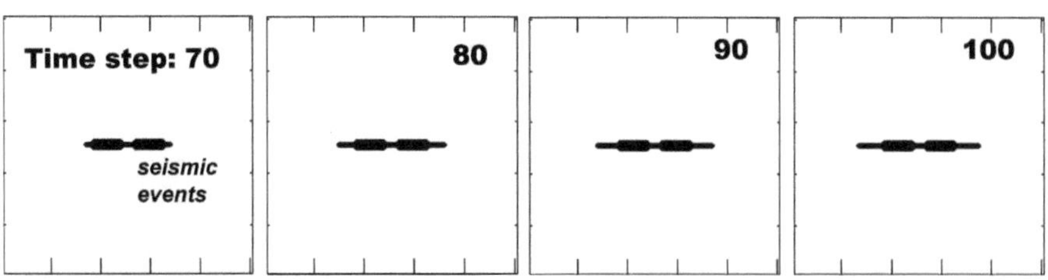

Fig. 2 Microseismicity predicted for hydrofracturing in homogeneous reservoir

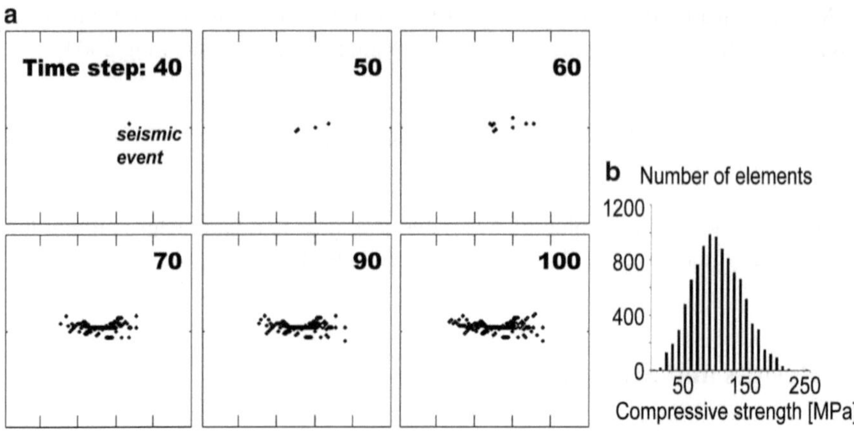

Fig. 3 Microseismicity predicted for hydrofracturing in the heterogeneous reservoir. The graph on the right shows the distribution of heterogeneity for the compressive strength

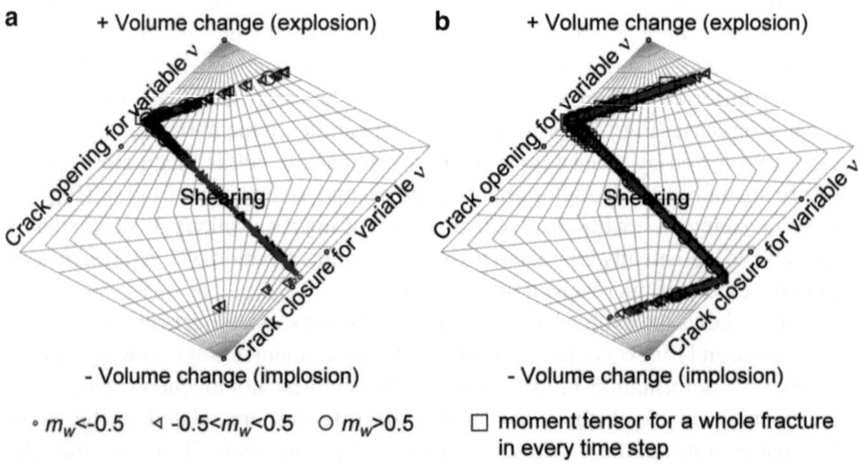

Fig. 4 Predicted moment tensors in the graph of Hudson et al. (1989) for (**a**) homogeneous and (**b**) heterogeneous cases

same; however the magnitudes of the deformation modes are distributed differently. In the first case, the events with biggest magnitude would be related to fracture opening, whereas in the second case they are spread over all observed deformation modes. The seismic moment tensors can be calculated for the total size of fracture (Eq. 7) in every time step (Fig. 4). Then, the deformation mode is a tensile crack opening for every time step in the homogeneous case and very close to tensile crack for the heterogeneous case. This is a very important result because it shows that the microseismicity reflects the tensile crack opening on a smaller and more variable scale.

In a summary, our model successfully predicts the orientation of the hydrofracture parallel to the maximum stress orientation in agreement with the general observations. For relatively homogeneous rocks (e.g. massive shale), the observed microseismicity forms a narrow belt whereas for heterogeneous rocks (e.g. turbidite fan) the belt is wide. Our modeling shows that on the scale of observed microseismicity, seismic moment tensors are naturally variable as a result of stress redistribution. On a scale of a whole fracture, the moment tensors reflect the fracture opening as a tensile crack. If a reservoir can be characterized in terms of statistical distribution of pre-existing fractures, then this technique can predict microseismicity a priori for in-situ hydrofracturing.

References

K. Aki, P. Richards, *Quantitative Seismology* (University Science Books, Sausalito, CA, 2002)

A. Baig, T. Urbancic, Microseismic moment tensors: a path to understanding frac growth. *Leading Edge* **29**, 320–324 (Mar 2010)

J.F. Hazzard, R.P. Young, Moment tensors and micromechanical models. Tectonophysics **356**, 181–197 (2002)

J.A. Hudson, R.G. Pearce, R.M. Rogers, Source type plot for inversion of the moment tensors. J. Geophys. Res. **94**, 765–774 (1989)

A. Settari, R. B. Sullivan, D.A. Walters, 3-D analysis and prediction of microseismicity in fracturing by coupled geomechanical modeling, SPE 75714 (2002)

C.A. Tang, P.K. Kaiser, Numerical simulation of cumulative damage and seismic energy release during brittle rock failure – Part I: fundamentals. Int. J. Rock Mech. Min. Sci. **35**, 113–121 (1998)

T. Wong, R.H.C. Wong, K.T. Chau, C.A. Tang, Microcrack statistics, Weibull distribution and micromechanical modeling of compressive failure in rock. Mech. Mater. **38**, 664–681 (2006). doi:10.1016/j.mechmat.2005.12.002

W.C. Zhu, C.A. Tang, Micromechanical model for simulating the fracture process of rock. Rock Mech. Rock Eng. **37**, 25–56 (2004)

Polydisperse Segregation Down Inclines: Towards Degradation Models of Granular Avalanches

Benjy Marks, Itai Einav, and Pierre Rognon

Abstract Segregation is a well known yet poorly understood phenomenon in granular flows. Whenever disparate particles flow together they separate by size, density and shape. If we wish to know how to separate particles more efficiently, or even how to keep them mixed together, we require a good understanding of both the phenomenology of the flow, and a quantitative analysis of the evolving particle size distribution towards a steady state. This chapter outlines the continuing effort towards this end, and provides a clue as to the future direction of our research.

Keywords Granular materials • Segregation • Avalanches • Kinetic sieving • Grain size distribution

Many forms of segregation occur in different geometries, with varying granular materials and flow regimes (Jaeger et al. 1996; Knight et al. 1993; Makse et al. 1997). A summary of such segregative phenomena is described in Hutter and Rajagopal (1994) and Ottino and Khakhar (2000). We deal here with the segregation due to size and density differences in particles undergoing shallow free surface flow down an inclined plane.

This type of segregation occurs in natural avalanches, causing small particles to migrate to the base of the flow. These small particles create a lubrication layer which speeds up the flowing layers above it, creating higher shear rates and more crushing. To be able to model an avalanche, we first need to understand how a polydisperse flow segregates before we can investigate the role of degradation in the flow.

In general, research on segregation is studied in a bi-mixture framework. This is a useful path to capture the underlying mechanisms related to industrial separation processes, which often tend to deal with only two grainsizes. Nevertheless, it should

B. Marks (✉) · I. Einav · P. Rognon
School of Civil Engineering, The University of Sydney, Sydney, NSW 2006, Australia
e-mail: benjy.marks@sydney.edu.au; itai.einav@sydney.edu.au; pierre.rognon@sydney.edu.au

S. Bonelli et al. (eds.), *Advances in Bifurcation and Degradation in Geomaterials*, Springer Series in Geomechanics and Geoengineering 11,
DOI 10.1007/978-94-007-1421-2_19, © Springer Science+Business Media B.V. 2011

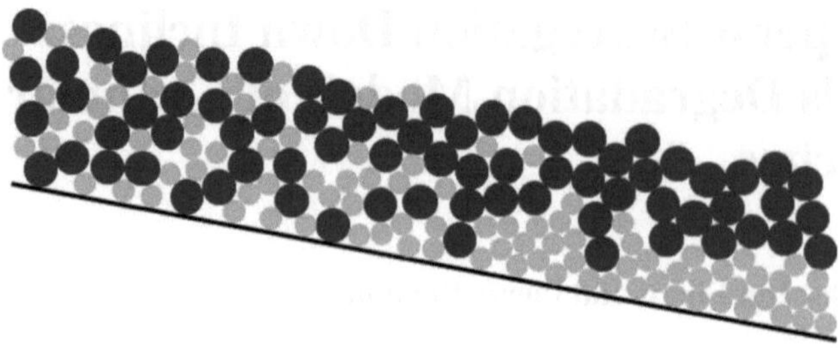

Fig. 1 Effect of kinetic sieving on particles flowing down an inclined slope from *left* to *right*. At the *left*, the particle size distribution is constant with height, while at the *right* it varies with height. This arrangement of sizes, with large particles at the top and smaller particles at the base of the flow is termed inverse grading

be emphasised that even in such problems the idea of identifying only two species is restrictive, since this depends on the tolerance. In some industrial scenarios, this tolerance may end up being of the order of the grainsize contrast between the two species. In other cases, for example in most natural granular flows, there is a broad range of sizes over many orders of magnitudes. An interesting open question, which was not necessarily answered clearly by existing theories, is how continuous polydispersity can affect the time required for complete segregation (Fig. 1).

Segregation along inclined planes is thought to be a result of a mechanism known as kinetic sieving (Savage and Lun 1988). As the particles roll down-slope, void spaces are created by collisions. Since a small particle is more likely to fit into a void than a large particle, we observe a net movement of small particles downwards through the bulk and a corresponding net movement of large particles upwards. The speed of this movement is termed the percolation velocity, w.

In the field of kinetic sieving, two main theories are those described by Savage and Lun (1988) and Gray and Thornton (2005). They both describe thin, rapidly flowing avalanches of bidisperse mixtures down an inclined plane. Savage and Lun use an information entropy argument to develop a solution which describes the concentration profile. Gray and Thornton usé a binary mixture theory to find concentration shocks which define their solutions.

Both theories include a mean segregation velocity that dictates the rate at which segregation occurs. Gray and Thornton have defined the mean segregation velocity as $q^{GT} = \pm\frac{B}{c} g\cos(\theta)$, where θ is the inclination of the slope, c is the interparticle drag and B is a dimensionless parameter. Savage and Lun proposed a different, but related equation of the form $q^{SL} = D_a\dot{\gamma}(\bar{q}_b - \bar{q}_a)$, where D_a is the diameter of the large particles, $\dot{\gamma}$ is the shear strain rate and $\bar{q}_a$, $\bar{q}_b$ are non-dimensional volume averaged velocities of the two constituents in the down-slope direction.

Gray and Thornton use a dimensionless parameter to indicate the specific initial conditions of the problem, the segregation number $S_r = \frac{q^{GT}L}{UH}$, where L, H and U are the typical avalanche length, thickness and down-slope velocity magnitudes respectively.

The percolation velocity of a particular species with volumetric concentration ϕ is defined by w, and for the Gray and Thornton solution is described by

$$w^{GT} = \pm \frac{B}{c} g \cos(\theta)(1 - \phi)$$

More recently, May et al. (2010) have investigated the problem by combining the segregation velocities proposed by Gray and Thornton (2005) and Savage and Lun (1988) and using

$$w^{May} = \dot{\gamma}(z) w^{GT}$$

where $\dot{\gamma}(z)$ is a function of the height only and has to be assumed within the context of May et al's theory. Here we have used the specific form of Gray, Thornton and May's proposed convex flux $f(\phi) = \phi(\phi - 1)$ (Gray and Thornton 2005; May et al. 2010).

Khakhar et al. (1999) investigate not only the role of differing particle size, but also particle density between species, again in a bi-mixture. By comparing their analytic work with numerical simulations, they make progress towards unifying a theory between two modes of segregation in natural inclined granular flows.

The above theories are restricted to bi-mixtures with moderate grainsize contrast, and neglect the effect of spontaneous percolation, whereby particles much smaller than the average can sink rapidly to the base of a flow. Most theories do not provide insight on how the size contrast between the species should affect the segregation time. An exception is given by Savage and Lun (1988), which motivates a mathematical form where the size contrast tends to enhance the segregation speed. We believe that a generalised polydisperse segregation theory can resolve this question in the most insightful way.

The solution we will present (only briefly) has the advantage of resolving $\dot{\gamma}$ as a function of height and time towards a steady state (Marks et al., Unpublished manuscript). The inclusion of shear rate in the percolation velocity is chosen to account for the rate of creation of voids which dictate the kinetic sieving mechanism. For inclined plane flow, a Bagnold velocity profile of the form $\dot{\gamma} \propto \sqrt{H - z}$ is representative of the equivalent mono-disperse system (MiDi 2004). A more realistic assumption for two species is a modified Bagnold velocity profile such as in Rognon et al. (2007).

The form of the flow equations following the method first described by Gray and Thornton (2005) are invariably some form of hyperbolic partial differential equation. These equations are notoriously difficult to solve both numerically and analytically (LeVeque 2002). We expect discontinuities (or in this case concentration shocks) to form in our solutions, but we are hampered considerably by a lack of simple solution tools to evaluate new theories.

One rather successful method of investigation has been to describe the problem as a cellular automaton (Marks and Einav 2011). This very rudimentary model can resolve a first order solution to the problem while retaining the essential physics. This method is attractive as it handles discontinuities naturally and without qualms. It has been used to create the images for this chapter. With increasing computational time, these stochastic simulations approach the smooth behaviour of the partial differential equations.

For a conservation equation to model polydisperse segregation, a number of improvements are needed to be implemented. Firstly, the idea of a continuous internal variable (Ramkrishna 2000) representing the grainsize s has to replace the hard-wired two phases in the existing models. This allows us to represent the solid fraction of a particular phase as $\phi(s)$. Secondly, each grainsize has to have its own density $\rho(s)$. Most importantly, conservation of momentum has to be used to derive a percolation velocity that explicitly includes shear rate $\dot{\gamma}$ and some function of s and $\rho(s)$ that accounts for the actual particle size and density. In Marks et al. (Unpublished manuscript) we arrive at:

$$w(z,s,t) = \dot{\gamma}(z,t)\left(\frac{s}{\bar{s}(z,t)} - \frac{\rho(s)}{\bar{\rho}(z,t)}\right)\frac{g\cos\theta}{C}$$

This form of the percolation velocity w reduces our number of fitting parameters down to one, namely C the co-efficient of interparticle friction (note that it is now dimensionless).

It was shown in Rognon et al. (2007) that there is a strong dependence of $\dot{\gamma}$ on the grain size distribution. We quantify this by including a shear rate dependent constitutive equation such as that described in MiDi (2004), concluding that an appropriate shear rate can most simply be described by

$$\dot{\gamma}(z,t) = \frac{\sqrt{3g\cos\theta}(\tan\theta - \mu_c)}{\sqrt{\bar{s^2}(z,t)4\pi}}\sqrt{H-z}$$

Here μ_c is the critical shear parameter at which the onset of flow occurs. Results are shown for this model in a bidisperse case in Fig. 4, and a polydisperse case with a size difference of four times between the smallest and largest particle sizes in Fig. 5.

Figure 2 shows how, using w^{GT}, the time for complete segregation does not vary with the initial concentration of small particles. Figure 3 shows how using w^{May}, large particles move faster than the slower particles, as described in Marks and Einav (2011) as being a result of the depth dependent shear rate.

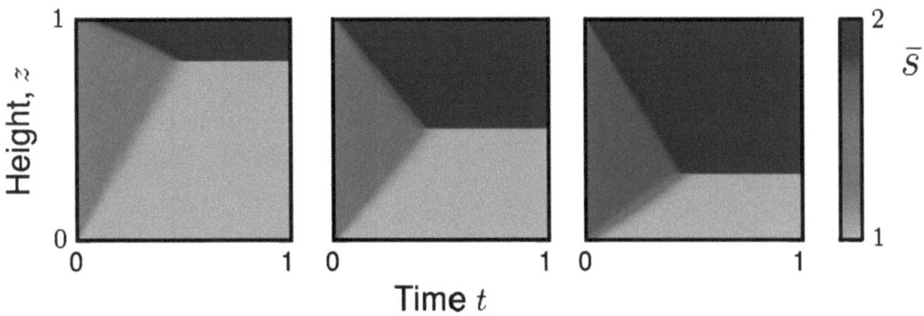

Fig. 2 Time evolution for inclined plane flow in one spatial dimension using w^{GT}. The system is initially filled with a bidisperse mixture with 30%, 50% and 80% (*left* to *right*) large concentration by volume. *Colourbar* represents the large particle concentration ϕ

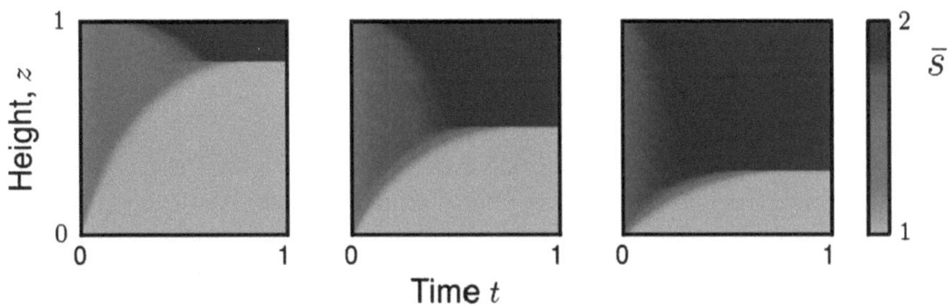

Fig. 3 Time evolution for inclined plane flow in one spatial dimension using w^{May} and $\dot{\gamma} = \sqrt{H - z}$. The system is initially filled with a bidisperse mixture with 30%, 50% and 80% (*left* to *right*) large concentration by volume. *Colourbar* represents the large particle concentration ϕ

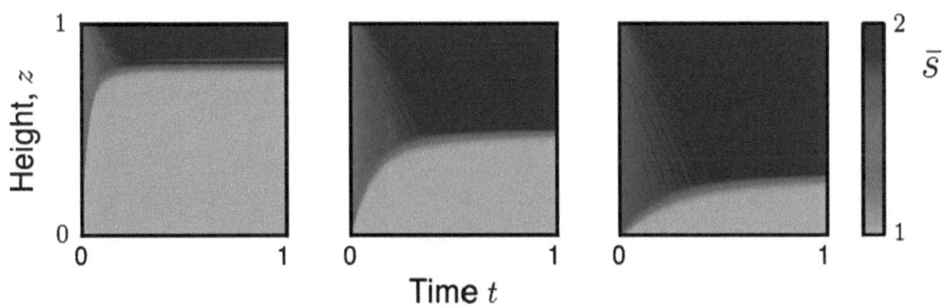

Fig. 4 Time evolution for inclined plane flow in one spatial dimension using $w(z, s, t)$ and $\dot{\gamma}(z, t)$. The system is initially filled with a bidisperse mixture with 30%, 50% and 80% (*left* to *right*) of size $s = 2$. The remainder has size $s = 1$. *Colourbar* represents the average particle size $\bar{s}$

As seen in Fig. 4, the time for complete segregation is now distinctly a function of the initial particle size distribution, even for a bi-mixture. Increasing the concentration of small particles decreases the time for segregation. For a polydisperse mixture, the time for segregation actually varies with size range. This is encapsulated in the percolation velocity, w, which now is different for each particle size.

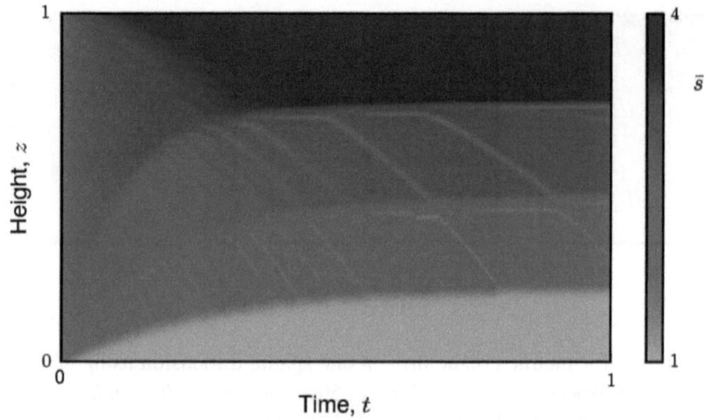

Fig. 5 Time evolution for inclined plane flow in one spatial dimension using $w(z, s, t)$ and $\dot{\gamma}(z, t)$. The system is initially filled with a polydisperse mixture of sizes varying from $s = 1$ to $s = 4$ and has volume fraction spread uniformly between these two values. *Colourbar* represents the average particle size $\bar{s}$

Each grainsize creates its own shock wave as shown in Fig. 5. These waves split the solution into regions of differing, uniform, particle size. At steady state, the particles are stacked in descending order of size so that the smallest particle is at the bottom of the flow, and the largest at the top.

The polydisperse framework captures the behaviour of particles with quantified size and density contrast, and describes the evolution of the grain size distribution at any point in space. These initial 1D numerical solutions already show the significance of the polydisperse description in terms of the grain size dependent shear strain rate, the time for complete segregation and its dependence on the initial grain size distribution.

With 2D simulations, natural avalanches and industrial mixing processes may be understood, modelled and optimised in a cohesive framework for the first time.

Acknowledgements We would like to thank the members of the Particles, Grains and Complex Fluids Group at the University of Sydney, in particular Dr. Bjornar Sandnes for useful discussions. IE acknowledges grant DP0986876 from the Australian Research Council.

References

J. Gray, A. Thornton, A theory for particle size segregation in shallow granular free-surface flows. Proc. R. Soc. A **461**(2057), 1447–1473 (2005)

K. Hutter, K. Rajagopal, On flows of granular materials. Continuum Mech. Thermodyn. **6**(2), 81–139 (1994)

H. Jaeger, S. Nagel, R. Behringer, Granular solids, liquids, and gases. Rev. Mod. Phys. **68**(4), 1259–1273 (1996)

D. Khakhar, J. McCarthy, J. Ottino, Mixing and segregation of granular materials in chute flows. Chaos **9**, 594 (1999)

J. Knight, H. Jaeger, S. Nagel, Vibration-induced size separation in granular media: the convection connection. Phys. Rev. Lett. **70**(24), 3728–3731 (1993)

R. LeVeque, *Finite Volume Methods for Hyperbolic Problems* (Cambridge University Press, Cambridge, 2002)

H. Makse, S. Havlin, P. King, H. Stanley, Spontaneous stratification in granular mixtures. Nature **386**(6623), 379–382 (1997)

B. Marks, I. Einav, A cellular automaton for segregation during granular avalanches, Granular Matter 13, Springer Berlin/Heidelberg, 211–214 (2011)

B. Marks, I. Einav, P. Rognon, A polydisperse framework for size segregation in inclined chute flow. Unpublished manuscript

L. May, M. Shearer, K. Daniels, Scalar conservation laws with nonconstant coefficients with application to particle size segregation in granular flow. J. Nonlinear Sci. **20**, 1–19 (2010)

G. MiDi, On dense granular flows. Eur. Phys. J. E **14**(4), 341–365 (2004)

J. Ottino, D. Khakhar, Mixing and segregation of granular materials. Ann. Rev. Fluid Mech. **32**(1), 55–91 (2000)

D. Ramkrishna, *Population Balances: Theory and Applications to Particulate Systems in Engineering* (Academic, San Diego, 2000)

P. Rognon, J. Roux, M. Naaim, F. Chevoir, Dense flows of bidisperse assemblies of disks down an inclined plane. Phys. Fluids **19**, 059101 (2007)

S. Savage, C. Lun, Particle size segregation in inclined chute flow of dry cohesionless granular solids. J. Fluid Mech. **189**, 311–335 (1988)

Numerical Modelling of Interaction Between Snow Avalanche and Protective Structures

M. Wawra, Y. Wang, and W. Wu

Abstract A numerical study of the interaction between granular flow and an obstacle on an inclined plane is presented. A depth-averaged, two-dimensional Savage-Hutter-type model is used. The underlying differential equations are usually solved by finite difference. The model, which has been proved to perform well for steady granular flow, turns out to be inappropriate for the interaction between granular flow and obstacles. In this paper, the quality of the numerical solution is significantly improved by grid refinement. We make use of the Adaptive Mesh Refinement, where only local grid refinement around the obstacle is needed.

Keywords Granular flow • Obstacle • Savage-Hutter model • Finite differences • Grid refinement

1 Introduction

Snow avalanches are a major natural hazard in alpine regions. Rational design of protective structures requires understanding of the dynamics of snow avalanches. Numerical modelling of such protection systems can help to understand the dynamics. In the last two decades the Savage-Hutter theory (Savage and Hutter 1989, 1991) with its extensions for three dimensions (Gray et al. 1999; Greve et al. 1994; Hutter et al. 1993; Pudasaini and Hutter 2003) for gravity-driven, free-surface granular flows has become an attractive approach to describe the behaviour of dense granular flows. It consists of depth-integrated balance laws of mass and momentum and has a similar mathematical structure as the shallow-water equation, known from hydrodynamics. The granular material is considered

M. Wawra (✉) · Y. Wang · W. Wu
Institut für Geotechnik, Universität für Bodenkultur, Vienna 1180, Austria

S. Bonelli et al. (eds.), *Advances in Bifurcation and Degradation in Geomaterials*, Springer Series in Geomechanics and Geoengineering 11,
DOI 10.1007/978-94-007-1421-2_20,

incompressible and obeys a dry Coulomb-type friction law with a constant internal friction angle. Nonlinear earth pressure coefficients are introduced to describe the ratio of overburden pressure and normal pressure in down-flow- and cross-flow-direction. Altogether, the internal friction angle and the bed friction angle, which measures the friction between the moving granular mass and the bed surface, enter the model as only material parameters. Several numerical methods were applied to solve the Savage-Hutter equations. We use the NOC (Non-Oscillatory Central Differencing) scheme (Nessyahu and Tadmor 1990, 1998) with a Minmod TVD (Total Variation Diminishing) limiter (Harten 1983), which provides the best performance (Wang et al. 2004). The model is further extended by Adaptive Mesh Refinement (AMR) (Berger and Colella 1989; Berger and Oliger 1984). AMR is a well known technique for local grid refinement. There exists little work in the literature on the interaction of avalanches and obstructions. Experimental and numerical studies of the interaction of dense-flow snow avalanches with obstacles are scarce. Gray et al. (2003) used a generalized hydraulic theory to study uniform flows around obstacles. Tai et al. (2001) used the Savage-Hutter theory to analyze the perturbation of a tetrahedral obstruction in a steady dense flow. Chiou (2006) and Chiou et al. (2005) carried out laboratory experiments of finite mass granular avalanches interacting with tetrahedral and cuboid obstructions. In this paper some numerical results obtained with AMR are presented.

2 Governing Equations

The basic Savage-Hutter model (1989) is two-dimensional, but has been extended to three dimensions by Hutter et al. (1993), Greve et al. (1994), Gray et al. (1999) and Pudasaini and Hutter (2003). The granular mass is assumed to be dry, cohesionless, volume preserving, isothermal and shallow, which allows depth integration. These reasonable simplifications lead to an elegant mathematical formulation. Some features of this model were already known from shallow-water equations. To model a complex basal topography a curvilinear coordinate system is used. The coordinate system follows the so called reference surface. On this reference surface a function is defined that accounts for the local differences of the real basal topography and the reference surface. This function will be referred to as elevation function. The elevation function has to be small in comparison with flow depth. This requires a well chosen reference surface. In the following the reference surface is assumed to follow the mean down-slope bed topography. An orthogonal curvilinear coordinate system $oxyz$ is defined by setting the z-coordinate normal to the reference surface, while x and y are tangential. The x-axis is assumed to follow the thalweg, the y-axis is normal in cross-slope direction. The inclination angle ζ is used to define the reference surface as function of the down-slope coordinate x, but is independent of the cross-slope coordinate y. In the curvilinear coordinate system described above, the leading-order, dimensionless, depth-integrated mass and momentum balance equations can be written as

$$\frac{\partial \mathbf{w}}{\partial t} + \frac{\partial \mathbf{f}(\mathbf{w})}{\partial x} + \frac{\partial \mathbf{g}(\mathbf{w})}{\partial y} = \mathbf{s}(\mathbf{w}) \tag{1}$$

where $\mathbf{w}$ denotes the vector of conservative variables, $\mathbf{f}$ and $\mathbf{g}$ represent the transport fluxes in x-, respectively y-direction and $\mathbf{s}$ denotes the source term. They are

$$\mathbf{w} = \begin{pmatrix} h \\ hu \\ hv \end{pmatrix}, \mathbf{f} = \begin{pmatrix} hu \\ hu^2 + \beta_x h^2/2 \\ huv \end{pmatrix}, \mathbf{g} = \begin{pmatrix} hv \\ huv \\ hv^2 + \beta_y h^2/2 \end{pmatrix}, \mathbf{s} = \begin{pmatrix} 0 \\ hs_x \\ hs_y \end{pmatrix}, \tag{2}$$

where

$$\begin{aligned} \beta_x &= \varepsilon \cos\zeta K_x^b, \\ \beta_y &= \varepsilon \cos\zeta K_y^b, \\ s_x &= \sin\zeta - \frac{u}{|\mathbf{u}|} \tan\delta(\cos\zeta + \lambda\kappa u^2) - \varepsilon\cos\zeta\frac{\partial z_b}{\partial x}, \\ s_y &= -\frac{v}{|\mathbf{u}|} \tan\delta(\cos\zeta + \lambda\kappa u^2) - \varepsilon\cos\zeta\frac{\partial z_b}{\partial y}. \end{aligned} \tag{3}$$

h is the avalanche thickness, $\mathbf{u} = (u,\ v)$ is the depth averaged velocity vector, with components u and v in down- and cross-slope-direction, δ is the bed friction angle, $\kappa = -\partial\zeta/\partial x$ is the local curvature, z_b is the above mentioned elevation function, ε is the aspect ratio of the height and the horizontal extent of the avalanche and K_x and K_y are called earth pressure coefficients and describe the ratio of normal pressure in down-, cross-slope direction to the overburden pressure. Two different functions are used for the earth pressure coefficients depending on whether the motion is dilatational or compressional:

$$K^{x_{act/pass}} = 2\left(1\mu\sqrt{1-\cos^2\varphi/\cos^2\delta}\right)\Big/\cos^2\varphi - 1, \tag{4}$$

$$K^{x_{act/pass}}_{y_{act/pass}} = \frac{1}{2}\left(K^{x_{act/pass}} + 1\mu\sqrt{(K^{x_{act/pass}}-1)^2 + 4\tan^2\delta}\right), \tag{5}$$

where subscripts *act* and *pass* stand for active (dillitational) and passive (compressional). All Eqs. 1–5 are presented in non-dimensional form. The physical counterparts are given by

$$\begin{aligned} (x, y, h, z_b)_{\text{dim}} &= (Lx, Ly, Hh, Hz_b)_{\text{nondim}}, \\ (t, u, v, k)_{\text{dim}} &= \left(\sqrt{L/g}t, \sqrt{Lg}u, \sqrt{Lg}v, k/R\right)_{\text{nondim}}, \end{aligned} \tag{6}$$

where g is the constant of gravitational acceleration, L and H are the typical length and thickness of the avalanche and R is the radius of curvature of the reference surface.

3 Numerical Results

The Savage-Hutter equations for rapidly moving granular flows are hyperbolic equations. An appropriate numerical scheme must be able to handle shock waves, occurring if the velocity of the flow changes from supercritical to subcritical. This can be observed in experiment when the material reaches the run-out zone or obstructions are encountered in the slope. The mass decelerates abruptly and a shock wave propagates against the flow direction. In the past decades various numerical techniques were applied to solve the Savage-Hutter equations. Wang and Hutter (2001) analyzed various traditional high-order numerical integration techniques. The non-oscillatory central scheme (NOC) (Lie and Noelle 2003; Nessyahu and Tadmor 1990, 1998) with a Minmod Total Variation Diminishing (TVD) limiter (Harten 1983) for cell reconstruction has shown the best performance (Wang et al. 2004).

Our numerical simulations are based on laboratory experiments performed and described by Chiou (2006). Chiou used a Plexiglas chute, on which different objects, such as tetrahedral wedges, were fixed and interacted with granular flows, consisting of different sorts of sand, which were released from a holding cap on the top of the chute.

For demonstration (see Fig. 1), we use a 30 dimensionless units long and 20 dimensionless units wide chute. We consider chute flow down an inclined slope. The slope consists of three sections, namely an inclined section, where the granular mass accelerates, a transition zone and a horizontal run-out zone. The inclined section is 20 units long, followed by 4 units long transition zone and 6 units long run-out zone. The inclination angle of the chute is $\zeta = 40°$. A tetrahedra is placed in the center of the slope. The distance between the lower edge of the tetrahedra and the upper edge of the slope is chosen to be 17. The base of the tetrahedral is an equilateral triangle with the length of 4 units. The height of the tetrahedra is 1.2 units. Sand with the internal friction angle $\varphi = 35°$ and the basal friction angle $\delta = 30°$ is used in our calculations. Initially the sand mass is contained in an elliptical cap, which is 4 units long, 2 units wide and 1 unit high. The cap is placed in the center of the chute. The whole chute is covered by a rectangular basic grid with 301×201 grid points. The numerical calculations with different grids showed that the quality of the solution is largely dependent on the grid size. Most significant has proven to be the area around the obstruction, especially at the front side where granular flow interacts with the obstruction. To improve the quality of the solution the area around the obstacle is refined locally. Adaptive Mesh Refinement (AMR) (Berger and Colella 1989; Berger and Oliger 1984) is applied, which is a well known technique for local grid refinement. Selected cells of the basic grid are overlaid by new grids,

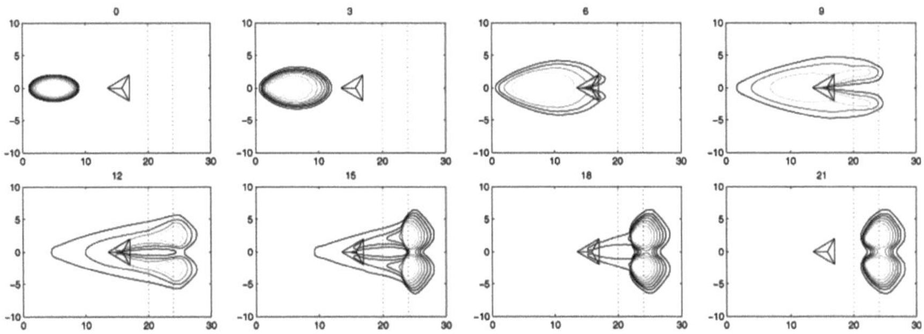

Fig. 1 Ground view of the interaction between granular flow and tetrahedral obstruction at dimensionless time $t = 0, 3, 6, 9, 12, 15, 18$ and 21. The contour lines indicate the depth of the mass. The first line contours a depth of 0.02. The *dotted line* at $x = 20$ marks the end of the slope of the chute, followed by the transition zone. The *dotted line* at $x = 24$ marks the beginning of the run-out zone

which are two times as fine as the basic grid. These new grids build a first level of refinement. The cells within these new grids can be overlaid by still new grids to form a second level of refinement and so on. For the simulation shown in Fig. 1, two levels of refinement are applied. The second level grid covers a rectangular array, which overlaps the basal triangle of the tetrahedral by a minimal distance of 1 unit in each direction. The first level grid overlaps the second level grid by 5 grid points in each direction. The numerical simulation shown in Fig. 1 estimates the impact of tetrahedral obstructions on granular flows very well. The granular mass is split and deflected by the wedge. In the run-out zone two humps are deposited. Also the shock waves at the fronts of the depositions are captured well, as seen in the pictures at times $t = 15$ to 21.

4 Conclusions

The depth-averaged, two-dimensional, Savage-Hutter model has proven to simulate unimpeded shallow granular flow appropriately in the past. However, if topographies with large gradients are present, as is the case when obstructions are considered, the numerical solution becomes rather poor. For the tetrahedral obstructions, the solution quality can be significantly improved by grid refinement. It is sufficient to refine the grid locally around the obstruction, which is accomplished with the Adaptive Mesh Refinement (AMR) method. Well instrumented experiments of tetrahedral obstructions are needed to validate the model. Obstructions with steep, plane fronts, like walls or dams, are more difficult to model, since the mass is not just deflected to the sides as for the tetrahedral wedge. In this case, some granular material will be held back by the wall, which provides some challenge for the numerical simulation. The successive mass moves along the surface of the deposition. This complex behaviour cannot be fully captured by a depth-averaged model.

Acknowledgements We thank the Austrian Science Fund (FWF) for the financial support. We thank S. P. Pudasaini for fruitful discussions.

References

M.J. Berger, P. Colella, Local adaptive mesh refinement for shock hydrodynamics. J. Comput. Phys. **82**, 64–84 (1989)

M.J. Berger, J. Oliger, Adaptive mesh refinement for hyperbolic partial differential equations. J. Comput. Phys. **53**, 484–512 (1984)

M.-C. Chiou, Modelling dry granular avalanches past different obstructions. Ph.D. thesis, University of Darmstadt, Darmstadt, 2006

M.-C. Chiou, Y. Wang, K. Hutter, Influence of obstacles on rapid granular flows. Acta Mech. **175**, 105–122 (2005)

J.M.N.T. Gray, M. Wieland, K. Hutter, Gravity driven free surface flow of granular avalanches over complex basal topography. Proc. R. Soc. Lond. A **455**, 1841–1874 (1999)

J.M.N.T. Gray, Y.-C. Tai, S. Noelle, Shock waves, dead zones and particle free regions in rapid granular free-surface flows. J. Fluid Mech. **491**, 161–181 (2003)

R. Greve, T. Koch, K. Hutter, Unconfined flow of granular avalanches along a partly curved surface. Part I: theory. Proc. R. Soc. Lond. A **445**, 399–413 (1994)

A. Harten, High resolution schemes for hyperbolic conservation laws. J. Comput. Phys. **49**, 357–393 (1983)

K. Hutter, M. Siegel, S.B. Savage, Y. Nohguchi, Two-dimensional spreading of a granular avalanche down an inclined plane. Part I: theory. Acta Mech. **100**, 37–68 (1993)

K.A. Lie, S. Noelle, An improved quadrature rule for the flux-computation in staggered central difference schemes in multidimensions. J. Sci. Comp. **18**, 69–80 (2003)

H. Nessyahu, E. Tadmor, Non-oscillatory central differencing for hyperbolic conservation laws. J. Comput. Phys. **87**, 408–463 (1990)

H. Nessyahu, E. Tadmor, Nonoscillatory central for multidimensional hyperbolic conservation laws. SIAM J. Sci. Comput. **19**, 1892–1917 (1998)

S.P. Pudasaini, K. Hutter, Rapid shear flows of dry granular masses down curved and twisted channels. J. Fluid Mech. **495**, 193–208 (2003)

S.B. Savage, K. Hutter, The motion of a finite mass of granular material down a rough incline. J. Fluid Mech. **199**, 177–215 (1989)

S.B. Savage, K. Hutter, The dynamics of avalanches of granular materials from initiation to runout. I. Analysis. Acta Mech. **86**, 201–223 (1991)

Y.-C. Tai, J.M.N.T. Gray, K. Hutter, S. Noelle, Flow of dense avalanches past obstructions. Ann. Glaciol. **32**, 281–284 (2001)

Y. Wang, K. Hutter, Comparison of numerical methods with respect to convectively-dominated problems. Int. J. Numer. Meth. Fluids **37**, 721–745 (2001)

Y. Wang, K. Hutter, S.P. Pudasaini, The Savage-Hutter theory: a system of partial differential equations for avalanche flows of snow, debris and mud. J. Appl. Math. Mech. **84**, 507–527 (2004)

Crack Propagations on the Rock Face at Glacier Point of the Yosemite National Park After the 1999 Rockfall

K.T. Chau and P. Lin

Abstract This study applies finite element analysis to examine the mechanism of crack propagations observed at the Glacier Point of the Yosemite National Park following the June 13, 1999 rockfall event. Extensive crack propagations were observed on a rectangular area of about 18 m by 23 m of rock exfoliation sheet on a rock face about 550 m above the Yosemite Valley 54 days following the rockfall by helicopter flights and by photographs taken by from the valley. This paper models the crack pattern on the hook-shaped rock sheet by using a plane stress analysis subjected to various ratios of vertical to horizontal strain increments. The present finite element analysis uses a code called RFPA that adopts a reduced modulus approach to model the damage process of rock. A mesh of 160 by 160 elements was used to model a rock face of about 35 m by 35 m, and a Weibull distribution is used to model the heterogeneity of the rock face. We found that when the vertical to horizontal strain ratio is smaller than one, the crack pattern closely resembles that observed at the Glacier Point. Thus, the present paper provides a plausible mechanism for the observed crack propagation.

Keywords Rockfall • Progressive crack propagation • Yosemite National Park • Finite element analysis • Damage

---

K.T. Chau (✉)
Department of Civil and Structural Engineering, The Hong Kong Polytechnic University, Kowloon, Hong Kong, China
e-mail: cektchau@polyu.edu.hk

P. Lin
Department of Hydraulic Engineering, Tsinghua University, Beijing, China
e-mail: celinpe@tsinghua.edu.cn

S. Bonelli et al. (eds.), *Advances in Bifurcation and Degradation in Geomaterials*, Springer Series in Geomechanics and Geoengineering 11,
DOI 10.1007/978-94-007-1421-2_21,

1 Introduction

Yosemite Valley is within the Yosemite National Park, one of the most crowded and popular park in the US National Park system. The valley is about 1 km deep, glacially carved canyon in granite located in the central Sierra Nevada. Since 1857, various types of landslides, particularly rockfalls, rock slides, debris slides, and debris flows, have been recognized in the Yosemite Valley. A more systematic reporting of rockfall began in 1916 in the monthly National Park Service (NPS) Superintendent's reports. Between 1857 and 2004, 14 people have been killed and at least 62 injured by more than 541 landslides that have been documented in Yosemite National Park (Wieczorek and Snyder 2004; Guzzetti et al. 2003). Therefore, rockfall, rock slide and rock avalanche appear to be a continuous natural process of mountain formation in the Yosemite Valley. Chau et al. (2003, 2004) shown that the rockfall frequency at the Yosemite Valley is comparable to that of Hong Kong.

Starting from 1998, a series of rockfalls from an elevation of about 1,800 m on the north-facing wall of Glacier Point have threatened employees and visitors to Camp Curry, which was established in 1899. One of the major rockfalls in this series occurred on June 13, 1999 (see Fig. 1a). Following the June 13 rockfall event, the National Park Service (NPS) and U.S. Geological Survey (USGS) began monitoring the release area shown in Fig. 1b. The sequence of crack development at the release area was observed from June 13 to August 9, 1999 (Wieczorek and Snyder 1999) by helicopter and from the valley using a telescope. A major part of cracking was occurred on June 15 to June 16, showing a series of new extensional fractures and fine hairline cracks in fresh rock. The exposed thickness of exfoliation sheets at this site suggests that the thickness of the unstable mass could range from 1 to 2 m thick.

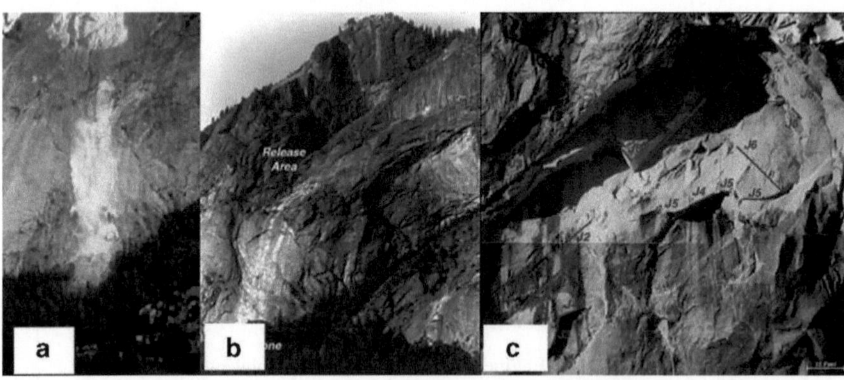

Fig. 1 (**a**) Rockfall on June 13, 1999 at Glacier Point at Yosemite National Park (Photograph by Lloyd De Forrest) (Guzzetti et al. 2003); (**b**) Release area at Glacier Point at Yosemite National Park, showing all joint sets; (**c**) close-up of release area (Wieczorek and Snyder 1999)

This cracking observation provides a rare opportunity in monitoring the process of cracking that led to rock dislodgement and subsequent rockfall event. In addition, it provides valuable data of field experiments that can be used for calibration of mechanics or mathematical modeling of weathering and degradation.

2 Finite Element Analysis

In this section, we will describe the finite element method (FEM) RFPA that we use. The acronym RFPA is for "rock failure process analysis". Comparing to Fig. 2b of Wieczorek and Snyder (1999), the window area becomes apparent only after the rockfall of 1999.

2.1 *Brief Description of RFPA*

In the present study, RFPA is used and it uses a equivalent elastic finite analysis is used to model failure of heterogeneous rock. In essence, in this RFPA elastic moduli and yield strength are assigned to each element whereas these elastic and strength parameters may vary from one element to another following a Weibull distribution for strength parameter σ and its cumulative probability:

$$f(\sigma) = \frac{m}{\sigma_0}\left(\frac{\sigma}{\sigma_0}\right)^{m-1} \exp\left[-\left(\frac{\sigma}{\sigma_0}\right)^m\right]; \quad P(\sigma) = 1 - \exp\left[-\left(\frac{\sigma}{\sigma_0}\right)^m\right] \tag{1}$$

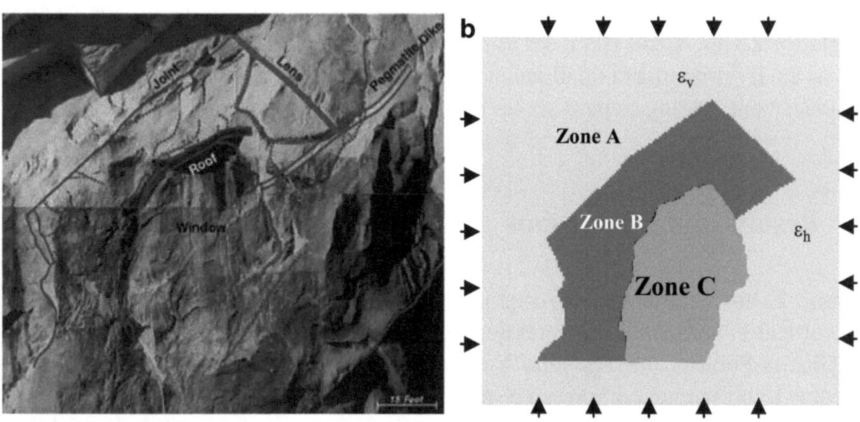

Fig. 2 (**a**) Crack propagation at the hook (*dotted lines*) appears on the north face of Glacier Point at Yosemite National Park (Wieczorek and Snyder 1999); (**b**) Finite Element Modeling of the release area at Glacier Point of the Yosemite National Park

where σ_0 is the mean strength parameter and m is a dimensionless parameter characterizing the degree of heterogeneity. A high m value corresponds to a homogeneous solid, whereas a low m value represents a heterogeneous solid. Once the prescribed yield strength is exceeded, reduced elastic moduli will be assigned to the "failed" element (Tang 1997). One major drawback of such approach is the arbitrariness of the choice of Weibull parameters. To alleviate this problem, Wong et al. (2006) formulated the microstructural basis for the Weibull statistical parameters for strength. The degradation factor for characterizing the ratio between the residual and peak strength in an element after damage is set to 0.1, as used by Tang (1997). The reduced modulus approach will account approximately for element failure and thus in turn will simulate the stress re-distribution and crack propagation process.

2.2 *FEM Model*

The June 13 rockfall apparently created a "window" on the rock face as an elliptical shape exfoliation sheet fell, as shown in Fig. 2a. This paper models the cracking pattern on the hook-shaped rock sheet by assuming a plane stress analysis subjected to various ratios of vertical to horizontal strain increments. A mesh of 160×160 element is used. A square domain of 35 m×35 m on the rock face is modeled by three material zones: Zone A—the rigid zone, Zone B—the cracking zone, and Zone C—the cavity zone (see Fig. 2b). Therefore, Zones A, B and C model the undamaged rock, the cracking zone and the fallen "elliptical window" zone respectively. In particular, the m values of Zones A, B, and pre-existing joint are assumed as 1,000, 5 and 2 respectively (for both elastic and strength distributions). The Young's moduli of Zones A, B, and pre-existing joint are 3 MPa, 60 kPa and 0.6 kPa respectively, whereas the cohesions of them are 3 kPa, 40 Pa and 4 Pa respectively. The friction angle for Zones A and B are set as a fixed value of 45°. Properties of Zone C are set as cavity properties (small values that do not cause numerical instability). The displacement loading steps is set to 0.2 mm.

3 Results and Discussions

Figure 3 shows the cracking patterns (shown as dark elements) for various ratios of vertical to horizontal compression (1/0, 2/1, 1/1, and 1/2) for the release area at Glacier Point of the Yosemite National Park. Figures 3a–d are for the cases of vertical to horizontal compressive strain of 1:0, 2:1, 1:1, and 1:2 respectively. The results shown in Fig. 3 is for those at an equivalent strain of 6×10^{-4}. The folding pattern of rocks above the release area shown in Fig. 2 suggests that the local stress or strain in the release area right above the ledge is generated by a complicated regional stress, and is not geostatic. The case of $\varepsilon_h/\varepsilon_v = 0$ is shown in Fig. 3a,

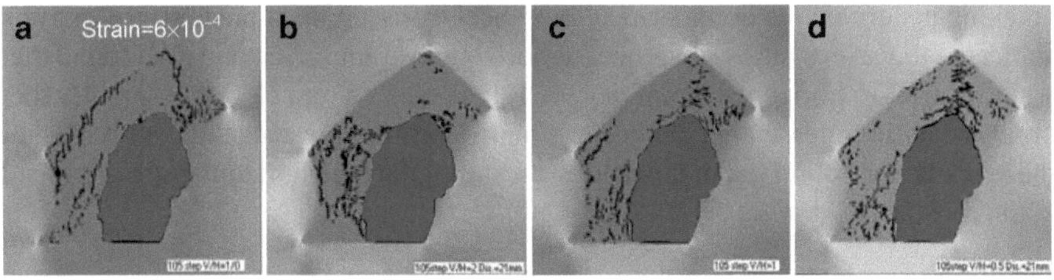

Fig. 3 Cracking pattern of various ratios of vertical to horizontal compression (1/0, 2/1, 1/1, and 1/2) for the release area at Glacier Point of the Yosemite National Park

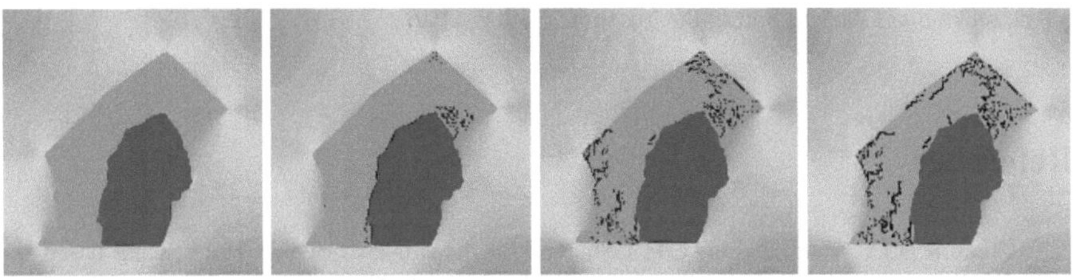

ε_v=1.2×10^{-4} ε_v=3.6×10^{-4} ε_v=6.0×10^{-4} ε_v=10.8 ×10^{-4}

Fig. 4 Cracking pattern for various vertical strain levels with vertical to horizontal compression ratio being 1/1 within the release area at Glacier Point of the Yosemite National Park

characterizing vertical cracking along the upper edge of the hook and in the region close to the tip of the upper hook area. Figure 3b shows the crack pattern for the case of $\varepsilon_v/\varepsilon_h = 2$. The cracking is concentrated at the lower part of the hook whereas the vertical cracks along the upper edge of the hook and the tip of hook shown in Fig. 3a disappears. When $\varepsilon_v/\varepsilon_h = 1$, two regions of distributed cracks appears one at the top of the hook and another at the bottom of the hook as shown in Fig. 3c. Figure 3d shows the crack pattern for $\varepsilon_v/\varepsilon_h = 0.5$. Figure 3 reveals that the crack pattern observed in the field compares more favorably with that of higher horizontal strain.

Figure 4 plots the cracking pattern at various levels of vertical strain with a vertical to horizontal strain ratio being 1/1. At a strain level of 0.012% strain, there is no crack initiation, at 0.035% strain cracks primarily initiate at the upper part of the "hook", at 0.06% more cracks initiate at both the lower and the upper parts of the hook, and eventually extensive cracking occurs in most of the "hook" at 0.108% strain.

4 Conclusions

This study uses a elastic damaged FEM analysis to examine the regional stress effect on the cracking pattern at the release area at Glacier Point of the Yosemite National Park, following the June 13, 1999 rockfall event. This paper models the cracking pattern on the hook-shaped rock sheet by assuming a plane stress analysis subjected to various ratios of vertical to horizontal strain increments. We found that when vertical to horizontal strain ratio < 1, the crack pattern closely resembles that observed at the release area at Glacier Point. Thus, the present study suggests that the crack pattern observed in the field is a result of the high horizontal stress trapped at the rock face.

Acknowledgements This study is fully supported by the Hong Kong Polytechnic University (Project No. 1-BBZF of PolyU) and by "Special Foundation for public welfare industry of the ministry of water resources (2010001035)" of China. The author is grateful to Prof. C.A. Tang in providing the computer program RFPA and Dr. Wieczorek of USGS in providing related materials on the rockfall at Glacier Point of the Yosemite National Park.

References

K.T. Chau, R.H.C. Wong, J. Lui, C.F. Lee, Rockfall hazard analysis for Hong Kong based on rockfall inventory. Rock Mech. Rock Eng. **36**(5), 383–408 (2003)

K.T. Chau, Y.F. Tang, R.H.C. Wong, GIS based rockfall hazard map for Hong Kong. Int. J. Rock Mech. Min. Sci. **41**(3), 530 (2004)

F. Guzzetti, P. Reichenbach, G.F. Wieczorek, Rockfall hazard and risk assessment in the Yosemite Valley, California, USA. Nat. Hazards Earth Syst. Sci. **3**, 491–503 (2003)

C.A. Tang, Numerical simulation on progressive failure leading to collapse and associated seismicity. Int. J. Rock Mech. Min. Sci. **34**, 249–262 (1997)

G.F. Wieczorek, J.B. Snyder, Rock falls from Glacier Point above Camp Curry, Yosemite National Park, California: U.S. Geological Survey Open-File Report 99–385, 22 p (1999)

G.F. Wieczorek, J.B. Snyder, Historical rock falls in Yosemite National Park: U.S. Geological Survey Open-File Report 03-491, 11 p (2004)

T.-F. Wong, R.H.C. Wong, K.T. Chau, C.A. Tang, Microcrack statistics, Weibull distribution and micromechanical modeling of compressive failure in rock. Mech. Mater. **38**(7), 664–681 (2006)

Soil – Machine Interaction: Simulation and Testing

Mustafa Alsaleh

Abstract Researchers at Caterpillar have been using Finite Element Analysis or Method (FEA or FEM), Mesh Free Models (MFM) and Discrete Element Models (DEM) extensively to model different earthmoving operations. Multi-body dynamics models using both flexible and rigid body have been used to model the machine dynamics. The proper soil and machine models along with the operator model can be coupled to numerically model an earthmoving operation. The soil – machine interaction phenomenon has been a challenging matter for many researchers. Different approaches, such as FEA, MFM and DEM are available nowadays to model the dynamic soil behavior; each of these approaches has its own limitations and applications. To apply FEA, MFM or DEM for analyzing earthmoving operations the model must reproduce the mechanical behavior of the granular material. In practice this macro level mechanical behavior is not achieved by modeling the exact physics of the microfabric structure but rather by approximating the macrophysics; that is using continuum mechanics or/and micromechanics, which uses length scales, that are larger than the physical grain size. Different numerical approaches developed by Caterpillar Inc. researchers will be presented and discussed.

Keywords Soil • Machine • Soil dynamics • DEM • FEM • MFM • Analytic methods • Machine testing

---

M. Alsaleh (✉)
Virtual Product Development Technology, Caterpillar Inc.; Mossville, IL 61616, USA
e-mail: alsaleh_mustafa_i@cat.com

S. Bonelli et al. (eds.), *Advances in Bifurcation and Degradation in Geomaterials*, Springer Series in Geomechanics and Geoengineering 11,
DOI 10.1007/978-94-007-1421-2_22,

1 Introduction

Geomaterials usually are composed of individual particles that range in size, shape and hardness. Such a structure causes the material to exhibit complex behavior when subjected to machine loading. Building robust virtual machine models is very important for Caterpillar Inc. to be able to understand machine performance and predict structural loads while the machine operates in such environments. To achieve these goals, the Caterpillar Inc. applied research development team has been developing different techniques to model these phenomena. The reason is essentially to be able to model the different machine operations in different surrounding environments. The different techniques and models are being used to model the soil – machine interaction at different levels of fidelity when the soil is subjected to different loading levels. The level of fidelity is essentially decided by the purpose of the modeling or simulation under consideration. For instance, if the purpose of the analysis is load prediction then the model has to be accurate enough to capture such forces, which mean a high fidelity model is needed. On the other hand, if the purpose is debugging some automation algorithms then a real low fidelity model will do the job. In some cases real-time models are needed; therefore, simplified equations can be used and implemented to run in a real time environment, in this case force prediction accuracy is not a concern. Finite element method is well developed and mature to be used for soil modeling in application that involve small to large deformation level, where server fragmentation is not experienced. Examples, but not limited to, are tire and track mobility. The lowest fidelity method is the classic soil mechanics or analytic – based approach. Such method is basically dependent on our understanding for fundamentals in soil mechanics and soil dynamics. Then we try to derive simple equations to describe the force-velocity response for the soil mass while subjected to machine dynamic loads through one or multiple machine implements or tools.

The second approach is the FEM, which is in general, a numerical approximation that represents the solution for systems of partial differential equations. Using FEM to model soil masses usually have several challenges such as; forming the proper equations that represent the real problem while maintaining numerical stability. The different FEM – based techniques developed by Caterpillar Inc. are usually linked to the commercial well-known FEM package, ABAQUS. In such techniques, the material models are in-house developed to suit the material type being dealt with. ABAQUS pre-processor, solver and post-processor are used to handle the problem in hand.

The third approach, which is potentially a higher fidelity approach, is DEM. This method is also called a distinct element method. DEM is the family of numerical methods used to compute the motion of a large number of particles with micro-scale size and above. DEM has been becoming more mature and widely accepted as a robust method to treat engineering problems in granular and discontinuous materials, especially in granular material flows, pharmaceutical applications, rock and powder mechanics. The various branches of the DEM family are the distinct element method proposed by Cundall in 1971, the generalized discrete element method proposed by Hocking, Williams and Mustoe in 1985. The theoretical basis of the method was established by Sir Isaac Newton in 1697. Williams, Hocking, and Mustoe in 1985 (Williams et al. 1985) showed that DEM could be viewed as a generalized finite element method. Its application to geomechanics problems is described in the book Numerical Modeling in Rock Mechanics, by Pande et al. (1990).

Caterpillar Inc. researchers have been developing and using a DEM code for the last 15 years; the code is called Rocks3DTD. It uses a very computationally efficient contact detection algorithm and can deal with any particle shape specified by the user. The contact frictional and normal forces are computed using the well-known Hertzian contact model for cohesionless materials. Additional algorithms are implemented to treat cohesive-like bonds when modeling fine-grained materials and rocks. The cohesion is modeled using cohesive pillars that bond neighboring particles together; this pillar can be strained until a strain threshold is reached and then the bond is broken (Gingold and Monaghan 1977). Recently, Cosserat rotation has been added to the code kinematics along with particle shape indices described by Libersky and Petschek (1991). This additional degree of freedom enabled the code to capture more of the micro-structural properties for the material being modeled (angularity, size, sphericity etc.). While Rocks3DTD is used to model particulate force responses and material flows, it is capable to link to full machine models using in-house built codes for modeling machine dynamics, tire-ground interaction, machine hydraulics, etc.

Rocks3DTD is also capable of interacting with tracked-type tractors to pass proper forces to the machine through the track shoes. The machine tools can be treated as either rigid and/or flexible bodies. The code had been parallelized to take advantage of mutli-threaded processors. It has been benchmarked against other commercial codes; to-date, Rocks3DTD usability, simulation speeds and accuracy have been found more encouraging. As acknowledged by many researchers, it is very challenging to obtain DEM model parameters that best represent real materials, specially, when dealing with fine-grained materials. Rocks3DTD developers have been successful in defining an engineered procedure to map these micro quantities to some material physical and macro quantities. Both small scale laboratory testing and full machine testing are being utilized to develop micro-macro parameter mapping functions. The particle size for instance, a very important DEM parameter, must be chosen carefully. Choosing the particle size for a given model will always has a great deal of trade-off between simulation accuracy and computational cost. Special attention had been given to this matter; the particle size distribution is established for a given model in a way to ensure highest simulation accuracy at the lowest computational cost. This way, the model parameters (micro-mechanical properties such as friction, stiffness etc) can be linked to macro properties to achieve better physical representation.

Rocks3DTD can predict the dynamic forces and flows of different discrete systems geometries under dynamic loading. As mentioned earlier, the contact parameters are micromechanical parameters that are very difficult to physically measure, and it is very challenging to evaluate because of the fact the that it is almost impossible to represent the actual shape and size of real materials. A real material is very complex to mimic in terms of shape, size, and size distributions.

The fourth approach is the MFM, which is still in a development phase. Gross distortion and eventual fragmentation of soil, which generally occur during earthmoving operations such as dozing and excavation, pose significant computational challenges to simulation by conventional Finite Element Method (FEM) [Joes report]. The main focus of this approach is to develop a 3-D earthmoving simulation code based in the use of the Mesh Free Method (MFM). This discretization method is seen as ideally suited for the prediction of implement forces and overall soil motion resulting from earthmoving operations in a fragmenting medium such as fine-grained cohesive soil. It is here, for simulations involving gross deformation and eventual fragmentation that the absence of fixed connectivity (or "mesh" as the name implies) gives MFM great flexibility, while still retaining the highly desirable characteristics of a continuum mechanics based formulation. MFM, a continuum dynamics based numerical method, is seen as ideally suited for the prediction of implement forces and overall soil motion resulting from earthmoving operations in a fragmenting medium such as fine-grained cohesive soils.

The classes of interpolation function-based discretization methods, which do not rely on fixed connectivity to describe the field variables and the instantaneous topology of the domain, have come to be known collectively as meshless methods or Mesh Free Methods (MFM). First invented in 1977, by Lucy (1977) and at the same time by Gingold and Monaghan (Gingold and Monaghan 1977), the then "smoothed particle hydrodynamics method" (now called standard-SPH) was originally applied to astrophysical and cosmological problems such as star and galaxy formation. Libersky et al. (Libersky and Petschek 1991; Libersky et al. 1993) extended the method to treat high-velocity dynamic response of solids and later Randles and Libersky (1996) proposed significant improvements to address some of the shortcomings of standard-SPH. Since (Lucy 1977; Gingold and Monaghan 1977), perhaps over twenty such methods have appeared in the literature.

For the discretization of partial differential equations (PDEs) that describe a deforming medium and in particular, for problems involving gross deformation and eventual fragmentation, the absence of fixed connectivity (or mesh) is probably the most attractive general characteristic of the MFMs. These methods may be divided into two main categories based on how they discretize the equation for balance of linear momentum; those that employ a variational (or weak formulation) and those that employ a collocation (or strong formulation). This research effort focuses on one of the collocation methods – the method of Corrected Smooth Particle Hydrodynamics (CSPH) (Bonet and Lok 1999; Bonet and Kulasegaram 2001) and how it may be adapted and applied to solving the partial differential equations that describe a deforming (and ultimately a fragmenting) medium.

The soil lab that exists in one of Caterpillar Inc. facilities has several soil bins that are being used to run scaled implement performance tests as well to collect data to validate the above mentioned numerical techniques. There are some occasions that numerical models lack the ability to handle certain operation or phenomenon. The soil lab with scaled tool size would be the alternative to resolve such issues. It is worth mentioning here that in any soil bin test, the known rules of scaling will be always applied to satisfy the physics of the problem.

2 Sample Applications

There are several applications that can be discussed in this section to clarify the applicability of the different numerical tools discussed above. We will in this section of the paper discuss the following application:

2.1 Bucket Loading Application

There are several CAT machines that would often require studying their bucket loading capabilities (Wheel loaders, Hydraulic Excavators, Backhoe loaders, Compact track and multi-terrain loaders, Industrial loaders, etc.). In such applications and to understand the overall machine performance, structural loads and fuel efficiency, it is essentially required to build a robust full machine virtual model. This model is potentially used to eliminate or reduce the amount of full machine field testing, which can be time consuming and quite expensive. Rocks3D^{TD} has proven that it is capable of modeling this phenomenon when coupled with the machine system model digging in fine to coarse – grained materials. Figures 1–3 show two examples on bucket loading applications.

2.2 Blading Operations

There are several of Caterpillar machines (wheeled and tracked) that have the blading functionality and would often need to investigate the performance of their

Fig. 1 Hydraulic excavator bucket loading – dumping operation (granular material) using Rocks3D^{TD}

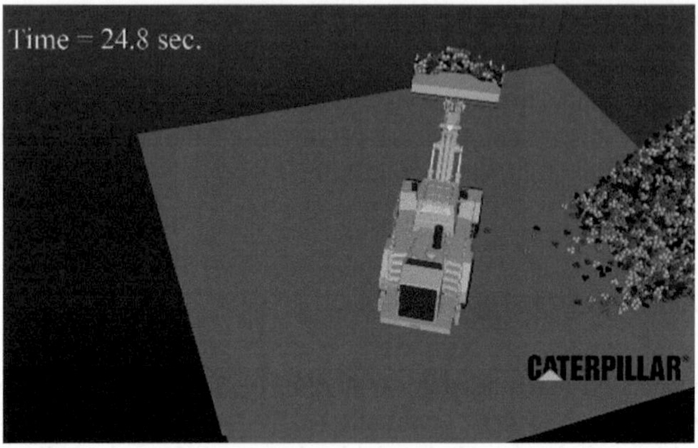

Fig. 2 Large wheel loader bucket loading – dumping operation (cohesive material) using Rocks3D^{TD}

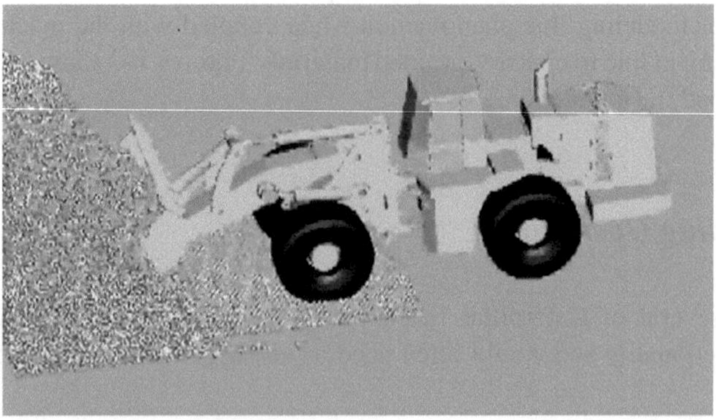

Fig. 3 Medium wheel loader bucket loading – dumping operation (granular material) using Rocks3D^{TD}

blades utilizing the existing numerical tools described earlier in this paper. Wheel tractor scrapers, tracked – type tractors, Motorgraders and wheeled dozers are some of the example machines that can be listed here. Figures 4–8 are some examples.

2.3 Ripping Operations

Caterpillar produces different types and sizes of ripper-equipped machines, these rippers essentially can operate in different types of soils and rocks. Machine

Fig. 4 D11 Track –type tractor dozing operation in cohesive soil using Rocks3D^{TD}

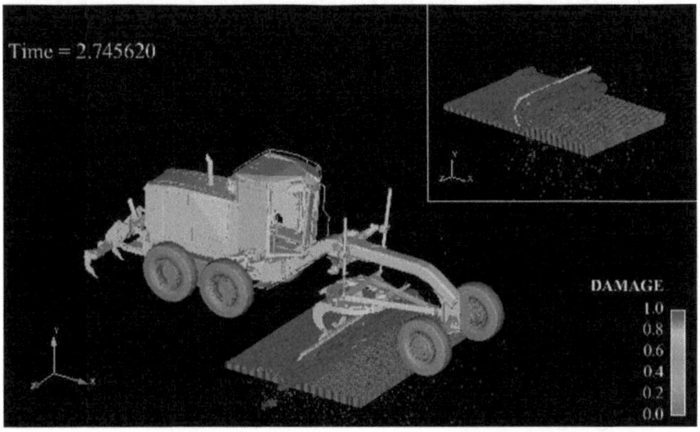

Fig. 5 Motor grader M160 fine dozing operation in cohesive material using MFM

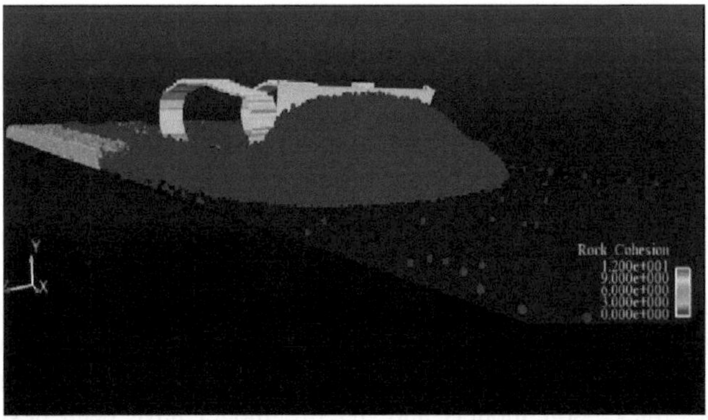

Fig. 6 D7 Track – type tractor dozing operation in granular material using Rocks3D

Fig. 7 Wheel tractor scraper 627 loading – unloading operation In Rocks3D^{TD} and analytic cutting edge forces model

Fig. 8 Motor grader M160 dozing operation in cohesive materials using Rocks3D^{TD} and analytic approach for the cutting edge forces

performance and ripper structural life are usually the main purpose of a ripping virtual model, for those goal to be achieved, ripping forces need to be accurately predicted. Figures 9 and 10 show two examples where Rocks3D^{TD} and MFM have been used to predict ripping forces.

2.4 Tire/Wheel– and Track Mobility

The different machines produced by Caterpillar are even tracked or wheeled machines, in either case, mobility models are needed to be able to capture their interaction with ground given the soil conditions and terrain topography. For this purpose, track –soil models and 3D tire model have been developed and implemented within the machine multi-body dynamics code (developed and owned

Fig. 9 D7 Track – type tractor ripping operation in a bedrock using Rocks3D^{TD}

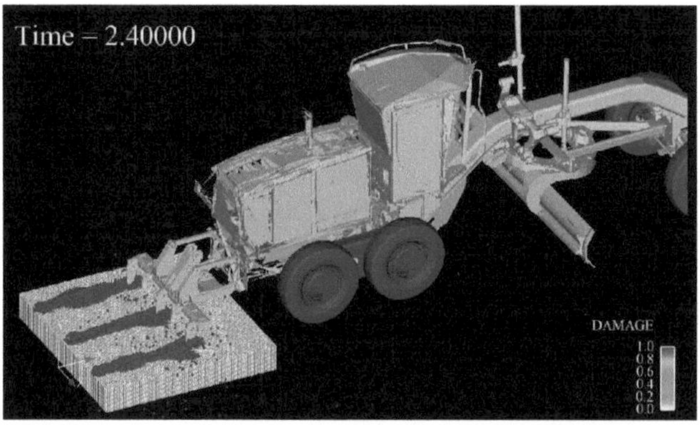

Fig. 10 Motor grader M160 3-shank ripping operation in cohesive soil using MFM

by Caterpillar). Moreover, compaction operations are usually needed to be modeled to understand soil and landfill compaction efficiency when these machines are used. The following examples show the applications of these models. All of the above shown wheeled machine simulations use the 3D tire model (Figs. 11 and 12).

2.5 Soil Lab Testing and Model Validation

The soil lab located at the Caterpillar Inc. Technical Center Facility in Mossville, IL has different soil bins. These soil bins are designed and equipped to be able run scaled tests for most of the above mentioned operations. The purpose of these scaled tests is usually to cover the gaps that some of the numerical tools have and

Fig. 11 Landfill compaction simulation using ABAQUS with an in-house developed UMAT

Fig. 12 Track mobility simulation for a D7 tractor on an Irregular terrain

to obtain data for the sake of model validation. Moreover, conventional triaxial tests are sometimes used to obtain full machine model parameters. Figures 13 and 14 below show some examples on both CAT soil lab tests and triaxial test validation.

3 Conclusions

Different numerical tools can be used to model the dynamic behavior of Geomaterials. FEM; DEM; MFM and Analytical Methods have been used by Caterpillar to model the different earthmoving operations. To date, reasonable results have been

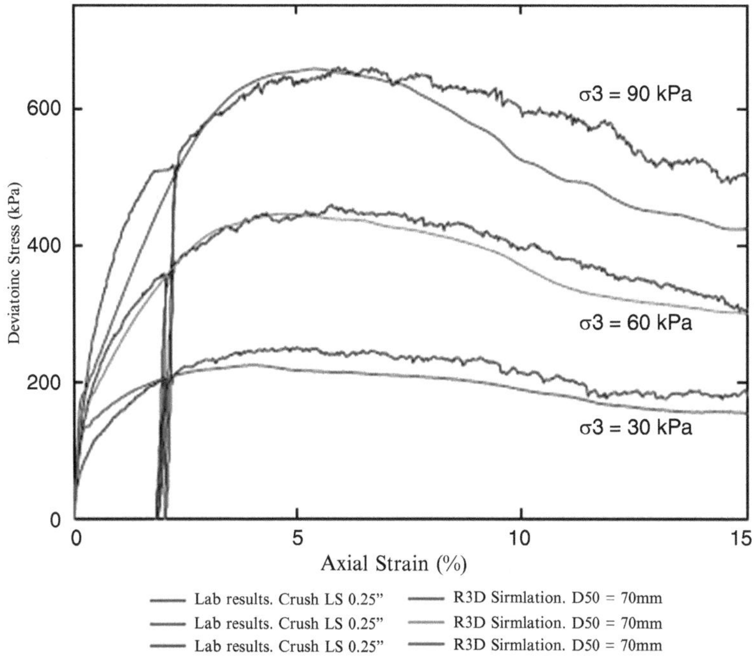

Fig. 13 Comparison between Rocks3D^{TD} and laboratory triaxial tests

Fig. 14 Comparison between scaled blade dozing operation in the soil lab and the virtual dozing operation in Rocks3D^{TD}

obtained and helped in resolving some design issues. However, lower cost and more accurate methods are always needed.

Acknowledgments The author would like to acknowledge the applied research management team for their continuous support for this area of research.

References

J. Bonet, S. Kulasegaram, Correction and stabilization of smooth particle hydrodynamics methods with applications in metal forming. IJNME **52**, 1203–1220 (2001)

J. Bonet, T.-S.L. Lok, Variational and momentum preserving aspects of smooth particle hydrodynamic formulations. Comput. Meth. Appl. Mech. Eng. **180**, 97–115 (1999)

P.A. Cundall, A computer model for simulating progressive large scale movements in blocky rock systems, in *Proceedings of the Symposium International Society Rock Mechanics*, Nancy, Paper II-8 (1971)

R.A. Gingold, J.J. Monaghan, Smoothed particle hydrodynamics: theory and applications to non-spherical stars, Mon. Not. R. Astr. Soc. **181**, 375 (1977)

L.D. Libersky, A.G. Petschek, Smoothed particle hydrodynamics with strength of materials, in *Proceedings, The Next Free Lagrange Conference*, vol. 395, ed. by H. Trease, J. Fritts, W. Crowley (Springer, New York, 1991), pp. 248–257

L.D. Libersky, A.G. Petschek, T.C. Carney, J.R. Hipp, F.A. Allahdadi, High strain Lagrangian hydrodynamics. J. Comput. Phys. **109**, 67–75 (1993)

L.B. Lucy, A Numerical approach to the testing of the fission hypothesis. Astron. J. **82**, 1013 (1977)

G. Pande, G. Beer, J.R. Williams, *Numerical Methods in Rock Mechanics* (Wiley, Chichester, 1990)

P.W. Randles, L.D. Libersky, Smoothed particle hydrodynamics: some recent improvements and applications. Comput. Meth. Appl. Mech. Eng. **139**, 375–408 (1996)

J.R. Williams, G. Hocking, G.G.W. Mustoe, The theoretical basis of the discrete element method, in *NUMETA 1985, Numerical Methods of Engineering, Theory and Applications* (A.A. Balkema, Rotterdam, Jan 1985)

J.R. Williams, G. Hocking, G.G.W. Mustoe, The theoretical basis of the discrete element method, in *NUMETA '85 Conference*, Sansea, 1985

Part III
Numerical Modeling of Failure and Localized Deformation in Geomaterials

Multiscale Semi-Lagrangian Reproducing Kernel Particle Method for Modeling Damage Evolution in Geomaterials

J.S. Chen, P.C. Guan, S.W. Chi, X. Ren, M.J. Roth, T.R. Slawson, and M. Alsaleh

Abstract Damage processes in geomaterials typically involve moving strong and weak discontinuities, multiscale phenomena, excessive deformation, and multi-body contact that cannot be effectively modeled by a single-scale Lagrangian finite element formulation. In this work, we introduce a semi-Lagrangian Reproducing Kernel Particle Method (RKPM) which allows flexible adjustment of locality, continuity, polynomial reproducibility, and h- and p-adaptivity as the computational framework for modeling complex damage processes in geomaterials. Under this work, we consider damage in the continua as the homogenization of micro-cracks in the microstructures. Bridging between the cracked microstructure and the damaged continuum is facilitated by the equivalence of Helmholtz free energy between the two scales. As such, damage in the continua, represented by the degradation of continua, can be characterized from the Helmholtz free energy. Under this framework, a unified approach for numerical characterization of a class of damage evolution functions has been proposed. An implicit gradient operator is embedded in the reproduction kernel approximation as a regularization of ill-posedness in strain localization. Demonstration problems include numerical simulation of fragment-impact of concrete materials.

Keywords Reproducing kernel • Semi-Lagrangian • Micro-cracks informed damage model • Geomaterials

J.S. Chen (✉) · P.C. Guan · S.W. Chi · X. Ren
Civil & Environmental Engineering, University of California, Los Angeles, CA, USA
e-mail: jschen@seas.ucla.edu

M.J. Roth · T.R. Slawson
U. S. Army Engineer Research and Development Center, Vicksburg, MS, USA

M. Alsaleh
Caterpillar Inc., Mossville, IL, USA
e-mail: alsaleh_mustafa_i@cat.com

S. Bonelli et al. (eds.), *Advances in Bifurcation and Degradation in Geomaterials*, Springer Series in Geomechanics and Geoengineering 11,
DOI 10.1007/978-94-007-1421-2_23,

1 Semi-Lagrangian Reproducing Kernel Approximation

Lagrangian formulation and discretization of conservation laws breaks down when mapping between the initial and current configurations is no longer one-to-one. This occurs under conditions such as new free surface formation (i.e. material separation) or free surface closure, which commonly exist in damage processes of geomaterials. Chen and Wu (2007) and Guan et al. (2009, 2011) proposed a semi-Lagrangian formulation to overcome the shortcoming of the Lagrangian formulation. The semi-Lagrangian reproducing kernel (RK) shape function formulated in the current configuration is expressed as:

$$\Psi_I(\mathbf{x}) = \mathbf{H}^T(\mathbf{x} - \mathbf{x}(\mathbf{X}_I, t))\,\mathbf{b}(\mathbf{x})\,\varphi_a(\mathbf{x} - \mathbf{x}(\mathbf{X}_I, t)) \tag{1}$$

where $\mathbf{x}$ is the spatial coordinate, $\mathbf{X}_I$ is the material coordinate of point I, and $\mathbf{H}$ is the vector containing the monomial basis functions:

$$\begin{aligned} &\mathbf{H}^T(\mathbf{x} - \mathbf{x}(\mathbf{X}_I, t)) \\ &\quad = \left[\, 1 \;\; x_1 - x_1(\mathbf{X}_I, t) \;\; x_2 - x_2(\mathbf{X}_I, t) \;\cdots\; (x_3 - x_3(\mathbf{X}_I, t))^n \,\right] \end{aligned} \tag{2}$$

The order n in the basis functions determines the order of completeness in the RK approximation. The function φ_a is called the kernel function which determines the locality characterized by the support measure "a", that is, $\varphi_a(\mathbf{x} - \mathbf{x}_I) \geq 0$, $\|\mathbf{x} - \mathbf{x}_I\|/a \leq 1$, and $\varphi_a(\mathbf{x} - \mathbf{x}_I) = 0$ otherwise, where $\mathbf{x}_I \equiv \mathbf{x}(\mathbf{X}_I, t)$. Further, φ_a determines the order of continuity of the RK approximation, for example, the B-spline kernel function has C^2 continuity. The coefficient vector $\mathbf{b}$ in (1) is obtained by the following reproducing kernel conditions:

$$\sum_{I=1}^{NP} \Psi_I(\mathbf{x})\, x_1(\mathbf{X}_I, t)^i\, x_2(\mathbf{X}_I, t)^j\, x_3(\mathbf{X}_I, t)^k = x_1^i x_2^j x_3^k,\; i + j + k = 0, 1, \ldots, n \tag{3}$$

Solving $\mathbf{b}(\mathbf{x})$ using Eq. 3 and substituting it into Eq. 1 yields the semi-Lagrangian RK shape function. Figure 1 shows the comparison of RK discretization with circular support and a FEM triangular mesh using the same set of points, with shaded areas as the domains of influence of the same point. In the semi-Lagrangian formulation, the RKPM points follow the material motion, while the distance measure $\|\mathbf{x} - \mathbf{x}(\mathbf{X}_I, t)\|$ in the kernel function $\varphi_a(\mathbf{x} - \mathbf{x}(\mathbf{X}_I, t))$ is defined in the deformed configuration. Under this construction, the kernel support of the semi-Lagrangian kernel function does not deform with the material motion. The comparison of the supports of Lagrangian and semi-Lagrangian kernels at undeformed and deformed states is shown in Fig. 2. The Lagrangian kernel supports cover the same group of material nodes before and after deformation, while the semi-Lagrangian kernel supports cover a different group of nodes after deformation. RKPM for large

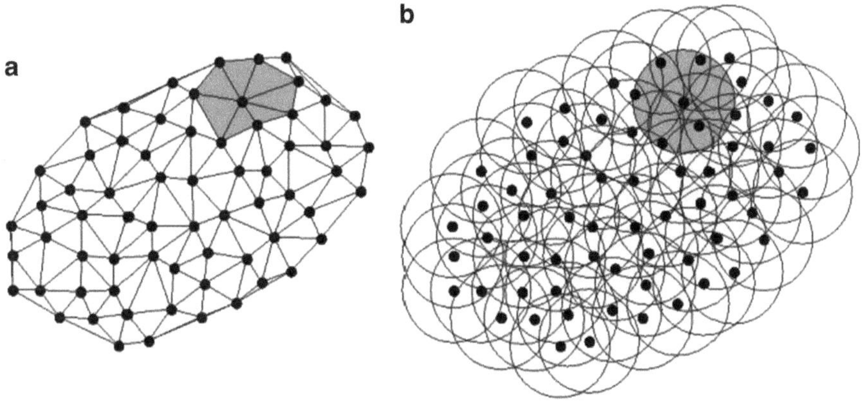

Fig. 1 Comparison of FEM and RKPM discretization and domains of influence: (**a**) FEM Discretization and (**b**) RKPM Discretization. The domain of influence of one node is marked in grey

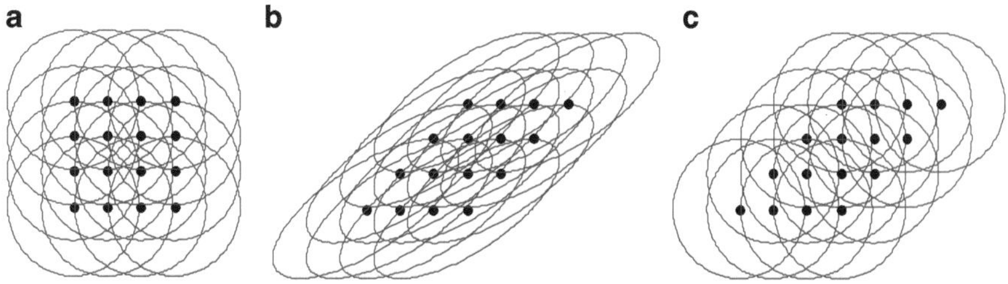

Fig. 2 Lagrangian and semi-Lagrangian kernel supports in undeformed and deformed configurations: (**a**) undeformed configuration, (**b**) Lagrangian kernel deformed with the material in the deformed configuration, and (**c**) semi-Lagrangian kernel in the deformed configuration

deformation analysis with stabilized nodal integration can be found in (Chen et al. 1996, 2002; Chen and Wu 2007), stability analysis of semi-Lagrangian RKPM formulation have been studied (Chen and Wu 2007), implicit gradient embedded in RKPM for localization problems (Chen et al. 2000, 2004), and adaptive RKPM (Lu and Chen 2002) have also been developed.

2 Multiscale Damage Model

A micro-cracks informed damage model for describing the softening behavior of brittle solids is proposed, in which damage evolution is treated as a consequence of micro-crack propagation. The homogenized stress-strain relation in the cracked microscopic cell defines the degradation tensor, which can be obtained by the equivalence between the averaged Helmholtz free energy (HFE) of the microscopic

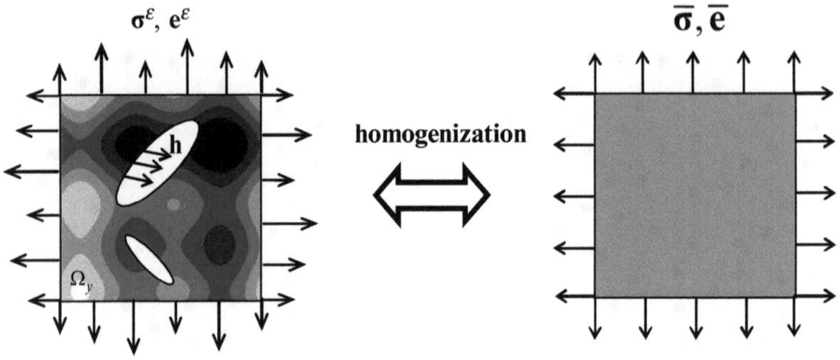

Fig. 3 Homogenization of microscopic cell with fluctuating fields

cell and the HFE of damaged continuum, $\overline{\Psi}$, as follows (with reference to Fig. 3) (Ren et al. 2003):

$$\overline{\Psi} = \frac{1}{V_y}\left(\int_{\Omega_y} \Psi^{\varepsilon} d\Omega + \frac{1}{2}\oint_{\Gamma_c} \mathbf{u}^{\varepsilon} \cdot \mathbf{h} ds\right) \tag{4}$$

Here superscript "ε" is used for the fine scale variables, $\boldsymbol{u}^{\varepsilon}$ is the displacement field in the microscopic cell, $\boldsymbol{h}$ is the cohesive traction acting on the crack surface Γ_c, Ψ^{ε} is the HFE of the microscopic cell, and V_y is the volume of the microscopic cell with domain Ω_y. This energy equivalence relationship serves as an energy bridging vehicle between the damaged continuum and the cracked microstructure.

By introducing bridging based on HFE, the damage tensor for continuum can be obtained. Recall (4), and let $\overline{\Psi}$ computed from the microscopic cell be related to the damaged continuum by

$$\overline{\Psi} = \frac{1}{2}\overline{\mathbf{e}} : (\mathbf{I}\text{-}\mathbf{D}) : \overline{\mathbf{C}}_0 : \overline{\mathbf{e}} \tag{5}$$

where $\overline{\mathbf{e}}$, $\overline{\mathbf{C}}_0$, and $\mathbf{D}$ are the homogenized strain tensor, undamaged material response tensor, and the continuum damage tensor, respectively. The damage energy release rate is defined as

$$\mathbf{Y} = -\frac{\partial\overline{\Psi}}{\partial\mathbf{D}} = \frac{1}{2}\overline{\mathbf{e}} : \overline{\mathbf{C}}_0 : \overline{\mathbf{e}} \tag{6}$$

By taking the derivative of the HFE of the microscopic cell in Eq. 4 with respect to the damage energy release rate, $\mathbf{Y}$, we obtain the damage tensor:

$$\mathbf{D} = I - \frac{\partial\overline{\Psi}}{\partial\mathbf{Y}} \approx I - \frac{\Delta\overline{\Psi}}{\Delta\mathbf{Y}} \tag{7}$$

Here a finite difference approach could be used in Eq. 7 to obtain the fourth order damage tensor using $\Delta\overline{\Psi}$ and $\Delta\mathbf{Y}$. This general approach can be reduced to a scalar or a bi-scalar damage models as discussed in (Ren et al. 2003).

3 Examples

The penetration experiments on a concrete panel were performed in the US Army Engineer Research and Development Center (ERDC) small-caliber ballistic test facility. Experimental measurements included the projectile impact and exit velocity, projectile mass loss, crater sizes in the concrete target, and projectile deformation after impact. The concrete material is modeled by the Advanced Fundamental Concrete (AFC) model (Adley et al. 2010) developed at ERDC. The AFC model is a 3-invariant plasticity model which incorporates damage evolution, strain rate effects, and a nonlinear pressure-volume relationship. The tensile damage evolution function in the AFC model is characterized by the multiscale damage model presented in Sect. 2, while the functions for compressive damage evolution remain unchanged. Tensile and compression damages were separated in the von Mises yield function, allowing the yield surfaces under tension and compression to evolve based on the separated scalar damage terms. Material properties used in the AFC model are: Young's modulus 200 GPa, Poisson's ratio 0.26, yield stress 3,000 MPa, hardening modulus 2,500 MPa, and density 7,806 $\mathrm{kg/m^3}$. The RKPM predicted progressive penetration states are plotted in Fig. 4. Comparison of damage patterns on the impact and exit faces between experimental measurement and computation is shown in Fig. 5, and the comparison of test data and simulation results is given in Table 1.

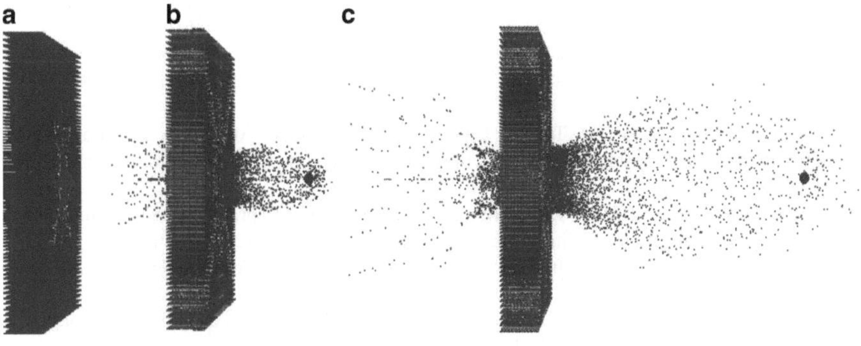

Fig. 4 Penetration process predicted by RKPM: (**a**) 50 μsec after impact, (**b**) 200 μsec after impact, (**c**) 500 μsec after impact

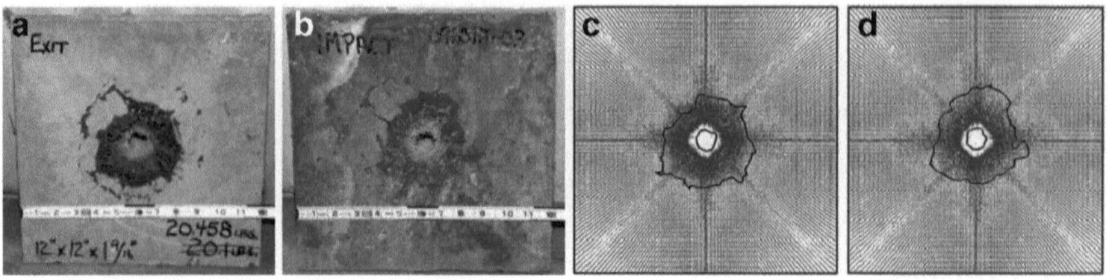

Fig. 5 Damage patterns: (**a**) Exp. impact face, (**b**) Exp. exit face, (**c**) RKPM impact face, (**d**) RKPM exit face

Table 1 Comparison of experimental and numerical results

	Impact velocity (m/s)	Exit velocity (m/s)	Velocity reduction (%)	Projectile mass loss (%)
Experiment	1,708	684	59.9	1
RKPM	1,708	672.3	60.6	0.6

References

M.D. Adley, A.O. Frank, K.T. Danielson, S.A. Akers, J.L. O'Daniel, The Advanced Fundamental Concrete (AFC) Model, Technical Report RDC/GSLTR-10-51, U.S. Army Engineer Research and Development Center, Vicksburg (2010)

J.S. Chen, Y. Wu, Stability in Lagrangian and semi-Lagrangian reproducing kernel discretizations using nodal integration in nonlinear solid mechanics, in *Computational Methods in Applied Sciences*, ed. by V.M.A. Leitao, C.J.S. Alves, C.A. Duarte (Springer, Berlin, 2007), pp. 55–77

J.S. Chen, C. Pan, C.T. Wu, W.K. Liu, Reproducing kernel particle methods for large deformation analysis of nonlinear structures. Comput. Methods Appl. Mech. Eng. **139**, 195–227 (1996)

J.S. Chen, C.T. Wu, T. Belytschko, Regularization of material instabilities by meshfree approximation with intrinsic length scales. Int. J. Numer. Methods Eng. **47**, 1303–1322 (2000)

J.S. Chen, C.T. Wu, S. Yoon, Y. You, A stabilized conforming nodal integration for Galerkin meshfree methods. Int. J. Numer. Methods Eng. **50**, 435–466 (2001)

J.S. Chen, C.T. Wu, S. Yoon, Y. You, Nonlinear version of stabilized conforming nodal integration for Galerkin meshfree methods. Int. J. Numer. Methods Eng. **53**, 2587–2615 (2002)

J.S. Chen, X. Zhang, T. Belytschko, An implicit gradient model by a reproducing kernel strain regularization in strain localization problems. Comput. Methods Appl. Mech. Eng. **193**, 2827–2844 (2004)

P.C. Guan, J.S. Chen, Y. Wu, H. Teng, J. Gaidos, K. Hofstetter, M. Alsaleh, A semi-Lagrangian reproducing kernel formulation for modelling earth moving operations. Mech. Mater. **41**, 670–683 (2009)

P.C. Guan, S.W. Chi, J.S. Chen, T.R. Slawson, M.J. Roth, Semi-Lagrangian reproducing kernel particle method for fragment-impact problems Int. J. Impact Eng. (2011) submitted

H. Lu, J.S. Chen, Adaptive meshfree particle method. Lect. Notes Comput. Sci. Eng. **26**, 251–267 (2002)

X. Ren, J.S. Chen, J. Li, T.R. Slawson, M.J. Roth, Micro-cracks informed damage model for brittle solids. Int. J. Solids Struct. **48**(10), 1560–1571 (2011)

Y. You, J.S. Chen, H. Lu, Filter, reproducing kernel, and adaptive meshfree methods. Comput. Mech. **31**, 316–326 (2003)

Tunnel Excavation Modeling with Micromechanical Approaches

Séverine Levasseur, Frédéric Collin, Robert Charlier, and Djimédo Kondo

Abstract A zone with significant irreversible deformations and significant changes in flow and transport properties is expected to be formed around underground excavation in the deep geological layers considered for the high level radioactive waste disposal. The present study concerns the modeling of this phenomena by a micromechanical damage model, based on a Mori-Tanaka homogenization on a cracked media. This anisotropic model is derived from Eshelby homogenized scheme, on which a coupling between damage and friction is taking into account on cracks. Compared to elastoplastic model on tunnel drilling modeling, micromechanical modeling seems very promising: both approaches provide similar EDZ sizes and shapes even if they do not have the same effects on perturbed mechanical behavior; micromechanical model also overcomes the elastoplastic one by a realistic characterization of crack processes.

Keywords Homogenization scheme • Multiscale behaviour laws • Damage models • Excavation • Damaged zone • Tunnel drilling

1 Introduction

Tunnel excavation in deep geological layers provides stress perturbations, which could lead to a significant increase of the hydromechanical properties, related to diffuse and/or localized microcracks propagation in the material, closed to

S. Levasseur (✉) · F. Collin (F.R.S-FNRS) · R. Charlier
Université de Liège, ArGEnCo – *Geo*3, Bat B52/3 chemin des 132 chevreuils 1, 4000 Liège, Belgium
e-mail: severine.levasseur@ulg.ac.be; f.collin@ulg.ac.be; robert.charlier@ulg.ac.be

D. Kondo
Laboratoire de Mécanique de Lille - Cité Scientifique, Université de Sciences et Technologies de Lille, 59655 Villeneuve d'Ascq Cedex, France
e-mail: djimedo.kondo@univ-lille1.fr

S. Bonelli et al. (eds.), *Advances in Bifurcation and Degradation in Geomaterials*, Springer Series in Geomechanics and Geoengineering 11,
DOI 10.1007/978-94-007-1421-2_24,

tunnel borders (Bernier et al. 2007). The creation of such zone, named Excavation Damaged Zone (or EDZ), is classically modeled by elastoplastic behavior laws, on which an analogy between damage and plasticity is proposed (Levasseur et al. 2010). However, recent developments in the field of micromechanically-based damage models, like homogenization methods, provide now physically and mathematically appropriate framework for the investigation of the behavior of micro-cracked media including the description of damage anisotropy-induced anisotropy, as well as cracks closure effects (Zhu et al. 2008; Dormieux et al. 2006). In the perspective of applications on geotechnical problems, the purpose of the present study is to evaluate some of these homogenization schemes, by an analysis of their assumptions and of the damaged macroscopic response that they predict, in comparison with elastoplastic approaches. More particularly in this paper, a micromechanical model, based on Mori-Tanaka homogenization scheme and incorporating damage coupled with friction, is proposed in Sect. 2 and applied to the modeling of a tunnel drilling in Callovo-Oxfordian Clay, with an assessment and a characterization of the excavation damaged zone, in Sect. 3.

2 Micromechanical Model

The micromechanical model is based on a thermodynamic potential Ψ consisting in purely elastic effects and inelastic effects dues to damage d^r, crack opening β^r and friction $\underline{\gamma}^r$ on each r-crack (Zhu et al. 2008):

$$\Psi = \frac{1}{2}(\underline{\underline{E}} - \underline{\underline{E}}^{pl}) : \mathbb{C}^s : (\underline{\underline{E}} - \underline{\underline{E}}^{pl}) + \frac{1}{2}\sum_{r=1}^{p} \frac{w^r}{d^r}\left(H_0 \beta^{r^2} + H_1 \underline{\gamma}^r . \underline{\gamma}^r\right) \tag{1}$$

where

- $\underline{\underline{E}}$ is the strain tensor and $\underline{\underline{E}}^{pl} = \sum_{r=1}^{p} w^r \left(\beta^r \underline{n}^r \otimes \underline{n}^r + \underline{\gamma}^r \overset{s}{\otimes} \underline{n}^r\right)$
- $\mathbb{C}^s = 3k^s \mathbb{J} + 2\mu^s \mathbb{K}$ is the elastic constitutive matrix
- H_0 and H_1 are material parameters, $H_0 = \dfrac{3E^s}{16(1-\nu^s)}$, $H_1 = H_0\left(1 - \dfrac{\nu^s}{2}\right)$

Thanks to Legendre-Fenchel transformation and by distinguishing open and closed cracks, the stress formulation of the potential, in case of Mori-Tanaka homogenization, can be written as:

$$\Psi^* = \frac{1}{2}\underline{\underline{\Sigma}} : \mathbb{S}^s : \underline{\underline{\Sigma}} - \frac{1}{2}\sum_{r=1}^{p_c} \frac{w^r}{d^r}\left(H_0 \beta^{r^2} + H_1 \underline{\gamma}^r . \underline{\gamma}^r\right)$$

$$+ \underline{\underline{\Sigma}} : \sum_{r=1}^{p_c} w^r \left(\beta^r \underline{n}^r \otimes \underline{n}^r + \underline{\gamma}^r \overset{s}{\otimes} \underline{n}^r\right)$$

$$+\frac{1}{2}\underline{\underline{\Sigma}}:\sum_{r=1}^{p_o} w^r\left(\frac{d^r}{H_0}\mathbb{E}^{2,r}+\frac{d^r}{2H_1}\mathbb{E}^{4,r}\right):\underline{\underline{\Sigma}} \tag{2}$$

where $\underline{\underline{\Sigma}}$ is the stress tensor, p_c is the number of closed cracks family and p_o the number of open cracks family, $\mathbb{E}^{2,r}$ and $\mathbb{E}^{4,r}$ tensors relative to the normal of crack $\underline{n}^r$, $\mathbb{S}^s$ the inverse of the elastic constitutive matrix. Then, the strain tensor is given by $\underline{\underline{E}} = \partial\Psi^*/\partial\underline{\underline{\Sigma}}$ and the first state law reads:

$$\underline{\underline{E}}=\mathbb{S}^s:\underline{\underline{\Sigma}}+\sum_{r=1}^{p_o} w^r\left(\frac{d^r}{H_0}\mathbb{E}^{2,r}+\frac{d^r}{2H_1}\mathbb{E}^{4,r}\right):\underline{\underline{\Sigma}}+\sum_{r=1}^{p_c} w^r\left(\beta^r\underline{n}^r\otimes\underline{n}^r+\underline{\gamma}^r\overset{s}{\otimes}\underline{n}^r\right) \tag{3}$$

as well as the thermodynamic force of damage $F^{d^r} = \partial\Psi^*/\partial d^r$:

$$F^{d^r}=\begin{cases}\dfrac{1}{2d^{r^2}}\left(H_0\beta^{r^2}+H_1\underline{\gamma}^r.\underline{\gamma}^r\right) & \text{closed cracks}\\[2ex] \dfrac{1}{2}\underline{\underline{\Sigma}}:\left(\dfrac{\mathbb{E}^{2,r}}{H_0}+\dfrac{\mathbb{E}^{4,r}}{2H_1}\right):\underline{\underline{\Sigma}} & \text{open cracks}\end{cases} \tag{4}$$

Considering the damage and friction criteria:

$$f^r = F^{d^r}-(c_0+c_1 d^r)\ \text{and}\ g^r = |\underline{F}^{\gamma^r}|+\mu_f F^{\beta^r} \tag{5}$$

with c_0 and c_1 damage resistance variables in J/m^2 and μ_f friction coefficient. The rates of d^r, β^r and $\underline{\gamma}^r$ follow the consistency relations $\dot{f}^r = 0$, $\dot{g}^r = 0$:

- for open cracks:
  $$\dot{\beta}^r = 0,\ \ \dot{\underline{\gamma}}^r = 0,\ \ \dot{d}^r = \frac{1}{c_1}\frac{\partial f^r}{\partial\underline{\underline{\Sigma}}}\dot{\underline{\underline{\Sigma}}}\ \text{with}\ \frac{\partial f^r}{\partial\underline{\underline{\Sigma}}}=\underline{\underline{\Sigma}}:\left(\frac{\mathbb{E}^{2,r}}{H_0}+\frac{\mathbb{E}^{4,r}}{2H_1}\right)$$
- for closed cracks:
  $$\dot{\beta}^r = \mu_f\dot{\lambda}^{\gamma^r},\ \ \dot{\underline{\gamma}}^r = \dot{\lambda}^{\gamma^r}\underline{v}^r,\ \ \dot{d}^r = \frac{1}{H_{\gamma\beta d}H_d^r}\left(\frac{H_1\underline{\gamma}^r.\underline{v}^r+\mu_f H_0\beta^r}{d^{r^2}}\right)\frac{\partial g^r}{\partial\underline{\underline{\Sigma}}}\dot{\underline{\underline{\Sigma}}}$$

  with $\underline{v}^r$ friction rate direction, $\dot{\lambda}^{\gamma^r} = \dfrac{1}{H^r_{\gamma\beta d}}\dfrac{\partial g^r}{\partial\underline{\underline{\Sigma}}}\dot{\underline{\underline{\Sigma}}}$, $\dfrac{\partial g^r}{\partial\underline{\underline{\Sigma}}} = \underline{v}^r\overset{s}{\otimes}\underline{n}^r+\mu_f\underline{n}^r\otimes\underline{n}^r$

$$H^r_{\gamma\beta d}=\frac{1}{d^r}(H_1+\mu_f^2 H_0)-\frac{1}{H_d^r}\left(\frac{H_1\underline{\gamma}^r.\underline{v}^r+\mu_f H_0\beta^r}{d^{r^2}}\right)^2$$
$$H_d^r=\frac{1}{d^{r^3}}\left(H_0\beta^{r^2}+H_1\underline{\gamma}^r.\underline{\gamma}^r\right)+c_1$$

The rate form of the coupled law is finally given by: $\dot{\underline{\underline{E}}} = \mathbb{S}_t^{hom}:\dot{\underline{\underline{\Sigma}}}$, with

$$\mathbb{S}_t^{hom} = \mathbb{S}^s + \sum_{r=1}^{p_c} w^r \frac{1}{H_{\gamma\beta d}^r} \frac{\partial g^r}{\partial \underline{\underline{\Sigma}}} \otimes \frac{\partial g^r}{\partial \underline{\underline{\Sigma}}} + \frac{1}{c_1} \sum_{r=1}^{p_o} w^r \frac{\partial f^r}{\partial \underline{\underline{\Sigma}}} \otimes \frac{\partial f^r}{\partial \underline{\underline{\Sigma}}} \tag{6}$$

3 Tunnel Excavation Modeling

3.1 Parameter Identification on Compression Test

The micromechanical model described in the previous section is first calibrated, by trial and error, on a compression test performed on argillite (Callovo-Oxfordien clay) by Chiarelli (see 2000). This calibration leads to the results presented on Fig. 1: $E^s = 4\text{GPa}$, $\nu^s = 0.2$, $\mu_f = 0.36$ (which corresponds to a friction angle $\varphi_f = 20°$), $c_0 = 20\,\text{kJ/m}^2$ and $c_1 = 60\,\text{kJ/m}^2$. These elastoplastic parameters are in agreement with the ones commonly used in elastoplastic models based on Van Eekelen criteria (Gerard et al. 2010, Van-Eekelen 1980) for the considered material (with a dilatancy angle $\psi = 0°$ and a hardened cohesion $c = 30\text{MPa}$).

3.2 Application to Tunnel Modeling

Starting from this calibrated micromechanical model, a synthetic modelling of a tunnel excavation drilled in CoX clay is proposed. Initial stresses are chosen isotropic. Tunnel excavation process consists to reduce the stress state to 0MPa.

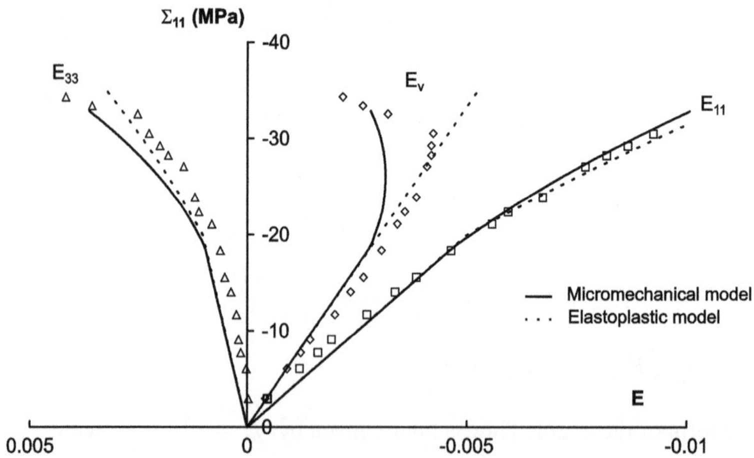

Fig. 1 Parameter calibration on compression test (*symbols*: experimental data)

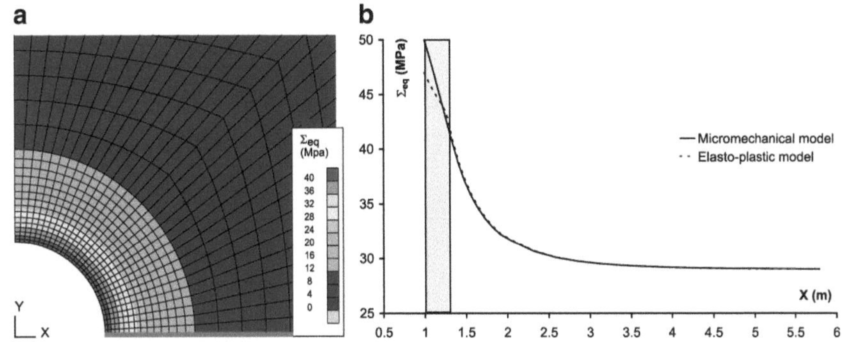

Fig. 2 Von Mises equivalent stress fields around the tunnel according to the micromechanical model; Stress distribution along radial section compared to elastoplastic model

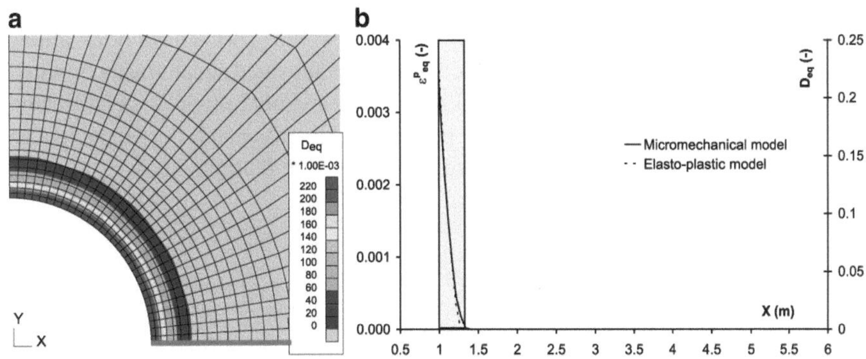

Fig. 3 Excavation damaged zone around the tunnel according to the micromechanical model (D_{eq}); Comparison along radial section with elastoplastic model E^p_{eq}

Compare to elastoplastic modeling results, Von Mises equivalent stress field Σ_{eq} predicted around tunnel by micromechanical model is realistic. Both models present very similar results excepted closed to tunnel border as shown in Fig. 2.

To characterize Excavation Damaged Zones (EDZ) by micromechanical model, an equivalent damage variable D_{eq} can be calculated from damage tensor $\underline{\underline{D}} = \sum_{r=1}^{p_c+p_o} w^r d^r \underline{n}^r \otimes \underline{n}^r$ by analogy with the equivalent stress Σ_{eq}. In the elastoplastic model, an analogy is commonly made between damage and plasticity. EDZ is then characterized by the equivalent plastic stain E^p_{eq}. According to these definitions, it results the damage distributions around tunnel represented on Fig. 3. These EDZ present similar sizes and shapes even if they do not have the same effects on perturbed mechanical behavior (cf. Fig. 2).

However, if micromechanical model is similar to elastoplastic model in term of stress–strain fields and EDZ shapes, micromechanical model overcomes the elastoplastic one through a characterization of crack processes. In EDZ, at each Gauss point of the finite element model, micromechanical model provides the

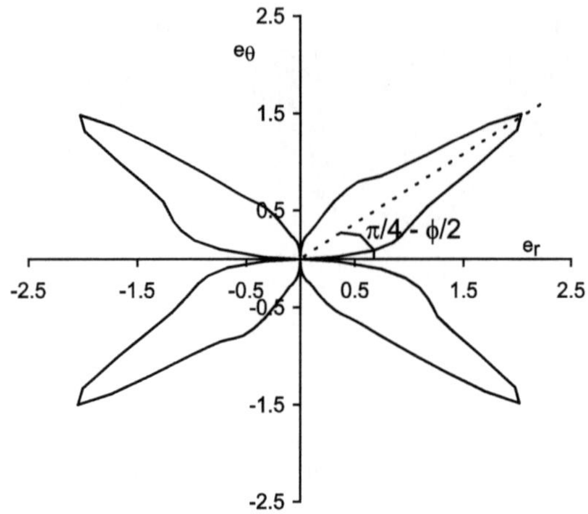

Fig. 4 Orientational distribution of the normal of damage in EDZ (in local axis)

orientational distribution of the normal of damage (cf. Fig. 4). These orientations can be linked to cracks: EDZ is mainly controled by activated shear cracks oriented at $\pi/4 + \varphi/2$ degrees. On these cracks, micromechanical model gives also the corresponding opening. This last point presents a big advantage in perspective to evaluate transport properties in EDZ, like permeability evolution.

4 Conclusions

The present study concerns the modeling of tunnel excavation by a micromechanical damage model based on Mori-Tanaka homogenized scheme and coupled with friction coefficient. From a calibration performed on compression test on Callovo-Oxfordian Clay, this micromechanical model, in comparison with an elastoplastic one, shows very promising results: both approaches provide similar EDZ sizes and shapes even if they do not have the same effects on perturbed mechanical behavior; micromechanical model also overcomes the elastoplastic one by a realistic characterization of crack processes and evolutions. Nevertheless, the micromechanical model still need to be improved by taking into account hydromechanical coupling as well as the dependency between initial stresses and damage-friction coupling like in Levasseur et al. (2010a, 2011).

References

F. Bernier, X.L. Li, W. Bastiaens, L. Ortiz, M. Van Geet, L. Wouters, B. Frieg, P. Blümling, J. Desrues, G. Viaggiani, C. Coll, S. Chanchole, V. De Greef, R. Hamza, L. Malinsky, A. Vervoort, Y. Vanbrabant, B. Debecker, J. Verstraelen, A. Govaerts, M. Wevers, V. Labiouse, S. Escoffier, J.F. Mathier, L. Gastaldo, Ch. Bühler, Fractures and self-healing within the

excavation disturbed zone in clays (SELFRAC). Final report, 5th EURATOM Framework Programme 1998–2002 (2007)
A.S. Chiarelli, Etude expérimentale et modélisation du comportement mécanique de l'argile de l'Est. Ph.D. Thesis, Université des Sciences et Technologies de Lille, 2000
L. Dormieux, D. Kondo, F.J. Ulm, *Microporomechanics* (Wiley, Chichester, 2006)
P. Gerard, J.P. Radu, J. Talandier, R. de La Vaissière, R. Charlier, F. Collin, Numerical modelling of the resaturation of swelling clay with gas injection, in *Proceedings of UNSAT 2010, Barcelona* (Balkema, Rotterdam, 2010)
S. Levasseur, R. Charlier, B. Frieg, F. Collin, Hydro-mechanical modelling of the Excavation Damaged Zone around an underground excavation at Mont Terri Rock Laboratory. Int. J. Rock Mech. Min. Sci. **47**(3), 414–425 (2010)
S. Levasseur, F. Collin, R. Charlier, D. Kondo, On a class of micromechanical damage models with initial stresses for geomaterials. Mech. Res. Commun. **37**, 38–41 (2010a)
S. Levasseur, F. Collin, R. Charlier, D. Kondo. A two scale anisotropic damage model accounting for initial stresses in microcracked materials. Eng. Fract. Mech. **78**, 1945–1956 (2011)
H.A.M. Van-Eekelen, Isotropic yield surfaces in three dimensions for use in soil mechanics. Int. J. Numer. Anal. Meth. Geomech. **4**, 98–101 (1980)
Q. Zhu, D. Kondo, J.F. Shao, V. Pensée, Micromechanical modelling of anisotropic damage in brittle rocks and application. Int. J. Rock Mech. Min. Sci. **45**(4), 467–477 (2008)

Coupling Discrete Elements and Micropolar Continuum Through an Overlapping Region in One Dimension

Richard A. Regueiro and Beichuan Yan

Abstract The paper presents recent progress in the development of a computational multiscale modeling approach for simulating the interfacial mechanics between dense dry granular materials and deformable solid bodies. Applications include soil-tire/track/tool/penetrometer interactions, geosynthetics- soil pull out strength, among others. The approach involves a bridging scale decomposition coupling between a three-dimensional ellipsoidal discrete element (DE) model and a finite strain pressure-sensitive micromorphic constitutive model implemented in a new multi-field coupled finite element (FE) method. The concept borrows from the atomistic-continuum bridging scale decomposition methods, except for the relevant differences for our problem in granular materials: (1) frictional, large relative motion of DE particles/grains upon shearing by deformable solid; (2) open window representation of DE region in contact with deformable solid; (3) overlapping finite strain micromorphic constitutive model for granular material with additional kinematics and higher order stresses; and (4) adaptivity of DE-FE region. The paper focusses on a simpler subset problem of topics (1–3): a one-dimensional glued elastic string of spherical DEs, overlapped partially with a one-dimensional micropolar continuum FE mesh. A numerical example is presented.

Keywords Overlap coupling • Finite element • Discrete element • One-dimension • Micropolar elasticity

---

R.A. Regueiro (✉)
Department of Civil, Environmental, and Architectural Engineering, University of Colorado at Boulder, 1111 Engineering Dr., Boulder, CO 80309, USA
e-mail: regueiro@colorado.edu

B. Yan
Community Surface Dynamics Modeling System, University of Colorado at Boulder, 3100 Marine St., Boulder, CO 80303, USA
e-mail: beichuan.yan@colorado.edu

S. Bonelli et al. (eds.), *Advances in Bifurcation and Degradation in Geomaterials*, Springer Series in Geomechanics and Geoengineering 11,
DOI 10.1007/978-94-007-1421-2_25,

1 Introduction

Interfacial mechanics between dense granular materials and deformable solid bodies involve large shear deformation and grain motion at the interface. To resolve such granular physics at the grain scale in contact with the deformable solid, but in a computationally tractable manner, a concurrent multiscale computational method is needed (Regueiro and Yan 2011).

As a simple problem to verify the method, a one-dimensional (1D) string of glued elastic spherical discrete elements is overlapped with a linear elastic micropolar continuum finite element implementation. The overlap coupling is enabled by the bridging scale decomposition method (Wagner and Liu 2003; Klein and Zimmerman 2006), but now with rotational degrees of freedom (dof) in addition to axial and transverse displacements. The paper presents the formulation of the 1D elastic glued discrete elements and linear elastic micropolar continuum reduced by Timoshenko beam kinematics with axial stretch, as well as a numerical example.

2 Discrete Element and Micropolar Continuum Finite Element Formulations in One Dimension

2.1 *Discrete Element Model of Glued Elastic Circular Disks*

Since the spherical particles are assumed glued and elastic, we employ Hertz-Mindlin interparticle nonlinear elastic constitutive behavior (Mindlin 1949). For eventual extension to frictional sliding, we can incorporate later the realistic particle geometry in three dimensions, with particle contact search and interparticle friction (Yan et al. 2010).

The balance of linear and angular momentum for a string of elastic spherical particles in contact may be written as

$$\boldsymbol{M}^Q \ddot{\boldsymbol{Q}} + \boldsymbol{F}^{INT,Q}(\boldsymbol{Q}) = \boldsymbol{F}^{EXT,Q} \tag{1}$$

$$\boldsymbol{M}^Q = \mathop{\mathsf{A}}_{\delta=1}^{N} \boldsymbol{m}_\delta^Q \; ; \quad \boldsymbol{m}_\delta^Q = \begin{bmatrix} m_\delta & 0 & 0 \\ 0 & m_\delta & 0 \\ 0 & 0 & m_\delta^\omega \end{bmatrix}$$

$$\boldsymbol{F}^{INT,Q} = \mathop{\mathsf{A}}_{\delta=1}^{N} \boldsymbol{f}_\delta^{INT,Q} \; ; \quad \boldsymbol{f}_\delta^{INT,Q} = \sum_{\varepsilon=1}^{n_c} \begin{bmatrix} \boldsymbol{f}^{\varepsilon,\delta} \\ \boldsymbol{f}_\omega^{\varepsilon,\delta} \end{bmatrix}$$

$$\boldsymbol{F}^{EXT,Q} = \mathop{\mathsf{A}}_{\delta=1}^{N} \boldsymbol{f}_\delta^{EXT,Q} \; ; \quad \boldsymbol{f}_\delta^{EXT,Q} = \begin{bmatrix} \boldsymbol{f}^{EXT,\delta} \\ \boldsymbol{\ell}^{EXT,\delta} \end{bmatrix}$$

where $\boldsymbol{M}^Q$ is the mass and rotary inertia matrix for a system of N particles, $\boldsymbol{m}_\delta^Q$ is the mass and rotary inertia matrix for particle δ, m_δ is the mass for

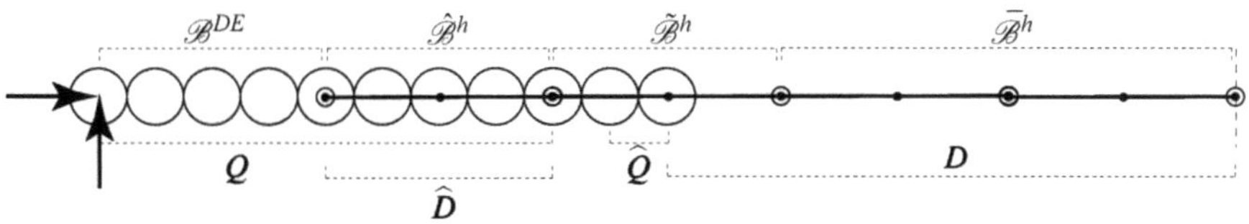

Fig. 1 1D string of 11 elastic spheres overlapped partially by 4 1D micropolar continuum FEs. Free particle dofs are indicated by $\boldsymbol{Q}$, ghost particle dofs by $\hat{\boldsymbol{Q}}$, prescribed FE nodal dofs by $\hat{\boldsymbol{D}}$, and free FE nodal dofs by $\boldsymbol{D}$

particle δ in x_1 and x_2 directions, m_δ^ω is the rotary inertia for particle δ about the x_3 axis, $\mathbf{A}_{\delta=1}^{N}$ is an assembly operator to obtain the system matrices from the individual particle matrices and contact vectors, $\boldsymbol{F}^{INT,Q}$ the internal force and moment vector associated with n_c particle contacts which will be a function of particle displacements and rotations when particles translate and rotate, $\boldsymbol{f}_\delta^{INT,Q}$ the resultant internal force and moment vector for particle δ, $\boldsymbol{f}^{\varepsilon,\delta}$ the internal force vector for particle δ at contact ε, $\boldsymbol{f}_\omega^{\varepsilon,\delta}$ the internal moment vector at the centroid of particle δ, $\boldsymbol{F}^{EXT,Q}$ the assembled external force and moment vector, $\boldsymbol{f}_\delta^{EXT,Q}$ the external body force and moment vector for particle δ, $\boldsymbol{f}^{EXT,\delta}$ the external body force vector at the centroid of particle δ, and $\boldsymbol{\ell}^{EXT,\delta}$ the external body moment vector at the centroid of particle δ. $\boldsymbol{Q}$ is the generalized dof vector for particle displacements and rotations

$$\boldsymbol{Q} = [\boldsymbol{q}_\delta, \boldsymbol{q}_\varepsilon, \ldots, \boldsymbol{q}_\eta, \boldsymbol{\omega}_\delta, \boldsymbol{\omega}_\varepsilon, \ldots, \boldsymbol{\omega}_\eta]^T \,, \quad \delta, \boldsymbol{\varepsilon}, \ldots, \eta \in \mathcal{A} \tag{2}$$

where $\boldsymbol{q}_\delta$ is the displacement vector of particle δ, $\boldsymbol{\omega}_\delta$ its rotation vector, and $\mathcal{A}$ is the set of free particles (see Fig. 1). In general, a superscript Q denotes a variable associated with particle motion, whereas a superscript D will denote a variable associated with continuum deformation.

2.2 *Finite Element Model of Elastic Micropolar Continuum with Reduced Kinematics*

Following the formulation of Eringen (1968), the balance equations for linear and angular momentum ignoring body forces and couples may be written as

$$\sigma_{lk,l} - \rho \dot{v}_k = 0 \tag{3}$$

$$m_{lk,l} + e_{kmn}\sigma_{mn} - \rho \dot{\beta}_k = 0 \tag{4}$$

where σ_{lk} is the unsymmetric Cauchy stress tensor, ρ the mass density, $v_k = \dot{u}_k$ the velocity vector, m_{lk} the unsymmetric couple stress, e_{kmn} the permutation operator, $\beta_k = j\dot{\varphi}_k$ the intrinsic spin per unit mass, j the micro-inertia, indices

$k, l, \dots = 1, 2, 3$, and $(\bullet)_{,l} = \partial(\bullet)/\partial x_l$ denotes partial differentiation with respect to the spatial coordinate x_l. Introducing w_k and η_k as weighting functions for the macro-displacement vector u_k and micro-rotation vector φ_k, respectively, we apply the Method of Weighted Residuals to formulate the partial differential equations in (3) and (4) into weak form. The weak, or variational, equations then result as

$$\int_{\mathcal{B}} \rho w_k \dot{v}_k dv + \int_{\mathcal{B}} w_{k,l} \sigma_{lk} dv = \int_{\Gamma_t} w_k t_k da \tag{5}$$

$$\int_{\mathcal{B}} \rho \eta_k \dot{\beta}_k dv + \int_{\mathcal{B}} \eta_{k,l} m_{lk} dv - \int_{\mathcal{B}} \eta_k e_{kmn} \sigma_{mn} dv = \int_{\Gamma_r} \eta_k r_k da \tag{6}$$

where $\mathcal{B}$ is the volume of the continuum body, $t_k = \sigma_{lk} n_l$ is the applied traction on the portion of the boundary Γ_t with outward normal vector n_l, and $r_k = m_{lk} n_l$ is the applied surface couple on the portion of the boundary Γ_r.

We introduce the following reduced kinematics for Timoshenko beam with stretch (Pinsky 2001) as

$$\boldsymbol{u} = \begin{bmatrix} u_1 \\ u_2 \\ u_3 \end{bmatrix} = \begin{bmatrix} u - x_2\theta \\ v \\ w \end{bmatrix}, \ \boldsymbol{\varphi} = \begin{bmatrix} 0 \\ 0 \\ \theta \end{bmatrix} \tag{7}$$

where u is the axial displacement, v the transverse displacement, θ the rotation. Note that the third component of $\boldsymbol{\varphi}$ is a rotation about the x_3 axis, thus a rotation in the $x_1 - x_2$ plane. We ignore the transverse displacement and stretch w in the x_3 direction, although we can solve for σ_{33} if need be (i.e., when we extend to pressure-sensitive plasticity for the micropolar or micromorphic continuum Regueiro (2009)). After substituting into the coupled weak form, assuming a mixed 1D finite element (quadratic in v, and linear in u and θ), details omitted, we arrive at the following coupled finite element matrix form as

$$\boldsymbol{M}^D \ddot{\boldsymbol{D}} + \boldsymbol{K}^D \boldsymbol{D} = \boldsymbol{F}^{EXT,D} \tag{8}$$

$$\boldsymbol{M}^D = \begin{bmatrix} \boldsymbol{M}^{uu} & \boldsymbol{0} & -\boldsymbol{M}^{u\theta} \\ \boldsymbol{0} & \boldsymbol{M}^{vv} & \boldsymbol{0} \\ -\boldsymbol{M}^{\theta u} & \boldsymbol{0} & \boldsymbol{M}^{\theta\theta} \end{bmatrix} \quad \boldsymbol{D} = \begin{bmatrix} \boldsymbol{d}_u \\ \boldsymbol{d}_v \\ \boldsymbol{d}_\theta \end{bmatrix}$$

$$\boldsymbol{K}^D = \begin{bmatrix} \boldsymbol{K}^{uu} & \boldsymbol{0} & -\boldsymbol{K}^{u\theta} \\ \boldsymbol{0} & \boldsymbol{K}^{vv} & -\boldsymbol{K}^{v\theta} \\ -\boldsymbol{K}^{\theta u} & -\boldsymbol{K}^{\theta v} & \boldsymbol{K}^{\theta\theta} \end{bmatrix} \quad \boldsymbol{F}^{EXT,D} = \begin{bmatrix} \boldsymbol{F}_F \\ \boldsymbol{F}_V \\ \boldsymbol{F}_M + \boldsymbol{F}_{M^\theta} \end{bmatrix} \tag{9}$$

where $\boldsymbol{M}^D$ is symmetric, but $\boldsymbol{K}^D$ is unsymmetric. The submatrices of $\boldsymbol{M}^D$ and $\boldsymbol{K}^D$ couple u, v, and θ.

3 Coupling Approach

The coupling implementation extends to particle mechanics and micropolar continuum (Regueiro and Yan 2011) the "bridging scale decomposition" proposed by Wagner and Liu (2003) and modifications thereof by Klein and Zimmerman (2006). The approach in one-dimension is illustrated in Fig. 1, where the leftmost particle has an applied axial and shear force representing contact with another solid.

The coupling approach involves two main features: (1) coupling the kinematics, and (2) partitioning the energy in the overlap region. As an example of kinematic coupling, consider a particle α, whose displacements can be interpolated from the finite element nodal dofs by evaluating the shape functions at the particle centroid $\boldsymbol{X}_\alpha$ as

$$\boldsymbol{q}_\alpha(t) = \boldsymbol{u}^h(\boldsymbol{X}_\alpha, t) = \sum_{a \in \check{\mathcal{N}}} N_a^u(\boldsymbol{X}_\alpha)\boldsymbol{d}_a(t) \quad \alpha \in \widehat{\mathcal{A}} \tag{10}$$

and likewise for the particle rotation

$$\omega_\alpha(t) = \varphi^h(\boldsymbol{X}_\alpha, t) = \sum_{b \in \check{\mathcal{M}}} N_b^\varphi(\boldsymbol{X}_\alpha)\theta_b \quad \alpha \in \widehat{\mathcal{A}} \tag{11}$$

The energy partitioning involves writing each system of equations for particles and continuum FE mesh in a Lagrangian equations form, such that kinetic and potential energy of the system can be added with scaling coefficients applied to the respective particle and continuum stiffnesses in the overlap region. Assuming statics for now, upon partitioning energy, the resulting nonlinear coupled equations written in residual form for solution by the Newton-Raphson method are

$$\begin{aligned} \boldsymbol{R}^Q(\boldsymbol{Q},\boldsymbol{D}) = {} & \boldsymbol{F}^{INT,Q}(\boldsymbol{Q}) + \boldsymbol{B}^T_{\widehat{Q}Q}\boldsymbol{F}^{INT,\widehat{Q}}(\widehat{\boldsymbol{Q}}) \\ & + \boldsymbol{B}^T_{\widehat{D}Q}\boldsymbol{F}^{INT,\widehat{D}}(\widehat{\boldsymbol{D}}) - \boldsymbol{F}^{EXT,Q} = \boldsymbol{0} \end{aligned} \tag{12}$$

$$\boldsymbol{R}^D(\boldsymbol{Q},\boldsymbol{D}) = \boldsymbol{B}^T_{\widehat{Q}D}\boldsymbol{F}^{INT,\widehat{Q}}(\widehat{\boldsymbol{Q}}) + \boldsymbol{F}^{INT,D}(\boldsymbol{D}) - \boldsymbol{F}^{EXT,D} = \boldsymbol{0} \tag{13}$$

The projection matrices $\boldsymbol{B}_{\widehat{Q}Q}$, $\boldsymbol{B}_{\widehat{Q}D}$, and $\boldsymbol{B}_{\widehat{D}Q}$ project the free dofs to the prescribed (or ghost) dofs, allowing the solution of the coupled equations only for the free dofs $\boldsymbol{Q}$ and $\boldsymbol{D}$.

4 Numerical Example

The one-dimensional overlapped coupled domain in Fig. 1 is compressed axially and sheared in the transverse direction as indicated in the figure. The particles are assumed to be isotropic elastic quartz ($E = 2.9\text{e}10$ Pa, $\nu = 0.25$), and likewise for

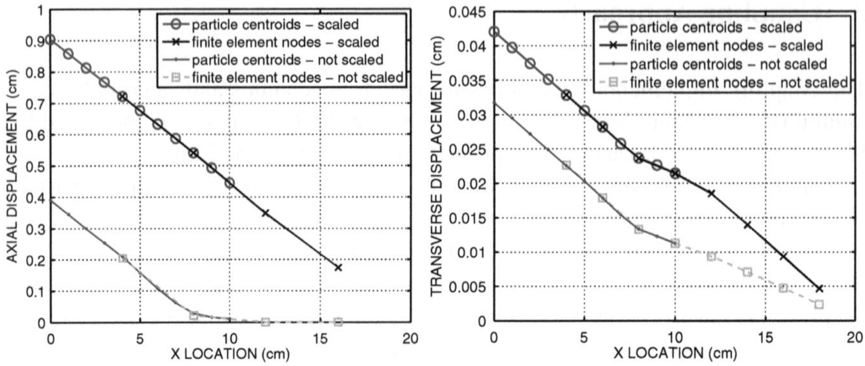

Fig. 2 (*left*) Axial displacement versus position along the 1D string of particles and FE mesh. (*right*) Transverse displacement

corresponding parameters of the 1D micropolar continuum (the additional elastic parameters are scaled to these parameters). Preliminary results of the coupling response is shown in Fig. 2. It can be seen that if the elastic stiffnesses of the 1D micropolar continuum rod finite element mesh are not scaled, the response is too stiff, and the partitioning of energy via a volume fraction in the overlapped region $\tilde{\mathcal{B}}^h$ will not lead to a smooth displacement field along the rod. When scaled, the homogeneous axial displacement along the bar can be achieved, however, the transverse displacement still demonstrates an artificial boundary effect for the transverse displacement and force when the stiffnesses are scaled. These are preliminary results, and we are in the process of rectifying the scaling for the transverse dof.

5 Conclusion

The paper presented the preliminaries of coupling DE regions and micropolar FE regions for eventual simulation of granular soil-tire/tool/ penetrometer applications, wherein three-dimensional DE formulation and finite strain micromorphic continuum FE is required (Regueiro 2009). A one-dimensional elastic example was shown for a string of glued elastic spheres experiencing Hertzian contact constitutive relations, overlapped by a 1D micropolar continuum finite element mesh. Results showed the stiffer response of the 1D micropolar continuum FE if the stiffness is not scaled. Proper scaling relations are being worked out.

Acknowledgments Funding for this research was provided by National Science Foundation grant CMMI-0700648, Army Research Office grant W911NF-09-1-0111, and the Army Research Laboratory. This funding is gratefully acknowledged.

References

A.C. Eringen, Theory of micropolar elasticity, in *Fracture, an Advanced Treatise*, ed. by H. Liebowitz (Academic, New York, 1968), pp. 622–729

P.A. Klein , J.A. Zimmerman, Coupled atomistic-continuum simulations using arbitrary overlapping domains. J. Comput. Phys. **213**(1), 86–116 (2006)

R.D. Mindlin, Compliance of elastic bodies in contact. J. Appl. Mech. **16**(3), 259–268 (1949)

P.M. Pinsky, Finite element structural analysis, Course notes, Stanford University, Stanford (2001)

R.A. Regueiro, Finite strain micromorphic pressure-sensitive plasticity. J. Eng. Mech. **135**, 178–191 (2009)

R.A. Regueiro, B. Yan, Concurrent multiscale computational modeling for dense dry granular materials interfacing deformable solid bodies. Bifurcations, Instabilities and Degradations in Geomaterials, *Springer Series in Geomechanics and Geoengineering*, ed. by R. Wan, M. Alsaleh, J. Labuz (Springer-Verlag, Berlin, 2011), pp. 251–273

G.J. Wagner, W.K. Liu, Coupling of atomistic and continuum simulations using a bridging scale decomposition. J. Comput. Phys. **190**(1), 249–274 (2003)

B. Yan, R.A. Regueiro, S. Sture, Three dimensional discrete element modeling of granular materials and its coupling with finite element facets. Eng. Comput. **27**(4), 519–550 (2010)

On the Numerical Implementation of a Multi-Mechanism Cyclic Plasticity Model Associated to a Dilation Second Gradient Model Aiming Strain Localization Mitigation

A. Foucault, F. Voldoire, and A. Modaressi

Abstract This paper deals with numerical simulations of granular soils, idealized by a cyclic multi-mechanism anisotropic non-associated plasticity constitutive model (Aubry et al. 1982; Hujeux 1985) coupled to a dilation second gradient model (Fernandes et al. 2008), under a finite elements method framework. This constitutive model is integrated through an implicit scheme into the Finite Element software *Code_Aster* (Foucault 2009). The use of this constitutive model, in case of strain localization, exhibits mesh sensitivity as for any strain-softening model. First we verify the ability of the dilation second gradient model to (1) circumvent the problem of mesh sensitivity with this model and to (2) describe the post-peak behaviour of the studied loading process. We studied a biaxial laboratory test, on Hostun sand, under drained conditions and monotonic loading. Second we develop a methodology to determine the values of the second gradient parameter – i.e. characteristic length – for different sets of soil model parameters. We have adapted a 1-D theoretical approach

A. Foucault (✉)
Laboratoire MSSMat, CNRS UMR 8579, École Centrale Paris, Grande voie des vignes, F-92290 Châtenay-Malabry, France

Laboratoire LaMSID, UMR EDF/CNRS 2832, 1, avenue du Général de Gaulle, F-92141 Clamart, France

Département Analyses Mécanique et Acoustique, EDF Recherche et Développement, 1, avenue du Général de Gaulle, F-92141 Clamart, France
e-mail: alexandre.foucault@edf.fr

F. Voldoire
Laboratoire LaMSID, UMR EDF/CNRS 2832, 1, avenue du Général de Gaulle, F-92141 Clamart, France

Département Analyses Mécanique et Acoustique, EDF Recherche et Développement, 1, avenue du Général de Gaulle, F-92141 Clamart, France

A. Modaressi
Laboratoire MSSMat, CNRS UMR 8579, École Centrale Paris, Grande voie des vignes, F-92290 Châtenay-Malabry, France

S. Bonelli et al. (eds.), *Advances in Bifurcation and Degradation in Geomaterials*, Springer Series in Geomechanics and Geoengineering 11,
DOI 10.1007/978-94-007-1421-2_26,

proposed by Chambon et al. (2001), applied to our constitutive model, whose comparison with our numerical results provides a characteristic length. To check the ability of the dilation second gradient model to ensure objective results with this soil constitutive model, we performed an extensive validation on a bearing capacity case with excavation.

Keywords Soil cyclic behavior model • Multi-mechanism plasticity • Softening • Strain localization • Second gradient dilation model

1 Introduction

In order to deal with the safety of embankment dams or other geo-structures submitted to seismic events, engineers require simulation tools having good properties of accuracy, robustness, and CPU performance, despite the known difficulties in such a range of problems. We have chosen a cyclic multi-mechanism constitutive model, developed in the early 1980s, namely due to its matureness and the availability of the parameters needed to describe the behaviour of many soils and materials implemented in many engineering geo-structures.

2 Overview on the Cyclic Multi-Mechanism Constitutive Model

The Hujeux's constitutive model (or ECP one) devoted to granular or clayey soils is written in terms of effective stress. Its mathematical formulation is rather complicated, motivated by the refined fitting of complex soils rheology. The representation of all irreversible phenomena is done by four coupled elementary plastic mechanisms: three plane-strain deviatoric plastic mechanisms in orthogonal planes and a volumetric one, to describe consolidation stages, see (Aubry et al. 1982; Hujeux 1985; Foucault 2009). Limit yield surfaces are based on Coulomb-type failure criteria. The critical state notion is integrated. Kinematic hardening on each mechanism is controlled by back-stress tensors. The consistency relations are used to determine the plastic flows, from the yield surfaces. The Roscoe's dilatancy law is used to obtain the volumetric plastic strain rate of each deviatoric mechanism. To take into account the cyclic behaviour a kinematical hardening based on the state variables at the last load reversal is used (by means of specific evanescent state variables). The soil deviatoric strength is described by the following yield function f_k^m, depending on an isotropic hardening variable r_k^m, the plastic volumetric strain ε_v^p, the mean pressure $p_k(\sigma)$ and the deviatoric stress $q_k(\sigma)$:

$$f_k^m\left(\sigma,\varepsilon_v^p,r_k^m\right) = q_k\left(\sigma\right) + p_k\left(\sigma\right) . F\left(p_k\left(\sigma\right),\varepsilon_v^p\right) . \left(r_k^m + r_{éla}^d\right) \leq 0 \quad (1)$$

with the following function parameterised by the internal friction angle ϕ_{pp} and the positive scalar b_h, depending on the critical pressure P_c of the plastic volumetric strain ε_v^p:

$$F\left(p_k(\sigma), \varepsilon_v^p\right) = \sin\phi_{pp}.\left(1 - b_h \ln\left|\frac{p_k(\sigma)}{P_c\left(\varepsilon_v^p\right)}\right|\right) \quad (2)$$

3 Numerical Implementation and Examples

We aim to get accuracy in computing of both balance equations and constitutive relations, during the appropriated steps in the nonlinear Newton-Raphson iterative procedure. That is why we choose an implicit integration scheme, as other authors promoted; see (Manzari and Prachathananukit 2001). The model has been implemented into *Code_Aster*, see (http://www.code-aster.org). At the local stage, the implicit integration consists in a small nonlinear system to be solved (constitutive relations, yield functions and flow equations), associated with a specific activated elastic-plastic mechanisms management procedure, giving a relevant prediction. Finally, the rate non-symmetric tangent stiffness operator is built to increase the performance of Newton iterations at the global stage, by updating at each iteration. Due to the rate-dependency of the tangent stiffness, some sub-stepping is introduced to increase robustness. Convergence criteria consist in dimensionless Euclidian norm of the residual.

For transient analyses, time integrators are commonly used, as Newmark (average acceleration method) or HHT (Hilber-Hughes-Taylor) schemes, the latter introducing a numerical damping, degrading the order of accuracy but reducing the detrimental spurious high frequencies stemming from the time-discretisation.

Two essential stages were done: (Aubry et al. 1982) the verification process, to check that the model implementation accurately represents the model specification, and (Hujeux 1985) the validation process, to check that model has accurate physical representation ability from the intended uses. For the first stage, we get pretty good results in a large set of loading situations through laboratory biaxial tests, both drained and undrained, monotonic and cyclic, see (Foucault 2010). Moreover, we performed embankment dam seismic loading analyses on the El Infiernillo benchmark as (Sica et al. 2008).

4 Dilation Second Gradient Enriched Kinematics

Hujeux's constitutive model exhibits high sensitivity to evolution directions on practical loading paths, coming from the multi-mechanism formulation, and also uniqueness loss. Moreover, before reaching the critical state, or incoming into liquefaction stage, this model can lead to softening. Both features induce numerical instabilities, at the material level, and divergence of Newton-Raphson iterative procedure, whatever the appropriate tangent operator management. It appears at a loading level depending on the finite element mesh used for the structure or the specimen. We observe strain localisation, in general, organised along finite thickness bands of highly concentrated evolving strains, like shear bands in perfect plasticity. To overcome this difficulty, we introduce spatial regularisation tool into the finite element formulation. It has been proved in (Fernandes et al. 2008), that the dilation second gradient model remedies to the mesh dependency on the localized zones width for Drücker-Prager constitutive model, by a good cost/efficiency compromise; that is why we decided to extend this approach to Hujeux's model of soil behaviour (Foucault 2010).

4.1 Theoretical Background

We briefly recall the theoretical presentation of the dilation second gradient model (Fernandes et al. 2008). This model can be seen as a constrained micromorphic dilation kinematics (Chambon et al. 2001). We assume the macro volumetric change equal to the microdilation: $\chi = \varepsilon_v$. We obtain the mixed augmented formulation of the virtual strain work, for any kinematically admissible field u_i^* and microdilation χ^* (Fernandes et al. 2008):

$$\forall u_i^* \in U_0^{ad},\ \forall \chi^* \in V_0^{ad},\ \forall \lambda^* \in L^{ad},$$

$$W_{int} = \int_\Omega \sigma_{ij}.\varepsilon_{ij}\left(u_i^*\right).d\Omega + \int_\Omega \Big(\left(S_j.\nabla \chi_j^*\right) - \lambda\left(\varepsilon_v^* - \chi^*\right) + \lambda^*\left(\varepsilon_v - \chi\right)$$

$$+r\left(\varepsilon_v^* - \chi^*\right).\left(\varepsilon_v - \chi\right)\Big)\, d\Omega$$

where regularisation comes from product $S_j.\nabla \chi_j^*$, associated to the relationship $\varepsilon_v = \chi$. We assume elastic constitutive equations linking double stress and second gradient tensors (Mindlin 1965), by means of a new parameter b_s defined by $[S_j] = b_s.\delta_j^i.\left[\dfrac{\partial \varepsilon_v}{\partial x_i}\right]$.

4.2 Application

Simulations of drained biaxial tests on laboratory samples ($10 \times 20\,\text{cm}^2$) show a dependence of shear bands thickness with respect to initial stress state and material properties.

A 1D-analytical problem of a dilative band is used to extract the key factors of this dependence, based on the example solved in (Chambon et al. 2001). These theoretical solutions are compared and validated to numerical responses issued from

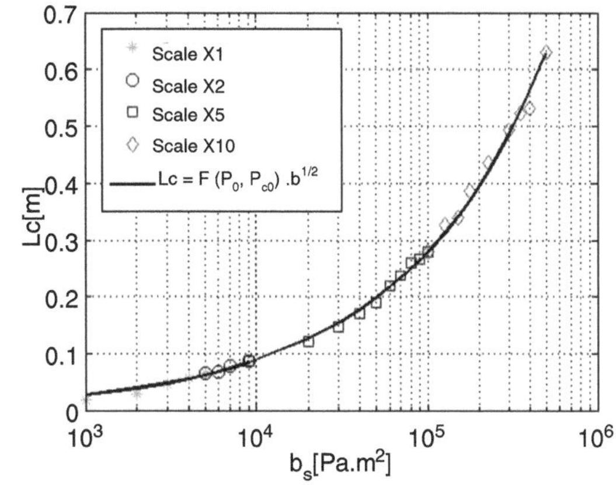

Fig. 1 Band thickness evolution function of b_s [Pa.m^2] with numerical simulation

Code_Aster software, which are in good accordance (Foucault 2010; Figs. 1 and 2) for Hostun sand material properties established by (Lopez-Caballero et al. 2003) and adapted to a simplified case of the Hujeux's model. The periodic non-trivial solutions of this analytical example involve a characteristic length depending on the tangent rigidity operator $\mathbf{H}^{ep}$:

$$\ell_c = 2\pi\sqrt{\frac{H^{ep}_{1212}b_s}{H^{ep}_{1211}H^{ep}_{1112} - H^{ep}_{1212}H^{ep}_{1111}}} = F\left(P, P_{c0}\right).\sqrt{b_s} \quad (3)$$

The numerical band thickness is numerically evaluated by considering the instantaneous plastic state at Gauss points at the bifurcation point of loading.

These results, extended to Hujeux's model, demonstrate the capacity of dilation second gradient model to mitigate the mesh sensitivity induced by softening. It appears clearly an independent characteristic length in simulations (Figs. 3 and 4) whatever the meshes (regular or not, coarse or fine). We also showed in (Foucault 2010) that the contribution of this model to the strain energy is finite and controlled, especially due to the critical state concept integrated in the Hujeux's model.

A bearing capacity problem with excavation was also solved to demonstrate the ability of the method on different meshes (Foucault 2010). We observed the main role of initial stress state on the shear band thickness in the context of the dilation second gradient model. Indeed, after a foundation settlement of 20 cm, the results show a band thickness enlargement in the vicinity of the excavation free surface, interpreted as a drawback in the methodology to determine the b_s values.

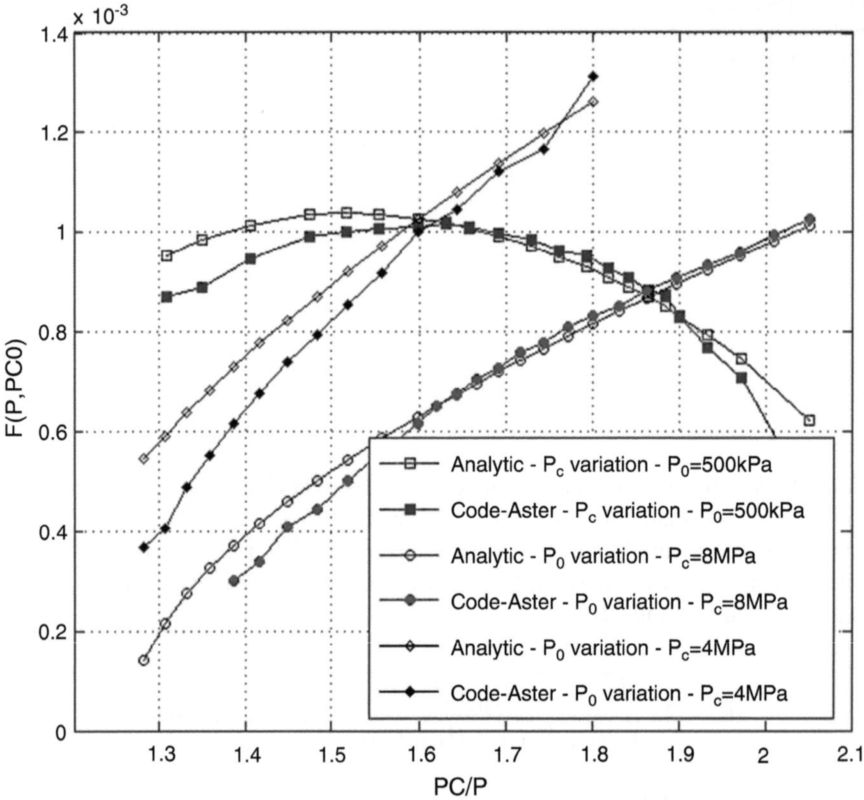

Fig. 2 $F(P, P_{c0})$'s evolution function of P_c/P for analytic and numeric approaches

Fig. 3 Drained biaxial test: force-displacement curve; mesh dependence

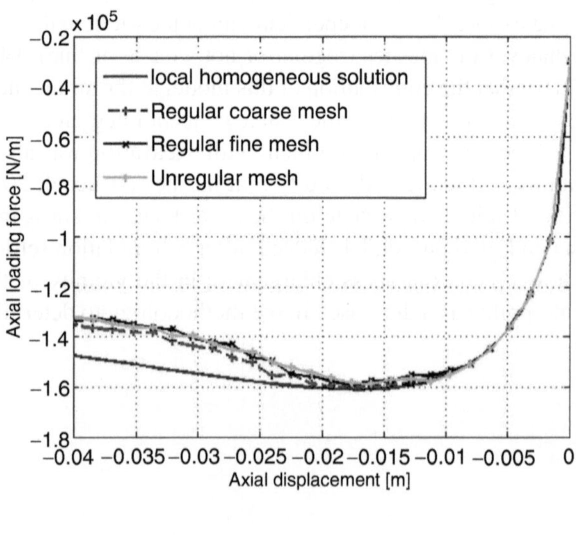

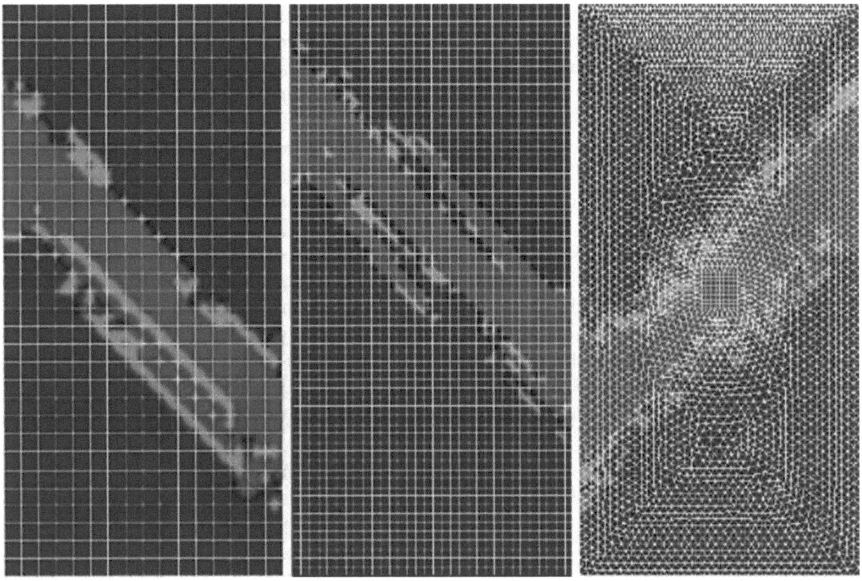

Fig. 4 Plots of instantaneous plastic state on Gauss points for the biaxial test after an axial displacement of 4 cm

5 Concluding Remarks and Forthcoming Research

From the numerical point of view, we can conclude the ability and the efficiency of the regularisation by second dilation model of geo-mechanical problem idealised by Hujeux's cyclic elastic-plastic constitutive model: the better selection of solution after critical point decreases the CPU time needed, bearing capacities and collapse modes are less dependent on the discretisation. Nevertheless we have to verify the same promising properties in dynamic problems: already, the effect of dilation second gradient model is shown (Foucault 2010) to be small, in the context of seismic wavelength. The b_s values evaluation needs also to be extended to any kind of load path using the second gradient dilation model on engineering geo-structures.

References

D. Aubry, J.C. Hujeux, F. Lassoudière, Y. Meimon, A double memory model with multiple mechanism for cyclic soil behaviour, in *International Symposium on Numerical Models in Geomechanics*, Zurich, 1982, pp. 3–13

R. Chambon, D. Caillerie, T. Matsuchima, Plastic continuum with microstructure, local second gradient theories for geomaterials: localization studies. Int. J. Solids Struct. **38**, 8503–8527 (2001)

Code_Aster, Open-Source FEM software. http://www.code-aster.org (2011)
R. Fernandes, C. Chavant, R. Chambon, A simplified second gradient model for dilatant materials: theory and numerical implementation. Int. J. Solids Struct. **45**, 5289–5307 (2008)
A. Foucault, R7.01.23: Loi de comportement cyclique de Hujeux pour les sols. Code_Aster documentation (2009), http://www.code-aster.org, 54 p
A. Foucault, Modélisation du comportement cyclique des ouvrages en terre intégrant des techniques de régularisation. Ph.D. thesis, École Centrale de Paris, Paris, 2010
J.C. Hujeux, Une loi de comportement pour le chargement cyclique des sols. Génie Parasismique, ed. by V. Davidovici (Presses ENPC, Paris, 1985) pp. 287–302
F. Lopez-Caballero, A. Modaressi, F. Elmi, Identification of an elasto-plastic model parameters using laboratory and in situ tests. *Deformation Characteristics of Geomaterials, IS Lyon*, eds. by Di Benedetto, T. Doanh, H. Geoffroy, C. Sauzéat (Balkema, Rotterdam, 2003), pp. 1183–1190
M.T. Manzari, R. Prachathananukit, On integration of a cyclic soil plasticity model. Int. J. Numer. Anal. Methods. Geomech. **25**, 525–549 (2001)
R.D. Mindlin, Second gradient of strain and surface tension in linear elasticity. Int. J. Solids Struct. **1**, 417–738 (1965)
S. Sica, L. Pagano, A. Modaressi, Influence of past loading history on the seismic response of earth dams. Comput. Geotech. **35**, 61–85 (2008)

2D and 3D Modelling of Geomaterials Using a Second Gradient Dilation Model

R. Fernandes, B. Ducoin, and R. Chambon

Abstract This paper deals with some recent results obtained with a simplified second gradient model to simulate localized patterns. The simplification is based on the use of the gradient of the volume variation only. It is first shown that this model is very efficient since it is less time consuming than a classical one. Consequently, such a model can be useful for 3-D computations. Finally some results show once more that even for 3-D computations, enhanced models such as second gradient ones do not restore the mathematical well posedness of the initial boundary value problem.

Keywords Second gradient • Dilation • 3D • FEM • Geomaterials

1 Introduction

Since some pioneering works of Ioannis Vardoulakis (see Vardoulakis and Sulem 1995 for a review of the first papers dealing with this topic), it is well known that classical continuum mechanics is not sufficient to model media undergoing strength degradation. Using second gradient models based on the media with microstructure theory (as defined by Germain and Mindlin) is a possible remedy to the mesh dependence of the results of the classical computations. The advantage of such second gradient theories is that it can give us objective extensions of any classical models.

R. Fernandes (✉) · B. Ducoin
LAMSID (EDF/R et D), 1 Avenue du Gnral De gaulle, 92140 Clamart, France
e-mail: romeo.fernandes@edf.fr

R. Chambon
Laboratoire 3S-R, University of Grenoble, BP 53X, 38041 Grenoble Cedex, France
e-mail: Rene.Chambon@hmg.inpg.fr

S. Bonelli et al. (eds.), *Advances in Bifurcation and Degradation in Geomaterials*, Springer Series in Geomechanics and Geoengineering 11,
DOI 10.1007/978-94-007-1421-2_27,

Here, we present first a simplified second gradient model using only the gradient of the volume variation. The details of the model can be seen in Fernandes et al. (2008). The classical part of the model is an elasto plastic Drucker Prager model involving a decreasing of the cohesion which induces softening behavior. Numerical implementation of this model for 2D and 3D problems is then detailed. Using 2D computations it is shown that best results are obtained by using both a field of Lagrange multiplier and a penalty method. This model moreover is able to deal with hydromechanical coupled problems. Some examples of the coupled modeling of the drilling of a borehole are detailed. A comparison of the performances of this simplified model, the classical second gradient model and the corresponding classical (without gradients) model is done. This shows clearly that this second gradient dilation model not only gives us objectives results but also is less expensive than the classical one as far as the CPU time is concerned.

Even if enhanced models are able to remedy to the mesh dependence of computations done with softening classical constitutive equations, the idea that they do not restore the uniqueness of the solutions of the boundary value problems is now emerging. This can be proved in very simple cases (Chambon et al. 1998) and this is shown clearly in some more complex 2D problems (Sieffert et al. 2009). Some examples of computation, given here, demonstrate that this is also true for 3D problems. Let us first clarify the distinction between lack of well posedness and mesh independence which are often misunderstood. A well posed problem has been defined by Hadamard has a problem for which there is a solution, this solution is unique and the output depends continuously on the input. When strength degradation is included in a classical constitutive model, several solutions are possible. Some of these solutions are clearly non realistic since they imply rupture without any energy consumption (see Kotronis et al. 2008 for instance). Numerical experiments show that in this case decreasing the mesh size exhibits a convergence of the numerical solutions towards such an unrealistic solution. On the other hand it is now clear that enhancing the model by introduction of an internal length does not restore the well posedness since we still encountered several solutions (see Chambon et al. 1998 and Sieffert et al. 2009 for instance). However even if there are several solutions these solutions involve an intrinsic internal length. It is then possible to prescribe some solution independently on the mesh and thus for a given solution to restore the mesh independence and the convergence towards one of the several solutions.

The second gradient dilation model is then a powerful and efficient tool to deal with softening materials in geomechanics engineering (3D) problems and is likely a good approach to study the non uniqueness of the solutions which seems to be unavoidable for softening materials.

2 Framework of the Second Gradient Dilatant Model

2.1 General Formulation

Let us recall first the principle of second gradient dilation model which can be traced back to the paper of Cowin and Nunziato (1983) in the elastic framework and a complete presentation of which has been made in Fernandes et al. (2008). In the case of a second gradient dilation model, the virtual work principle reads:

$$\int_{\Omega} \left(\sigma_{ij} \frac{\partial u_i^*}{\partial x_j} + S_j \frac{\partial^2 u_i^*}{\partial x_j \partial x_i} \right) dv = \int_{\partial \Omega} \left(p_i u_i^* + P_i D u_i^* \right) ds. \tag{1}$$

where σ_{ij} is the ordinary stress, S_j the double dilation stress, u_i^* the virtual displacement field, p_i the external force per unit area, P_i the additional double force, and D denotes the normal derivative. In this model kinematics of the continuum is defined by the displacement field but also a field of some micro volumetric strain. Moreover this micro volumetric strain is constrained to be equal to the classic macro one. In this case the virtual work principle involves the gradient of the virtual volumetric strain and it is then necessary to defined the dual corresponding quantity S_j defining then the double dilation stress. Stresses are given by proper constitutive equations. In this paper the classical stress is obtained by an elasto plastic Drucker-Prager model, for which the decreasing of the cohesion induces plastic softening. On the contrary the double stress is simply proportional to the gradient of the volume variation. The prescribed boundary conditions are either the displacement and the normal derivative of the displacement, or the static dual quantities just defined. More details can be seen in Fernandes et al. (2008).

2.2 Numerical Implementation

If Eq. 1 is used directly to develop a finite element method like done in Chambon et al. (1998) for a more general model in a one dimensional case, this implies the use of C1 elements. In this paper we prefer the use of C0 finite elements. This implies the use of two independent fields, namely the displacement and the volumetric strain. In order to constraint the volumetric strain to be equal to the divergence of the displacement, we use Lagrange multipliers or/and penalty coefficients. After some numerical experiment we chose for efficiency reason (see Fernandes et al. 2008), to use both Lagrange multipliers and penalty methods, however as it can be seen in Sect. 4.2 in some case the penalty term is not activated. One of the great advantages of such a choice is that it is not necessary to use a too much big value for the penalty coefficient. Finally equations which have to hold for any kinematically admissible virtual displacement field, for any virtual volume variation ε_v^* and for any u_i^* read:

$$\int_{\Omega}\left(\sigma_{ij}\frac{\partial u_i^*}{\partial x_j}+S_j\frac{\partial \chi^*}{\partial x_j}-\Lambda\left(\varepsilon_v^*-\chi^*\right)+\Lambda^*\left(\varepsilon_v-\chi\right)+r\left(\varepsilon_v^*-\chi^*\right)\left(\varepsilon_v-\chi\right)\right)dv$$
$$=\int_{\partial\Omega}\left(p_i u_i^*+P_i D u_i^*\right)ds, \tag{2}$$

where r is the penalty coefficient. These equations are discretized to yield a finite element method. In the following computations has been done with or without (which means $r = 0$) penalty term.

2.3 Extension to Hydromechanical Problems

Extension to hydromechanical (saturated) problems of the second gradient dilation model is straightforward, following the method detailed in Collin et al. (2006) for the more general second gradient model. First we use the Terzaghi's definition of effective stress, which means a relation between the total stress σ_{ij}, the effective stress σ'_{ij} and the fluid pressure p^w:

$$\sigma_{ij} = \sigma'_{ij} + p^w \delta_{ij}. \tag{3}$$

σ'_{ij} is given by the constitutive equation of the skeleton. Then, in addition to Eq. 2, the mass balance equation for the percolating water has to hold. Written in a variational form, this yields:

$$-\int_{\Omega^t}\left(\frac{\partial m^w}{\partial t}p^{w,\star}+M_i^w\frac{\partial p^{w,\star}}{\partial x_i^t}\right)d\Omega^t=\int_{\Gamma}\overline{q}p^{w,\star}d\Gamma^t, \tag{4}$$

where, m^w is the fluid mass inside the reference solid volume, t is the time, p^w is the fluid pressure, M_i^w the fluid mass flow and $\overline{q}$ the eventually prescribed mass flux at the boundaries. $\bullet^\star$ as usual denotes virtual quantities. In the case of hydromechanical problems finite difference in time are written instead of time derivative. Details of this classical way of solving the hydromechanical problems can be seen in Collin et al. (2006).

2.4 Finite Element Used

The elements used in the computation reported in this paper are triangles for 2-D hydromechanical problems and tetrahedron for 3-D mechanical problems.

The elements used are defined in the following manner. The displacements are assumed to be polynomial of the second order of the space, meaning that the nodal values are taken for the vertices and for the mid points of the edges. The volumetric strain is assumed to be linear which means that its nodal values are related with

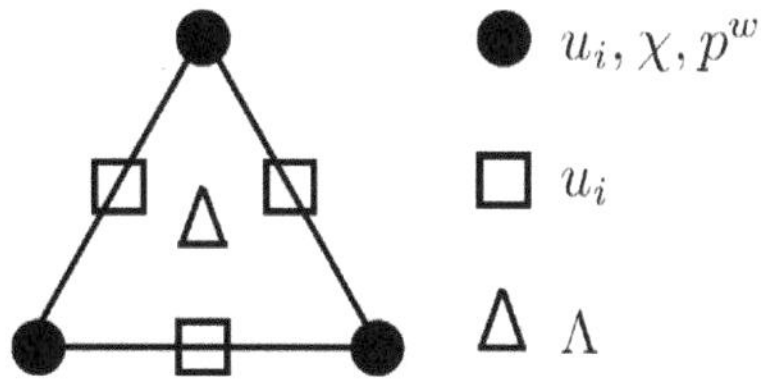

Fig. 1 Discretization of the ($u_i - P2$; $\chi - P1$; $p^w - P1$; $\Lambda - P0$) finite element to deal with coupled hydro-mechanical problems in the context of geomaterials

vertices only. The Lagrange multiplier are assumed to be constant over an element and in the case of hydromechanical problems, the fluid pressure is assumed to be linear with respect to the coordinates implying that the nodal values are related with the vertices of the elements.

When classical method and consequently classical elements are used, the displacement and the fluid pressure, which in this case are the only degrees of freedom, are defined in the same manner as for the second gradient dilation element. Figure 1 shows the triangle for the 2-D case, tetrahedrons for the 3-D case are built up in a similar manner except for the Lagrange multiplier which are linear.

3 2-D Numerical Experiments: Evaluation of Performance of the Second Gradient Dilation Model

This model has been checked and result shows clearly that localized solutions are mesh independent provided the mesh size is less than the implicit internal length involved in this model like in any second gradient model. The objective of the following computations is different. Even if it is known that classical computations are non objective and consequently mesh dependent, it is often thought that enhanced model are too costly and then that engineering use of such model is not possible. In the present section, the same computations have been done with a classical model and with the studied model. In order to do such a comparison the same problem with the same mesh (which clearly does not mean the same number of degrees of freedom) the same prescribed precision is computed with a classical model and with the corresponding (which means the same classical part) enhanced model. The computations have been done with the open source Code Aster.

3.1 The Problem Solved

The problem solved is the computation of the drilling of a vertical borehole. An anisotropic initial stress state is assumed. It is chosen to search only solutions symmetric with respect to the initial principal stress directions. This is why the computation is done on a quarter of the drilling. The aim is not here to study the non uniqueness of the solution like it as been extensively studied in Sieffert et al. (2009). Figure 2 shows the computed domain and the boundary conditions.

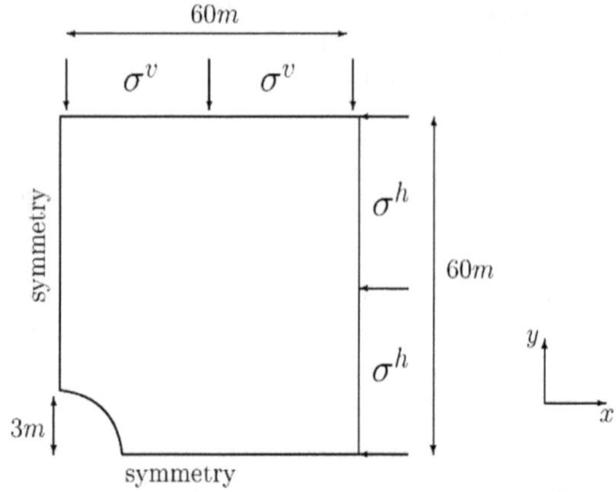

Fig. 2 The 2D problem solved

Table 1 Comparison of performance of a classical model and of the second gradient dilation model

Number of elements	D.O.F classical	CPU time classical	D.O.F second gradient	CPU time second gradient
18,867 triangles	86,540	26 h	114,985	10 h
30,471 triangles	139,028	3 days	184,909	19 h
78,164 triangles	355,056	7.5 days	472,606	2.5 days

The drilling is modeled by a linear decreasing (with respect to the time) of the stress of the mixture and of the pore pressure for the quarter of circle modeling the borehole. Zero values for both stresses are obtained after 17 days. Then these values remain constant and equal to zero during 15 months. All the details of the computation, namely the mechanical parameters, the hydraulic parameters such as the permeability or the Biot coefficient and the rate of the drilling can be seen in Fernandes et al. (2012) and are not given here.

3.2 *Results About the Performance of the Second Gradient Dilation Model*

Clearly the solutions obtained, using a classical model, are mesh dependent and in fact the CPU times mentioned in Table 1 do not correspond to the same solution, they can be seen in Fernandes et al. (2012). However all are properly converged. On the contrary the solutions obtained with the second gradient dilation model are the same since the size of the triangle close to the borehole are small enough with respect to the internal length even for the coarser mesh.

The code used namely Code Aster is based on implicit integration scheme for the constitutive equations and on the use of a consistent tangent stiffness matrix. It induces automatic decrease of the time step when the convergence becomes not sufficiently efficient. This explains why, despite the fact that second gradient dilation model involve more degrees of freedom than the classical model for the same mesh, the CPU time is very small for the enhanced model used. In fact on one hand time step can be longer and the number of iterations for a given time step can be smaller for the second gradient model. We did the same computations with the more classical second gradient model (see Chambon et al. 2001a and Matsushima et al. 2002) which proves that it induces CPU time similar to the classical models despite the fact that the number of degrees of freedom in this case is much greater. Moreover the solutions are also mesh independent.

3.3 Concluding Remarks

These results show clearly that second gradient models are usable for engineering problems. When a degradation of the strength is modeled, which is often the case for rock or soil like material, softening behavior is required. In this case it is now well known that classical models are unable to give objective solutions. Our study proves that the so called local (see Chambon et al. 2001a and Matsushima et al. 2002) second gradient model are not only a possible solution to restore the objectivity of the numerical results but also a very efficient solution. It is clear that similar studies have to be done for other enhanced models.

4 3-D Numerical Experiments

Engineering problems are almost always three dimensional ones. Since our model is less expensive than a classical one, it is obvious to develop it also for 3-D. In this section some preliminary results are presented. The development of the 3-D finite elements is not detailed; the principle of this new element has been given in Sect. 2.4. It has been checked that the objectivity of the numerical results holds also in 3-D. The consistency of the 3-D model with respect to the 2-D one has also been validated. The results presented hereafter are related with the question of uniqueness of the solutions. It has been extensively studied in the past for 2-D computations (see Chambon et al. 1998 and Sieffert et al. 2009 for instance), but it deserves studies in 3-D. In the following a very simple problem, namely a triaxial test with a defect is computed.

4.1 The Problem Solved

The problem solved is sketched in the left hand side of Fig. 3. Only mechanical effects are taken into account. Vertical displacements are assumed to be equal to

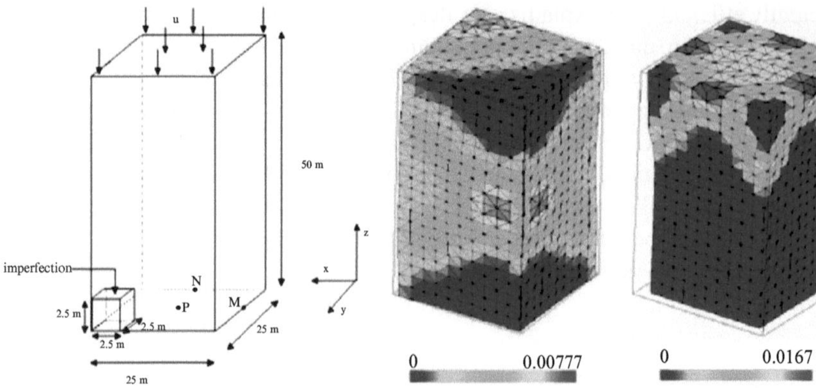

Fig. 3 The 3 D problem solved and two different obtained solutions

zero at the bottom of the domain and prescribed to the same value for the top of the modeled sample. Natural boundary conditions which mean that no forces are assumed for the other part of the boundary. Some point is assumed to be fixed in order to avoid rigid body motion. The parameters used are the same as the one used in Fernandes et al. (2008). In order to study the uniqueness of the solution an imperfection is put in the model (see Fig. 3, left). This imperfection is independent of the mesh used. In this area the initial cohesion is smaller than for the remain of the sample.

4.2 Main Results

For two dimensional problems it is now well known that for softening models the same numerical computations can give different solutions. This means that with the same mesh, the same time stepping, the same behavior parameters, which means the same physical parameters it is possible to obtain different solutions. Since computers are deterministic this is possible by changing only numerical parameters such as the first guess of the Newton-Raphson iterative procedures used to solve the equations corresponding to a given time step. These results have been obtained first for classical media (see Chambon et al. 2001b) and then extended to second gradient models (see Chambon et al. 1998 or Sieffert et al. 2009 for instance). We use a similar method here. Namely the same computations are performed using three different numerical inputs.

Since the localized modes are three dimensional it is difficult to have simply a clear view of these modes. We have chosen to plot external views of the same state, namely the final state of the computations. The color is related to the value of the plastic strain and the scale is given for the two given results. The results shown in the central part of Fig. 3 has been obtained by using a loading linked with the displacements for the degraded zone and without activation of the penalty method. The results shown in the right hand side of Fig. 3 is also obtained without penalty but in a more classical way, by using only the displacement of the top of the sample with respect to the displacement of the bottom. An other result giving a symmetric (i.e. localization in the bottom instead of the top) pattern as the previous one corresponds to the same way of prescribing the loading path but with adding the penalty term to the equations.

4.3 Comments and Concluding Remarks About the 3-D Computations

The 3-D second gradient dilation model is as efficient as the 2-D one. Moreover the same results hold. As soon as softening is activated, uniqueness of the solution of the corresponding initial boundary value problems is lost. Contrary to what is often believed, a defect is not able to induce uniqueness of the solution. Very different solutions can be obtained. The rupture mode seen on the right hand side of Fig. 3 is very similar to what has been observed in true triaxial apparatus some years ago (see Desrues et al. 1985) and it seems that the geometry of the sample has some influence on the found results.

5 Conclusion

The need of enhanced models for modeling softening is widely recognized. However it is strange that its use is not so widely spread. Our results, only related with local second gradient models and especially a particular case, namely the second gradient dilation model shows clearly that these models are easy to use since they do not induce too long computer time. It is necessary however to do the same numerical experiments for other enhanced models. It is not obvious that other enhancement (use of the gradient of some internal variables or the use of some non local strain or damage) yields similar good results. Clearly there is no good justification to do not use objective models. On the other hand, it is necessary to study carefully the non uniqueness of the solution found when using a constitutive equation modeling some softening. This point is may be more crucial since it implies a change in our view about modeling and computation. It has obviously to be studied in conjunction with the rarely studied problem of reproducibility of the experiments (see Sieffert et al. 2009 for instance).

Acknowledgments This paper is dedicated to the memory of Ioannis Vardoulakis. He suggested us some references during friendly discussions. He was the referee of the Ph. thesis of the first author. He supported by a very kind letter the scientific project, about second gradient dilation model allowing the third author to obtain a six months full time research position.

References

R. Chambon, D. Caillerie, N. El Hassan, One-dimensional localisation studied with a second grade model. Eur. J. Mech. A/Solids **17**, 637–656 (1998)

R. Chambon, D. Caillerie, T. Matsushima, Plastic continuum with microstructure, local second gradient theories for geomaterials: localization studies. Int. J. Solids Struct. **38**, 8503–8527 (2001a)

R. Chambon, S. Crochepeyre, R. Charlier, An algorithm and a method to search bifurcation points in non linear problems. Int. J. Numer. Methods Eng. **51**, 315–322 (2001b)

Code Aster, http://www.code-aster.org

F. Collin, R. Charlier, R. Chambon, A finite element method for poro mechanical modeling of geotechnical problems using local second gradient model. Int. J. Numer. Methods Eng. **65**, 1749–1772 (2006)

S.C. Cowin, J.W. Nunziato, Linear elastic materials with voids. J. Elast. **13**, 125–147 (1983)

J. Desrues, J. Lanier, P. Stutz, Localisation of the deformation in tests on sand sample. Eng. Fract. Mech. **21**, 909–921 (1985)

R. Fernandes, C. Chavant, R. Chambon, A simplified second gradient model for dilatant materials: theory and numerical implementation. Int. J. Solids Struct. **45**, 5289–5307 (2008)

R. Fernandes, C. Chavant, R. Chambon, Computations of the borehole stability problem using a classical model and a second gradient dilation model (in preparation) (2012)

P. Kotronis, S. Al-Holo, P. Besuelle, R. Chambon, Shear softening and localization: modeling the evolution of the shear zone. Acta Geotechnica **3**, 85–97 (2008)

T. Matsushima, R. Chambon, D. Caillerie, Large strain finite element analysis of local second gradient models, application to localization. Int. J. Numer. Methods Eng. **54**, 499–521 (2002)

Y. Sieffert, S. Al-Holo, R. Chambon, Loss of uniqueness of solutions of the borehole problem modeled with enhanced media. Int. J. Solids Struct. **46**, 3173–3197 (2009)

I. Vardoulakis, J. Sulem, *Bifurcation analysis in geomechanics* (Blackie Academic & Professional, London, 1995)

Local Second Gradient Models for Thermo-Hydro-Mechanical Coupling in Rock Like Materials

Y. Sieffert, F. Marinelli, and R. Chambon

Abstract In the design of nuclear waste disposals, an important topic concerns the evolution of an Excavated Damage Zone (EDZ) with a thermal exchanges. In this paper, a new model of local second gradient coupling with a thermo-hydro-mechanical is presented. As for monophasic case, the use of enhanced media induce the objective of the computation but not the uniqueness of the solution. Some classical engineering problems are presented which exhibit several solutions.

Keywords Thermo-hydro-mechanics • Localization • Second gradient model • Non uniqueness • Geomaterials

1 Introduction

It is well known that when localization can appear in a solution of an initial boundary value problem the classical continuum mechanics is no more available. It is then necessary to use enhanced continua in order to incorporate an objective internal length in the model. Several solutions are available. Local second gradient model is a good candidate for such a generalizing since it can be used with any classical model. Numerically it allows us to generalize the classical way of dealing with non linear problems. More precisely using a full Newton-Raphson method with a consistent tangent stiffness matrix for the stress point algorithm is easy to develop. Even if the price to pay is to develop the computer code, which increases the number of degrees of freedom, the results are very efficient computations which are in fact never more time consuming than classical (but not objective) computations.

Y. Sieffert (✉) · F. Marnelli · R. Chambon
Laboratoire 3S-R, Grenoble Université, Université Joseph Fourier BP 53 X Grenoble cedex 9
e-mail: yannick.sieffert@ujf-grenoble.fr; Rene.Chambon@hmg.inpg.fr

S. Bonelli et al. (eds.), *Advances in Bifurcation and Degradation in Geomaterials*, Springer Series in Geomechanics and Geoengineering 11,
DOI 10.1007/978-94-007-1421-2_28,

However geomaterials, especially natural ones, such as soils and rocks are never monophasic ones. Liquid (usually water) and often gas are percolating in the geomaterials. It is then crucial to take into account such couplings in order to be able to tackle engineer problems. On the other hand, for some problem, thermal effects have to be considered. Hydromechanical coupling for saturated situation has already been developed (see Collin et al. (2006) and Fernandes et al. (2008) for instance).

In this paper we give first an extension of the local second gradient model to thermo-hydro-mechanical problems. A new finite element is then detailed and validated. The non uniqueness of the solutions of initial boundary value problems is also studied. Similarly to what is observed in the monophasic case, several different solutions can be obtained in the case of biaxial tests as well as in the case of the borehole stability problem. It is proved that the different solutions are related with the internal length implicitly incorporated in the second gradient model and is not related with the coupling as soon as a pattern of localized zones is observed.

2 THM Constitutive Model

Geomaterials like soils, rock and concrete are porous media generally considered as a mixture of two continua (Coussy 1995) : the solid skeleton (grain assembly) and the fluid phases (water, air, oil, ...). For this study, our analysis is restricted to saturated conditions, but thermal impact is now added in the initial HM model developed by Collin et al. (2006).We introduce a dependence of the solid grain density, the liquid density and the viscosity of the fluid on the temperature. In derivative time form :

- $\dot{\rho}^{s,t} = -\rho^{s,t}\beta^{s}\dot{T}^{t}$
- $\dot{\rho}^{w,t} = \frac{\rho^{w,t}}{k^{w}}\dot{p}^{t} - \rho^{w,t}\beta^{w}\dot{T}^{t}$
- $\dot{\mu}^{t} = -\alpha^{w}\mu^{t}\dot{T}^{t}$

with $\rho^{s,t}$ the solid grain density, $\rho^{w,t}$ the water fluid density, β^{s} the solid thermal expansion coefficient, β^{w} the liquid thermal expansion coefficient, T the temperature, p the pore pressure, k^{w} the fluid bulk modulus, μ the fluid dynamic viscosity, α^{w} the liquid dynamic viscosity thermal coefficient.

2.1 Balance of Momentum of a Microstructured Porous Media

In the framework of microstructure continuum theory, a microkinematic gradient field v_{ij} is introduced to describe strain and rotation at the microscale. As usual in the local second gradient model, the assumption of the equality between the microkinematic gradient and the macrodeformation gradient F_{ij} is done :

$$v_{ij} = F_{ij} \tag{1}$$

For this study, we are assumed to the pore fluid and the temperature have no influence at the microstructure level. In a weak form we obtained:

$$\int_{\Omega^t}\left(\sigma^t_{ij}\frac{\partial u^\star_i}{\partial x^t_j}+\Sigma^t_{ijk}\frac{\partial^2 u^\star_i}{\partial x^t_j\partial x^t_k}\right)d\Omega^t=\int_{\Omega^t}\rho^{mix,t}g_i u^\star_i d\Omega^t + \int_{\Gamma^t_\sigma}\left(\bar{t}_i u^\star_i+\overline{T}_i Du^\star_i\right)d\Gamma^t, \tag{2}$$

with:

- Σ^t_{ijk} is the double stress
- $\rho^{mix,t}=\rho^{s,t}(1-\phi^t)+\rho^{w,t}\phi^t$ is the mass density of the mixture
- $\phi^t=\frac{\Omega^{w,t}}{\Omega^t}$ is the porosity
- $\bar{t}_i$ is the external (classical) forces per unit area
- $\overline{T}_i$ is an additional external (double) force per unit area. We assumed that the double forces are equal to zero
- D denotes the normal derivative of any quantity q ($Dq=(\frac{\partial q}{\partial x_k})n_k$)

2.1.1 Introducing Lagrange Multiplier Fields

The equation of the balance of momentum for the mixture in second gradient model media needs the use of C1 functions for the displacement field as second derivatives of the displacement are involved. In order to avoid such functions, the equalities between $v^\star_{ij}$ and $F^\star_{ij}$ and between v_{ij} and F_{ij} are introduced through a field of Lagrange multipliers λ_{ij} related to a weak form of the constraint (1) (Chambon et al. 1998):

$$\int_{\Omega^t}\left(\sigma^t_{ij}\frac{\partial u^\star_i}{\partial x^t_j}+\Sigma^t_{ijk}\frac{\partial v^\star_{ij}}{\partial x^t_k}\right)d\Omega^t-\int_{\Omega^t}\lambda^t_{ij}\left(\frac{\partial u^\star_i}{\partial x^t_j}-v^\star_{ij}\right)d\Omega^t$$
$$=\int_{\Omega^t}\rho^{mix,t}g_i u^\star_i d\Omega^t+\int_{\Gamma^t_\sigma}\left(\bar{t}_i u^\star_i+\overline{T}_i v^\star_{ik}n_k\right)d\Gamma^t, \tag{3}$$

$$\int_{\Omega^t}\lambda^\star_{ij}\left(\frac{\partial u_i}{\partial x^t_j}-v^t_{ij}\right)d\Omega^t=0, \tag{4}$$

2.2 *Mass Balance Equation for the Fluid*

In a week form, we obtained:

$$\int_{\Omega^t}\left(\dot{M}^{w,t}p^\star-m^{w,t}_i\frac{\partial p^\star}{\partial x^t_i}\right)d\Omega^t=\int_{\Omega^t}Q^{w,t}p^\star d\Omega^t-\int_{\Gamma^{w,t}_q}\overline{q}^{w,t}p^\star d\Gamma^t \tag{5}$$

with

- $M^{w,t}$ is the fluid mass in a unity of volume
- $m_i^{w,t}$ is the mass flow of the fluid
- $Q^{w,t}$ is a sink term
- $\Gamma_q^{w,t}$ is the part of the boundary where the input fluid mass per unit area $\overline{q}^{w,t}$ is prescribed.

2.2.1 The Fluid Mass $M^{w,t}$

In a volume Ω^t, the fluid mass $M^{w,t}$ is equal to

$$M_{\Omega^t}^{w,t} = \rho^{w,t}\phi^t\Omega^t$$
$$\dot{M}_{\Omega^t}^{w,t} = \dot{\rho}^{w,t}\phi^t\Omega + \rho^{w,t}\dot{\phi}^t\Omega^t + \rho^{w,t}\phi^t\dot{\Omega}^t$$

After some algebra, the variation of the fluid mass is considered:

$$\dot{M}^{w,t} = \rho^{w,t}\left(\frac{\dot{p}^t}{k^w}\phi^t - \beta^w\dot{T}^t\phi^t\right) + \rho^{w,t}\frac{\dot{\Omega}^t}{\Omega^t} - \rho^{w,t}\beta^s\dot{T}^t(1-\phi^t) \tag{6}$$

2.2.2 The Mass Flow of the Fluid $m_i^{w,t}$

The mass flow of the fluid $m_i^{w,t}$ is obtained with the Darcy's law :

$$m_i^{w,t} = -\rho^{w,t}\frac{\kappa}{\mu^t}\left(\frac{\partial p^t}{\partial x_i^t} + \rho^{w,t}g_i\right), \tag{7}$$

with κ is the intrinsic permeability.

2.3 *The Balance of Heat Equation*

In a weak form :

$$\int_{\Omega^t}\left(\dot{M}^{T,t}T^\star - m_i^{T,t}\frac{\partial T^\star}{\partial x_i^t}\right)d\Omega^t = \int_{\Omega^t} Q^{T,t}T^\star d\Omega^t - \int_{\Gamma_q^{T,t}}\overline{q}^{T,t}T^\star d\Gamma^t \tag{8}$$

- $M^{T,t}$ represents the stored heat quantity
- $m_i^{T,t}$ is the heat flux
- $Q^{T,t}$ is the volumetric heat source
- $\Gamma_q^{T,t}$ is the part of the boundary where the input heat per unit area $\overline{q}^{T,t}$ is prescribed.

2.3.1 The Stored Heat Quantity $M^{T,t}$

The stored heat quantity is:

$$M^{T,t} = H^{w,t} + H^{s,t} \tag{9}$$

- $H^{w,t}$ is the enthalpy of fluid phase

$$H^{w,t} = \phi^t \rho^{w,t} c_p^{w,t} (T^t - T^0) \tag{10}$$

  - $c_p^{w,t}$ is heat capacity of fluid

$$c_p^{w,t} = c_p^{w,0} + c_p^{w,0} C_w^T (T^t - T^0) \tag{11}$$

  - C_w^T is the heat variation of heat capacity of fluid

- $H^{s,t}$ is the enthalpy of solid phase

$$H^{s,t} = (1 - \phi^t) \rho^{s,t} c_p^{s,t} (T^t - T^0) \tag{12}$$

  - $c_p^{s,t}$ is heat capacity of solid

$$c_p^{s,t} = c_p^{s,0} + c_p^{s,0} C_s^T (T^t - T^0) \tag{13}$$

  - C_s^T is the heat variation of heat capacity of solid

In a volume Ω^t, the stored heat quantity is equal to

$$\begin{aligned} M_{\Omega^t}^{T,t} &= H^{w,t} \Omega^t + H^{s,t} \Omega^t \\ &= \phi^t \Omega^t \rho^{w,t} c_p^{w,t} (T^t - T^0) + (1 - \phi^t) \Omega^t \rho^{s,t} c_p^{s,t} (T^t - T^0) \end{aligned}$$

After some algebra :

$$\dot{M}^{T,t} = \left[\frac{\dot{\Omega}^t}{\Omega^t} \rho^{w,t} c_p^{w,t} + \beta^s \dot{T}^t (1 - \phi^t)(\rho^{s,t} c_p^{s,t} - \rho^{w,t} c_p^{w,t}) \right] (T^t - T^0) \tag{14}$$

$$\begin{aligned} &+ \phi^t \rho^{w,t} c_p^{w,t} \dot{T}^t + (1 - \phi^t) \rho^{s,t} c_p^{s,t} \dot{T}^t \\ &+ \phi^t \left[\rho^{w,t} \left(\frac{\dot{p}^t}{k^w} - \beta^w \dot{T}^t \right) c_p^{w,t} + \rho^{w,t} c_p^{w,t} C_w^T \dot{T}^t \right] (T^t - T^0) \\ &+ (1 - \phi^t) \left[-\rho^{s,t} \beta^s c_p^{s,t} \dot{T}^t + \rho^{s,t} c_p^{s,t} C_s^T \dot{T}^t \right] (T^t - T^0) \end{aligned} \tag{15}$$

2.3.2 The Heat Flux $m_i^{T,t}$

The heat flux is:

$$m_i^{T,t} = -\lambda^{m,t} \frac{\partial T^t}{\partial x_i^t} + c_p^{w,t} m_i^{w,t} (T^t - T^0) \tag{16}$$

- $\lambda^{m,t}$ is the thermal conduction of the mixture

$$\lambda^{m,t} = \phi^t \lambda^{w,t} + (1 - \phi^t)\lambda^{s,t} \tag{17}$$

- $\lambda^{w,t}$ is the thermal conduction of fluid

$$\lambda^{w,t} = \lambda^{w,0} + \lambda^{w,0} \Lambda_w^T (T^t - T^0) \tag{18}$$

  - Λ_w^T is the variation liquid thermal conductivity coefficient
- $\lambda^{s,t}$ is the thermal conduction of solid

$$\lambda^{s,t} = \lambda^{s,0} + \lambda^{s,0} \Lambda_s^T (T^t - T^0) \tag{19}$$

  - Λ_s^T is the variation solid thermal conductivity coefficient

2.4 A New Finite Element

A new element is developed to be associated with this THM constitutive law. Eight degrees of freedom is necessary to take into account the two displacements, the pore pressure, the temperature and the four microdisplacement.

2.5 The Numerical Problem Solved

Thermal loading is an important point for the design of storage nuclear waste in the underground. Indeed a radioactive waste disposal induced thermal fluxes in the host rock. In relation of the Timodaz EC project (Thermal Impact on the Damaged Zone Around a Radioactive Waste Disposal in Clay Host Rocks (http://www.timodaz.eu/), we have studied the evolution of the Excavation Damage Zone (EDZ) with an increase of the temperature on the cavity. It has been found that thermal loading does not substantially modify the evolution of the EDZ.

3 Non Uniqueness Study of the Solutions in Coupled Models

In the last 10 years, the loss of uniqueness has been clearly demonstrated in initial boundary value problems involving constitutive equations modeling the degradation of the strength of materials. In a one dimensional problem of traction in a bar where the analytical solutions can be developed (Chambon et al. 1998), it was demonstrated that it is possible to retrieve all the analytical solutions numerically, only by using different starting guesses in the Newtons method.

For some classical numerical tests in geomechanical the non uniqueness has been proved namely the biaxial test (Bésuelle et al. 2006) and the borehole problem (Sieffert et al. 2009). However in all these studies, the geomaterial is assumed to be monophasic.

In fact with a HM model, we can obtained different solutions of the same IBVP, in the case of biaxial tests as well as in the case of the borehole stability problem, as seen in the following.

Visualization of the different solutions is performed by observing the second invariant of the total strain (Fig. 1) and the loading index (Fig. 2). In the Fig. 2, a small square is plotted when a Gauss point undergoes a plastic loading. Otherwise which means that the corresponding Gauss point undergoes either elastic unloading or reloading, no tag is put on the picture.

It is proved that the different solutions are related with the internal length implicitly incorporated in the second gradient model and is not related with the coupling as soon as a pattern of localized zones is observed.

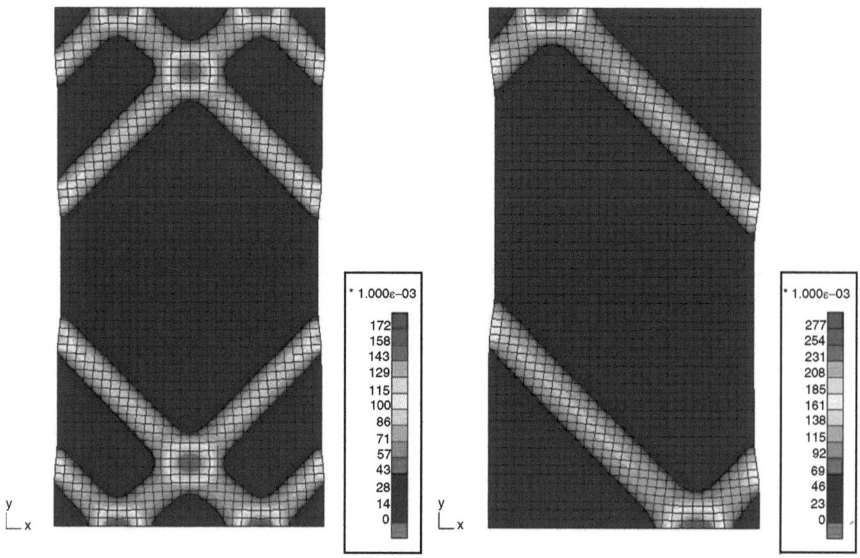

Fig. 1 Biaxial test : two solutions with HM model

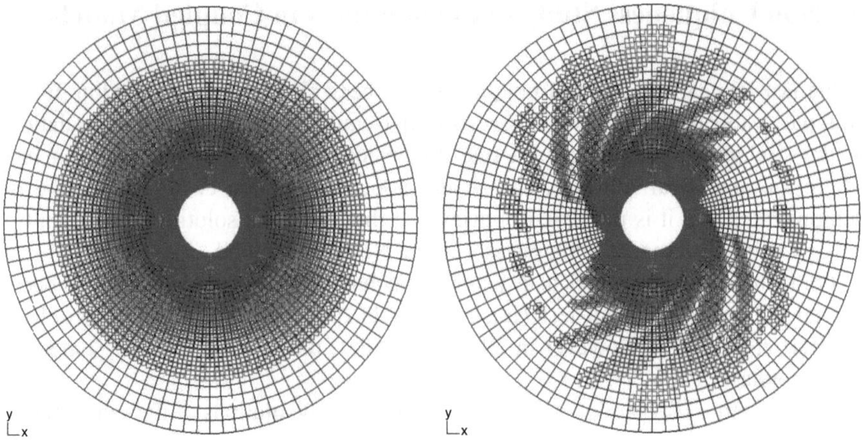

Fig. 2 Borehole problem : two solutions with HM model

4 Conclusions

In this paper a new THM model with second gradient media has been presented. A problem has been computed to validate this development. Similarly to the monophasic case, several different solutions has been obtained in the case of biaxial tests as well as in the case of the borehole stability problem.

Acknowledgments The authors would like to thank the Timodaz EC project for their financial support.

References

P. Bésuelle, R. Chambon, F. Collin, Switching deformation modes in post-localization solutions with a quasibrittle material. J. Mech. Mater Struct. **1**, 1115–1134 (2006)

R. Chambon, D. Caillerie, N. El Hassan, One-dimensional localisation studied with a second grade model. Eur. J. Mech. A/Solids. **17**, 637–656 (1998)

F. Collin, R. Charlier, R. Chambon, A finite element method for poro mechanical modeling of geotechnical problems using local second gradient model. Int. J. Num. Meth. Eng. **65**, 1749–1772 (2006)

O. Coussy, *Mechanics of Porous Continua* (Wiley, London, 1995)

R. Fernandes, C. Chavant, R. Chambon, A simplified second gradient model for dilatant materials: theory and numerical implementation. Int. J. Solids Struct. **45**, 5289–5307 (2008)

Y. Sieffert, S. Al-Holo, R. Chambon, Loss of uniqueness of solutions of the borehole problem modeled with enhanced media. Int. J. Solids Struct. **46**, 3173–3197 (2009)

Two Scale Model (FEM-DEM) For Granular Media

Michał Nitka, Gaël Combe, Cristian Dascalu, and Jacques Desrues

Abstract The macroscopic behavior of granular materials, as a consequence of the interactions of individual grains at the micro scale, is studied in this paper. A two scale numerical homogenization approach is developed. At the small-scale level, a granular structure is considered. The Representative Elementary Volume (REV) consists of a set of N polydisperse rigid discs (2D), with random radii. This system is simulated using the *Discrete Element Method* (DEM) – molecular dynamics with a third-order predictor-corrector scheme. Grain interactions are modeled by normal and tangential contact laws with friction (Coulomb's criterion).

At the macroscopic level, a numerical solution obtained with the Finite Element Method (FEM) is considered. For a given history of the deformation gradient, the global stress response of the REV is obtained. The macroscopic stress results from the Love (Cauchy-Poisson) average formula including contact forces and branch vectors joining the mass centers of two grains in contact.

The upscaling technique consists of using the DEM model at each Gauss point of the FEM mesh to derive numerically the constitutive response. In this process, a tangent operator is generated together with the stress increment corresponding to the given strain increment at the Gauss point. In order to get more insight into the consistency of the two-scale scheme, the determinant of the acoustic tensor associated with the tangent operator is computed. This quantity is known to be an indicator of a possible loss of uniqueness locally, at the macro scale, by strain localization in a shear band.

The results of different numerical studies are presented in the paper. Influence of number of grains in the REV cell, numerical parameters are studied. Finally, the two-scale (FEM-DEM) computations for simple samples are presented.

J. Desrues (✉) · M. Nitka · G. Combe · C. Dascalu
CNRS UMR 5521, Grenoble-INP, UJF-Grenoble 1, 3SR-Lab, Grenoble F-38041, France
e-mail: Jacques.Desrues@grenoble.cnrs.fr; micnitka@pg.gda.pl; cristian.dascalu@hmg.inpg.fr

S. Bonelli et al. (eds.), *Advances in Bifurcation and Degradation in Geomaterials*,
Springer Series in Geomechanics and Geoengineering 11,
DOI 10.1007/978-94-007-1421-2_29,

Keywords Granular materials • Computational homogenization • Discrete element method • Finite element method • 2D deformations

1 Introduction

The presented study considers a two-scale numerical scheme for the description of the behavior of granular materials. At the small-scale level, we consider that the granular structure consists of 2D round rigid grains, modeled by the discrete element method (DEM). At the macroscopic level, we consider a numerical solution obtained with the Finite Element Method (FEM). The link between scales is that of the computational homogenization, in which average REV stress response of the granular microstructure is obtained in each macroscopic Gauss point of the FEM mesh as the result of the macroscopic deformation history imposed to the REV. We also compute the tangent stiffness matrix, at the Gauss point, and the acoustic tensor, which is an indicator of possible unstable behaviors. The influence of different parameters on the stability of the macroscopic response is presented through the results of numerical tests. At the end, some results of two-scale computations are presented.

2 Macroscopic Modelling

For a given history of the deformation gradient, we compute the global stress response of the REV. The macroscopic stress results from the average formula $\sigma_{ij} = \frac{1}{S}\sum_{c=1}^{N_c} f_i^c \cdot l_j^c; \quad i,j \in \{x,y\}$, where S is the area of the sample, f_i^c and l_j^c are respectively the component i of the force acting in the contact c and the component j of the branch vector joining the mass centers of two grains in contact (Love 1927). Next, we convert the Cauchy stress into the Piola-Kirchhoff stress (Bonet and Wood 1997). The Piola-Kirchhoff stress is depended on the history of the *gradient of deformation* **F** (Bilbie et al. 2007; Bilbie et al. 2008)

$$\overline{\mathbf{P}}(t) = \Gamma^t\{\overline{\mathbf{F}}(\tau), \tau \in [0,t]\} \tag{1}$$

For any history of $\overline{\mathbf{F}}$, we assume that $\overline{\mathbf{P}}$ admits a right time derivative $\dot{\overline{\mathbf{P}}}$ with respect to t:

$$\dot{\overline{\mathbf{P}}} = lim_{\delta t \to 0} \frac{\overline{\mathbf{P}}(t+\delta t) - \overline{\mathbf{P}}(t)}{\delta t} \tag{2}$$

We also assume that, for given history of $\overline{\mathbf{F}}$ till time t, the right-sided derivative $\dot{\overline{\mathbf{P}}}$ depends only on the right time derivative $\dot{\overline{\mathbf{F}}}$, that is:

$$\dot{\overline{\mathbf{P}}} = \Theta(\dot{\overline{\mathbf{F}}}) \tag{3}$$

where the function Θ is generally non-linear with respect to its argument $\dot{\overline{\mathbf{F}}}$.

In what follows, we limit our study to the case when the history of $\overline{\mathbf{F}}$ is given by $\overline{\mathbf{F}} = \mathbf{I} + \alpha\mathbf{G^0}$ with $\mathbf{G^0}$ being a fixed tensor and α being a time-like loading parameter which runs monotonously from 0 to 1. In this case we get: $\dot{\overline{\mathbf{P}}} = \Xi(\alpha)$ and by differentiating with respect to α, we get that $\dot{\overline{\mathbf{F}}} = \mathbf{G^0}$ along the path. According to the definition of the function Θ we can write the approximate formula (Bilbie et al. 2007; Bilbie et al. 2008):

$$\Theta(\mathbf{G}^0) \approx \frac{\Xi(\alpha + \Delta\alpha) \; - \; \Xi(\alpha)}{\Delta\alpha} \tag{4}$$

The loss of uniqueness for the rate-type boundary value problems is analysed through the Rice approach (Rice 1976). Following this analysis we look for the rate of deformation gradient $\dot{\overline{\mathbf{F}}}$ which is discontinuous along the boundary of a localization band. It is known that such a discontinuity can be written as (Rice 1976):

$$\overline{\dot{F}^1_{kL}} = \overline{\dot{F}^0_{kL}} + q_k N_l \tag{5}$$

where $\mathbf{N}$ is the normal ($||\mathbf{N}|| = \mathbf{1}$) to the interface, $\overline{\dot{\mathbf{F}}^{\mathbf{1}}}$ is taken on the same side as $\mathbf{N}$ and $\overline{\dot{\mathbf{F}}^{\mathbf{0}}}$ on the opposite side. The stress vector has to be continuous across the interface:

$$\left(\overline{\dot{P}^1_{iJ}} \; - \; \overline{\dot{P}^0_{iJ}}\right) N_J \; = \; 0 \tag{6}$$

As $\overline{\dot{P}}^1_{iJ}$ and $\overline{\dot{P}^0_{iJ}}$ are linked to $\overline{\dot{F}^1_{iJ}}$ and, respectively, $\overline{\dot{F}^1_{iJ}}$, by Eq. 3, the unknowns $\mathbf{q}$ and $\mathbf{N}$ have to satisfy the equation

$$\left(\Theta_{iJ}\left(\overline{\dot{\mathbf{F}}^{\mathbf{0}}} + \mathbf{q} \otimes \mathbf{N}\right) - \Theta_{iJ}\left(\overline{\dot{\mathbf{F}}^{\mathbf{0}}}\right)\right) N_J = 0 \tag{7}$$

for given $\overline{\dot{\mathbf{F}}^{\mathbf{0}}}$.

In the considered macroscopic quasistatic deformation process, the question of loss of ellipticity therefore reduces to the determination of the value α for which Eq. 7 has a non-trivial solution $(\mathbf{q}, \mathbf{N})$, $\mathbf{q} \neq \mathbf{0}$.

In our case we restrict the search of non-trivial solutions to the case in which the tensor $\overline{\dot{\mathbf{F}}^{\mathbf{1}}}$ is closed to $\overline{\dot{\mathbf{F}}^{\mathbf{0}}}$. This leads to a continuous bifurcation mode in the sense of Rice (1976).

So, assuming that Θ is differentiable at $\overline{\dot{\mathbf{F}}^{\mathbf{0}}}$, Eq. 7 yields, after linearization:

$$B_{iJkL}\left(\overline{\dot{\mathbf{F}}^{\mathbf{0}}}\right) q_k N_L N_J = 0 \tag{8}$$

where $B_{iJkL}\left(\overline{\dot{\mathbf{F}}^{\mathbf{0}}}\right) = \frac{\partial \Theta_{iJ}}{\partial \overline{\dot{F}_{kl}}}|_{\overline{\dot{\mathbf{F}}} = \overline{\dot{\mathbf{F}}^{\mathbf{0}}}}$ It is clear that a non-trivial solution exists only if the so-called *acoustic tensor* $\mathbf{Q}$, defined by $Q_{ik} = B_{iJkL} N_L N_J$, is singular, that is only if:

$$det\, \mathbf{Q} = \mathbf{0} \tag{9}$$

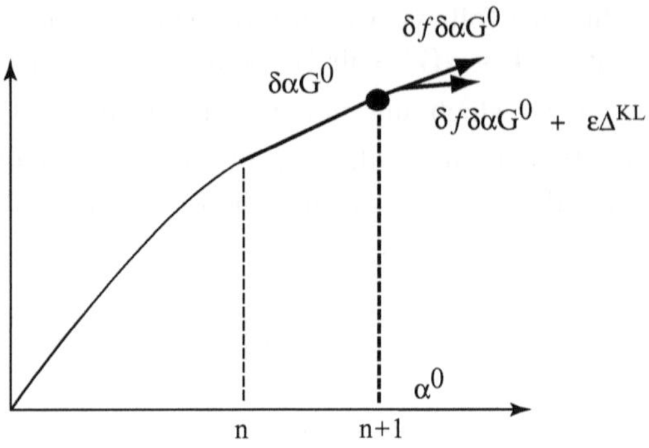

Fig. 1 Schematic representation of the computation of the tangent matrix

For this particular process considered here and given by $\overline{\mathbf{F}} = \mathbf{I} + \alpha\mathbf{G^0}$, we have seen that $\dot{\overline{\mathbf{F}}}$ is constant and equal to $\mathbf{G^0}$ and the function $\Theta(\mathbf{G^0})$ can be approximated by Eq. 4.

As to the derivation of Θ, it can be numerically approximated by finite differences:

$$B_{iJkL} = \frac{\Theta_{iJ}(\mathbf{G}^0 + \varepsilon\Delta^{kL}) - \Theta_{iJ}(\mathbf{G^0})}{\varepsilon} \tag{10}$$

where Δ^{kL} is a second-order tensor such that all its components are equal to 0 except the kL one which is equal to 1. In Fig. 1 we have represented the stress at the point $\delta f \Delta\alpha$ (in the same linear direction as point n) and stresses in points with perturbations $\varepsilon\Delta^{kL}$. We also computed the tangent matrix (Bilbie et al. 2007; Bilbie et al. 2008) as:

$$B_{iJkL} = \frac{P_{iJ}(\alpha^{n+1} + \delta f \Delta\alpha + \varepsilon\Delta^{kL}) \; - \; P_{iJ}(\alpha^{n+1} + \delta f \Delta\alpha)}{\delta f \;\; \varepsilon\Delta^{kL}} \tag{11}$$

where $\delta f \Delta\alpha$ is a small variation step in the main direction, $\varepsilon\Delta^{kL}$ is a small perturbation of the kL component.

3 Micro-Scale Modeling: DEM

The system consist of a set of N polydisperse discs, with the random radii homogeneously distributed between R_{min} and $R_{max} = 2.5R_{min}$. This system is simulated using a discrete element method – molecular dynamics with a third-order predictor-corrector scheme (Allen and Tildesley 1989). All grains interact via a linear elastic law and Coulomb friction when they are in contact (Cundall and Strack 1979). The normal contact force f_n is related to the normal apparent interpenetration δ of the contact as $f_n = k_n\delta$, where k_n is a normal stiffness coefficient ($\delta > 0$ if a contact is present, $\delta = 0$ if there is no contact). The tangential component f_t of the contact force is proportional to the tangential elastic relative displacement, with a tangential stiffness coefficient k_t. The Coulomb condition $|f_t| \le \mu f_n$ requires an incremental evaluation of f_t in each time step, which leads to some amount of slip each time one of the equalities $f_t = \pm\mu f_n$ is imposed. A normal viscous component opposing the relative normal motion of any pair of grains in contact is also added to the elastic force f_n to obtain a critical damping of the dynamics. As the boundary condition we considered *Periodic Limit Condition* (PLC).

4 Results

For the stability criterion the Rice (1976) criterion was chosen, which says that if determinant of acoustic tensor is equal 0 (**det Q** = **0**) for some angle θ, there may exist bifurcation.

The influence of the size of the sample and the numerical parameters: small variation step δf and perturbation ϵ will be studied in this section. Periodic limit condition is applied for those tests. Friction between grains is assumed at $\mu = 0.5$.

The influence of the sample size for shear test for stress is presented in Fig. 2 (to be more clear, the diagrams were moved up on y-axis). Stars represent instability zones, that correspond to the *det* **Q** < **0**. This test was made for $\delta f = 0.1$ and perturbation $\varepsilon\Delta^{kL} = 2 \cdot 10^{-5}$. Different number of grains (400, 1,024, 3,025 and 4,900 grains) were considered. We remark the diminution of the global number of potential instability points when the number of grains is increasing.

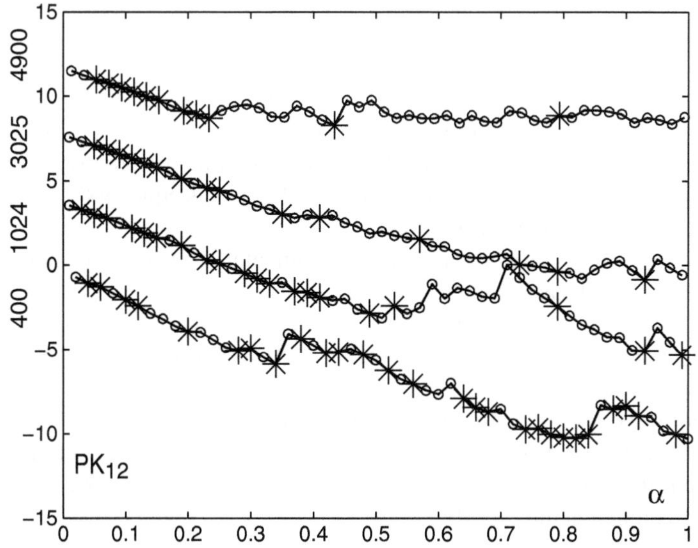

Fig. 2 Influence of the size of sample on macroscopic stability

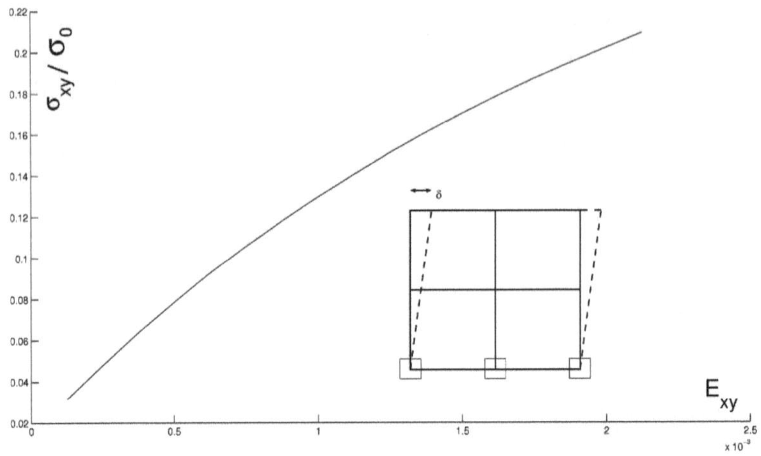

Fig. 3 DEM-FEM: Global stresses σ_{xy} for shear test with displacement $\delta = 0.00025$ imposed

We have also done tests to check the influence of the small variation step $\delta f \delta \alpha$ and the small perturbations $\varepsilon \Delta^{kL}$. In this case we have obtained larger stability zones for smaller values of small variation step $\delta f \delta \alpha$ and for smaller values of perturbation value. However, it is important to take the value of $\delta f \delta \alpha$ carefully to stay still in elasto-plasticy behaviour (not only elasticity).

To succeed in the macro–micro computations the numerical parameters should be chosen very carefully. In Discrete Element Method and also in Finite Element Method, there are different variations, which have significant influence on the convergence of the test.

For the FEM-DEM computations the open-source code 'FLagSHyP' written by J. Bonet is used. On FEM level, the two-dimensional quadratic elements with four Gauss points were chosen.

On the micro level, in DEM calculations, according to our study of instability, the parameters: small variation step δf and small perturbation ϵ were chosen as 0.1 and $2 \cdot 10^{-5}$, respectively. The boundary conditions was periodic (PLC). Friction between grains $\mu = 0.5$ was chosen. Number of grains is equal 400 (only, because calculations are very time consuming).

First, the shear test was done, where incremental shear displacement is equal $\delta = 0.00025$. Results are plotted in Fig. 3.

Next, biaxial tests with strain control with no volume changes were done. The incremental deformation is $\delta E_{yy} = -\delta E_{xx} = 0.00025$. In the Fig. 4a the stress–strain responses are presented.

The last tests were done for classical biaxial test. On the walls constant pressure is applied equal to the isotropic stress. On the top displacement incremental step is imposed (three different incremental steps equal 0.00001, 0.00005 and 0.0001). Results are plotted in the Fig. 4b.

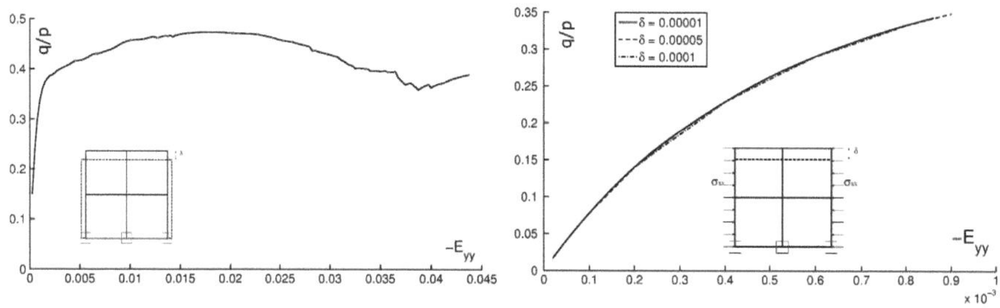

Fig. 4 DEM-FEM: Stress–strain response for (*left*) isochoric biaxial test (equivalent to undrained test) and (*right*) constant confining pressure biaxial test (drained). ($\delta = 0.00001$, $\delta = 0.00005$ or $\delta = 0.0001$)

Those tests show that two-scale approach is possible to compute. More complicated tests as classical biaxial or shear can be simulated. It proves that even for non-linear behavior, the two scale model can work well.

5 Conclusions

A two-scale numerical approach for granular materials has been proposed, combining DEM modeling of the granular micro structure with the FEM modeling of the overall response. We focused on the consistency of the discrete-to-continuous approach, through the identification of the numerically-induced macroscopic instabilities.

The sample size plays an important role for stability zones. It is very important to make the correct choice for the size of the sample (not too small). It is also very important to choose the correct small variation step δf of and small perturbation ϵ. Both of them (as well as the number of grains in the cell) have significant influence on the number of instability points. This two parameters should be chosen in respect to the problem we want to solve. Microscopic behaviour may be a true physical behaviour, linked to instability and shear banding. As such, it should not be rejected in general. But the occurrence of such events far from the failure regime of the sample, if not limited to a few points but spread over the sample in a significant number of points, may indicate non-relevant numerical behaviour. This is a challenge to deal with, maybe by exploring more changes in the numerical parameters. However, it is still possible to compute small two-scales geotechnical problems. After study of the influence of the small variation step and perturbations coefficient, we are able to avoid numerical instabilities. Since, the computations are extremely time consuming, all tests are done for small sample with 400 grains, what causes that incremental step should be very small. For bigger incremental steps there is a problem with convergence. However this problem should disappear, when bigger REV samples will be used.

References

M.P. Allen, D.J. Tildesley, *Computer Simulations of Liquids* (Clarendon Press, Oxford, 1989)

G. Bilbie, C. Dascalu, R. Chambon, D. Caillerie, Micro-fracture instabilities in granular solids, in *Bifurcations, Instabilities, Degradation in Geomechanics*, ed. by G. Exadaktylos, I. Vardoulakis (Springer, Berlin/Heidelberg, 2007), pp. 231–242

G. Bilbie, C. Dascalu, R. Chambon, D. Caillerie, Micro-fracture instabilities in granular solids. Acta Geotech. **3**, 25–35 (2008)

J. Bonet, R.D. Wood, *Nonlinear Continuum Mechanics for Finite Element Analysis* (Cambridge University Press, Cambridge, 1997)

P.A. Cundall, O.D.L. Strack, A discrete numerical model for granular assemblies. Géotechnique. **29**(1), 47–65 (1979)

A.E.H. Love, *A Treatise on the Mathematical Theory of Elasticity* (Cambridge University Press, Cambridge, 1927)

J.R. Rice, The localization of plastic deformation, in *Theoretical and Applied Mechanics*, ed. by W.T. Koiter (North-Holland Publishing Company, Amsterdam, 1976), pp. 207–220

Finite Element Modelling of Material Instability via an Enriched Elastoplastic Model

Richard Wan and M. Pinheiro

Abstract Geomaterials are dissipative particulate systems due to internal forces that arise from intergranular friction or viscosity. As such, their mechanical behaviour is highlighted by various forms of failure with either localization of deformations or diffuse deformations. Intriguingly, the latter failure mode, which does not involve any localized deformations or discontinuities, is normally observed well before plastic limit conditions are met. This work examines failure as a material (constitutive) instability phenomenon giving way to a bifurcation problem whereby a multiplicity of material response is possible for the same initial loading history. We use a rate-independent elastoplastic constitutive model with plastic strain softening and non-associativity of plastic flow through a micromechanically derived stress-dilatancy equation. The dependencies of the latter on density, stress, and fabric provide essential mathematical sources of material instability to promote the capturing of discontinuous response. An example problem involving diffuse and localization deformations in a water saturated sand sample as a boundary value problem is presented.

Keywords Second order work • Diffuse failure • Localized failure • Stress dilatancy • Non-associated flow rule

1 Introduction

This paper briefly touches upon the issue of instability in material behaviour and conditions under which it may occur, motivated by various forms of failure observed in geomechanics. For instance, a natural slope may undergo large movements as

R. Wan (✉)
Department of Civil Engineering, University of Calgary, Calgary AB, Canada, T2N 1N4
e-mail: wan@ucalgary.ca

M. Pinheiro
Thurber Engineering Ltd., Calgary, AB, Canada

S. Bonelli et al. (eds.), *Advances in Bifurcation and Degradation in Geomaterials*, Springer Series in Geomechanics and Geoengineering 11,
DOI 10.1007/978-94-007-1421-2_30,

a result of deformations either localizing into a shear band or developing in a diffuse manner with a chaotic displacement field throughout the entire mass. In both cases, the problem underlies a material (constitutive) instability phenomenon that originates in the small scale due to microstructural features of the geomaterial. At such a microscale, mechanisms of energy dissipation and inter-granular force transmission lead to local instabilities that reflect at the macroscopic scale as shear dilatancy, localized deformations, and liquefaction, among others.

Within the realm of localization, one finds distinctive forms of concentrated deformations such as shear bands, compaction bands, and dilation bands (e.g. Rudnicki and Rice 1975; Vardoulakis and Sulem 1995). These bands lead to an unstable response that is associated to a bifurcation at the material point level. On the other hand, the possibility of other types of instabilities appearing before plastic localized failure has also been evoked (Darve 1994, for instance). This means that other forms of failure may as well occur within the Mohr-Coulomb plastic limit surface. For example, when loose sand is sheared in undrained conditions, it may collapse as a result of a spontaneous loss in strength at stress levels far from the plastic limit surface. In this type of instability, coined as diffuse instability, no localization of deformations appears but a rather generalized failure sets in due to sufficiently large losses of inter-granular contact numbers with an accompanying chaotic deformational field. Such a diffuse instability may precede strain localization which can also be seen as a special case of the former. The implication is that stress states deemed to be safe with respect to a limit condition or localization can still be vulnerable to another type of instability as of the diffuse type.

This short paper presents a plausible model that can capture numerically both localized and diffuse instabilities. As an illustration, we show a triaxial test modelled as a boundary value problem in which both diffuse and localized instabilities are examined under axisymmetric conditions involving solid-liquid interaction. The order of precedence of failure modes is effectively demonstrated in that localized deformation comes on the heels of diffuse failure under the proper loading program. A density-stress-fabric dependent elasto-plastic model (Wan and Guo 2004) is used to represent the constitutive behaviour of the material.

2 Density-Stress-Fabric Dependent Constitutive Model

The ability of a constitutive model to describe soil behaviour within a high degree of fidelity is crucial in any failure analysis. For instance, a constitutive model for geomaterials must be able to capture essential behavioural features such as non-linearities, irreversibilities, and dependencies of the mechanical behaviour on stress level, density and fabric. The WG-model (Wan and Guo 2004) is a two-surface elastoplastic model that has the above attributes. Rather than advocating advanced concepts such as those found in micropolar plasticity (Tejchman and Bauer 1996), the model is founded on Rowe's 1962 stress-dilatancy theory for predicting volumetric changes under deviatoric loading. The latter theory has been further enriched to address density, stress level, and fabric dependencies as well as cyclic loading regime conditions; see Wan and Guo 2004 for details. A non-associated flow rule derived from the enriched stress-dilatancy theory is used to calculate the increment of plastic shear deformations. The updating of shear-yield and cap-yield surfaces is governed by two different hardening/softening laws.

3 Diffuse Against Localized Deformations

Material instability is strongly driven by particle forces which arise principally from externally applied boundary stresses transmitted along force chains that form within the solid skeleton. It is the stability of elementary force chains within the solid skeleton as revealed by experiment and computer simulation on granular media that gives rise to macroscopic instability through microstructural buckling. In the absence of buckling, a chaotic spatial state may prevail as in diffuse failure.

Diffuse instability has its origins in the loss of positive definiteness of the incremental constitutive relation (tensor $\mathbf{D}$) that relates incremental stresses ($d\boldsymbol{\sigma}$) to strains ($d\boldsymbol{\varepsilon}$) or a mixture of these under generalized loading. The connection with the vanishing of the second-order work, $W_2 := d\boldsymbol{\sigma} : d\boldsymbol{\varepsilon} = 0$, (Hill 1958) is well recognized. In fact, when the second order work is expressed in terms of $\mathbf{D}$, it yields a quadratic form that has to be positive for stability to prevail. If $\mathbf{D}$ is expressed in terms of its symmetric ($\mathbf{D}_{sym}$) and skew-symmetric ($\mathbf{D}_{skew}$) parts, it transpires that Hill's stability criterion is satisfied whenever $\det(\mathbf{D}_{sym}) > 0$. It can be shown that $\det(\mathbf{D}_{sym})$ always vanishes before that of $\mathbf{D}$ during loading history, which means that a loss of determinacy in incremental material response is possible well before plastic limit conditions for non-associated plastic geomaterials.

Furthermore, it can be readily verified that strain localization is a special case of the broader diffuse instability defined by the second order work. The demonstration is rather straightforward. Let the strain vector $d\boldsymbol{\varepsilon}$ compatible with a shear banding mechanism be written as $d\boldsymbol{\varepsilon} = (\mathbf{n} \otimes \mathbf{g} + \mathbf{g} \otimes \mathbf{n})/2$, where $\mathbf{n}$ is the normal to the shear band and $\mathbf{g}$ is a vector that refers to the jump in the gradient of the displacement field characterizing the shear band. As such, the writing of the second order work readily leads to the well-known localization criterion (Rudnicki and Rice 1975), i.e. $\det(\mathbf{n} : \mathbf{D} : \mathbf{n}) = 0$.

4 Finite Element Study of Localized and Diffuse Failure

The indeterminacy invoked in the previous section leads to a multiplicity of solutions for the underlying governing equations which become ill-posed, and hence represents a bifurcation problem. As for a lab experimental test, any post bifurcation (failure) state is revealed by the loss of homogeneity of initial field variables to

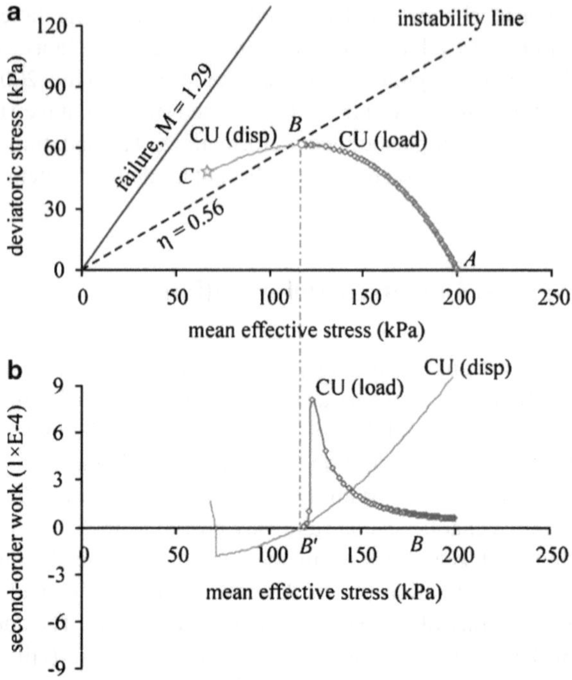

Fig. 1 Simulation of undrained compression test on loose sand showing diffuse failure and violation of second order work preceding localized failure in displacement controlled mode

give rise to either diffuse or localized deformations. These aspects can be modelled numerically in a mathematical boundary problem provided the constitutive model used has the basic requisites such as stress, density, fabric dependencies that lead to plastic strain-softening with non-associated plastic flow (Wan and Guo 2004).

4.1 Instabilities in Undrained Triaxial Test

A consolidated undrained (CU) compression test is simulated in both force and displacement controlled loading to illustrate the emergence of various forms of material instabilities. Figure 1 shows both tests giving rise to a vanishing second-order work at the peak of the effective stress path plotted in the $p'-q$ (mean effective stress versus deviatoric stress) space.

The vanishing of the second-order work is well inside the plastic limit surface as debated in the beginning of the paper. However, in the load controlled case (CU load), as the second-order work vanishes, the test cannot be controlled any further because the externally imposed loading is incompatible with the internal material response. As such, the computations break down, meaning collapse of the specimen.

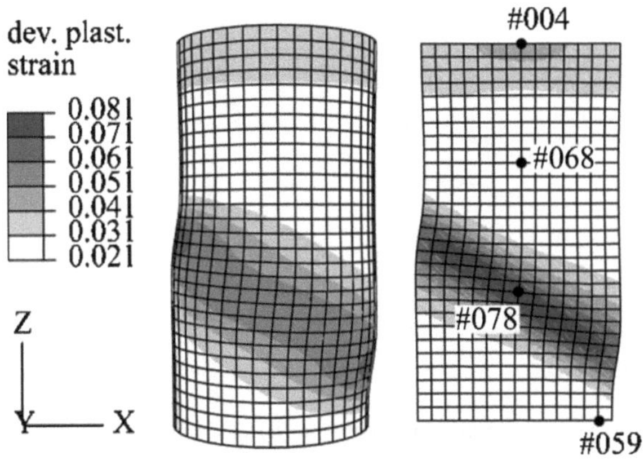

Fig. 2 Deviatoric plastic strain field on the final non-homogeneous deformed configuration

At this point, the specimen is found to remain homogeneous with spatially uniform void ratio, pore pressure, stress and strain fields, while the second-order work is zero throughout. This bifurcated state may be treated as one involving spatially chaotic states without any organization.

By contrast, the undrained test performed under strain controlled mode (CU disp) carries on past the peak B until it fails at point C near the plastic limit surface for lack of numerical convergence. This means that for diffuse failure with collapse to occur, it is not solely sufficient to have the second-order work criterion violated, but it is also necessary to have the proper control variables in place, here a force-controlled loading program.

It is next shown that beyond the point where the second-order work is violated, as the deformation history carries on in displacement control, the failure mode eventually evolves into a localized one as the stress path nears the plastic limit surface. This clearly confirms the chronology of failure modes discussed earlier in the paper. Figure 2 shows the deviatoric plastic strain field in the final non-homogeneous deformed configuration together with the locations of four strategic nodes to follow the field variable contrast inside the specimen.

The loss in homogeneity arising from bifurcation is reflected in all field variables, for example, excess pore water pressure, void ratio and deviatoric plastic strain fields as illustrated in Fig. 3 for the same four nodes mentioned above. As seen in Fig. 3a, there is a steady and uniform increase in pore pressures until when bifurcation ($W_2 = 0$) is first met. Thereafter, a distinct region of localized deformations eventually appears. Material points inside this region experience escalating pore water pressures while points outside this shear zone undergo a decrease in pore pressures. In line with the pore pressure field response, Fig. 3b indicates that the localized zones compact, leading to a decrease in void ratio and a concomitant increase in pore pressures. It is noted despite local volume changes, the overall sample's void ratio remains constant as required by the undrained test.

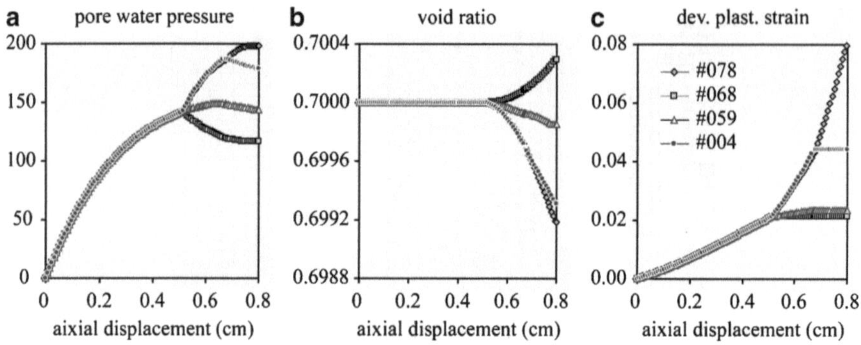

Fig. 3 Evolution of (**a**) excess pore water pressure, (**b**) void ratio and (**c**) deviatoric plastic strain at four selected nodes in the FE mesh

5 Conclusions

The failure of geomaterials with respect to diffuse and localized deformations has been presented within the framework of material instability and the theory of bifurcations. It is found that a non-associated elastoplastic constitutive equation enriched with density, pressure and fabric sensitivities as sources of material instability represents a plausible model for capturing both diffuse and localized failure modes and their respective order of precedence during deformation history. The success of such a simple model makes it a good starting point for further refinement to include large deformations, an internal length scale and other multiphysics phenomena based on the present theory.

Acknowledgments This work has been undertaken under funding provided by the Natural Science and Engineering Research Council of Canada. The first author is grateful to several colleagues, notably F. Darve, F. Nicot, and A. Daouadji, for various discussions of ideas on the central theme of second-order work. The finite element effort in this work originated from lab work performed by A. Daouadji.

References

F. Darve, Stability and uniqueness in geomaterials constitutive modeling, in *Localisation and Bifurcation Theory for Soils and Rocks*, ed. by R. Chambon, J. Desrues, I. Vardoulakis (Balkema, Rotterdam, 1994), pp. 73–88

R. Hill, A general theory of uniqueness and stability in elastic-plastic solids. J. Mech. Phys. Solids **6**, 236–249 (1958)

P.W. Rowe, The stress-dilatancy relation for static equilibrium of an assembly of particles in contact. Proc. R Soc. Lond. A Math. Phys. Sci. **269**(1339), 500–527 (1962)

J.W. Rudnicki, J.R. Rice, Conditions for the localization of deformation in pressure sensitive dilatant material. J. Mech. Phys. Solids **23**, 371–394 (1975)

J. Tejchman, E. Bauer, Numerical simulation of shear band formation with a polar hypoplastic constitutive model. Comput. Geotech. **19**(3), 221–44 (1996)
I. Vardoulakis, J. Sulem, *Bifurcation Analysis in Geomechanics* (Chapman & Hall, London, 1995)
R.G. Wan, P.J. Guo, Stress dilatancy and fabric dependencies on sand behavior. J. Eng. Mech. **130**(6), 635–645 (2004)

Comprehensive DEM Study of the Effects of Rolling Resistance On Strain Localization in Granular Materials

Marte Gutierrez and Abdalsalam Mohamed

Abstract This paper presents the results of a comprehensive study of the effects of rolling resistance in shear band formation in granular material using DEM (Discrete Element Modeling). The study used the two-dimensional Particle Flow Code (PFC) to simulate biaxial compression and simple shear tests in granular materials. A rolling resistance model was implemented as a user-defined model (UDM) in PFC. A series of parametric studies were performed to investigate the effects of different levels of rolling resistance parameters on the emergence and shear bands in granular materials. The results reinforce prior conclusions by Oda et al. (Mech. Mater. 1:269–283, 1982) on the importance of rolling resistance in promoting shear band formation in granular materials. It was shown that increased rolling resistance results in the development of columns of particles in granular soils during the strain hardening process. The buckling of these columns of particles then leads to the development of shear bands. It is concluded that the variation of rolling resistance parameters has a significant effect on the orientation, thickness and the onset of occurrence of shear bands. The PFC models were tested under a wide range of macro-mechanical parameters and boundary conditions, and how they influence shear band characteristics.

Keywords Discrete element modeling • Granular materials • Rolling résistance • Strain localization • Shear banding

1 Introduction

Granular materials are ensembles of particles which can interact at contact points. Each particle can move relative to neighboring particles by sliding and/or rolling

M. Gutierrez (✉) · A. Mohamed
Division of Engineering, Colorado School of Mines, Golden, CO, 80401, USA
e-mail: mgutierr@mines.edu

S. Bonelli et al. (eds.), *Advances in Bifurcation and Degradation in Geomaterials*, Springer Series in Geomechanics and Geoengineering 11,
DOI 10.1007/978-94-007-1421-2_31,

at contact points. However, in classical continuum mechanics, sliding is considered to play the dominant role in microscopic deformation of granular materials. For example, friction and dilatancy of granular soils are conventionally explained to be mainly due to pure sliding. Particles are assumed to be perfectly rounded, and to rotate freely and offer no resistance to rotation. In reality, particles are non-circular and have rough surface textures, and particle rolling should not occur freely when in contact with other particles. Rolling resistance can be defined as a force couple that is be transferred between particles through the contacts, and this force couple inhibits particle rotation. Rolling resistance has been used in the Distinct Element Method (DEM) as a simple way of including the effects of non-circular, irregular and rough particle shapes (Iwashita and Oda 1998).

Oda et al. (1982) showed that rolling resistance has a very important effect on the DEM simulation of shear band formation in granular materials. It was shown that rolling resistance promotes shear band formation. Due to the particle rotation gradients, columns of vertical chains of particles buckle along the shear band. As result of localized buckling of columns of particles, large shear strain concentrations are accompanied by the formation of large voids between particles inside the shear band. Oda and Kazama (1998) reported that realistic DEM modeling of high particle rotations and large voids inside shear bands can only be accomplished by considering rolling resistance at contact points between particles. A comprehensive list of references on the effects of rolling resistance on shear band formation in granular materials can be found in Mohamed and Gutierrez (2010).

The goal of this paper is to investigate in greater detail the roll of rolling resistance in shear band formation in granular materials using the Discrete Element Method, and the Particle Flow Code (PFC) developed by Itasca (2008). An extensive parametric study is performed by varying the magnitude of rolling resistance in two-dimensional DEM simulations of strain localization of granular materials under biaxial and simple shear loading conditions.

2 DEM Model Accounting for Rolling Resistance

In conventional DEM, particle rolling occurs without any resistance. The granular medium is assumed to be an assembly of discrete particles and the mechanical relationships between particles are determined by normal spring constant, tangential spring constant and friction parameters. However, in reality particles contact each other on surfaces with finite area and, thus, a moment of rolling resistance exists when a particle rolls over another. The first model for rolling resistance in DEM was introduced by (Iwashita and Oda 1998). In this model, a rotational spring and rolling dashpot are installed between the particles.

The interaction force F_i which represents the action of two particles at the i-th contact may be decomposed into a normal component N_i and a shear component T_i. Two types of kinematical behavior can occur for a contact: sliding and/or rolling. Sliding starts working when N_i and $T_i$satisfy the condition $T_i = \mu N_i$where μ is the contact frictional coefficient. The normal contact force $N_i$and shear force

T_i are calculated from the normal displacement $U_i^n$and shear displacement U_i^s, respectively, by $N_i = K^n U_i^n$ and $T_i = K_i^s U_i^s$ where K^n and $K^s$are the incremental normal and shear stiffnesses, respectively.

For particle rolling, the elastic rotational force-displacement relation is described by the following equation:

$$M_i = K_r \theta_r \tag{1}$$

where M_i is the rotational moment, $\theta_r$is the relative rotation between two particles and K_r is the rolling stiffness. Rolling starts when M_i exceeds the threshold value of ηN_i:

$$|M_i| > \eta N_i \tag{2}$$

where η is the rolling resistance. Once the resistance is exceeded, the particle begins to roll without mobilizing any further rolling resistance. The parameter η is the dimensional coefficient of rolling friction as noted by Sakaguchi et al. (1993), and its physical meaning is similar to the inter-particle friction μ . Note that $\eta = \beta B$, where B is equal to half of contact width between two particles. The interface/contact width B is dependent on the shape parameter δ and the common radius r between two particles, and therefore:

$$\eta = \delta \beta r = \alpha r \tag{3}$$

where α is combination of two dimensionless constant β and δ . The parameter α, which depends totally on the distribution of contact force at the contact surface, is called rolling friction coefficient. It controls how much rolling resistance can be generated at any contact point. Note that $\alpha = 0$ corresponds to free rotation, while $\alpha \geq 1$ corresponds to fixed rotation (i.e., particles are completely prevented from rotating). The rolling resistance model described above was implemented as a User Defined Model (UDM) in PFC. This rolling resistance model and the parameter α are used in the parametric study of shear band formation below.

The comprehensive investigation of the effects rolling resistance on strain localization characteristic in granular materials is performed under two-dimensional plane strain and simple shear conditions. To generate the granular assembly, a random particle generation procedure called the "expansion method" is adapted to achieve a desired sample particle size distribution and porosity. This method employs a constant factor which is expressed as a multiple of the particle radii. The factor is adjusted and the particle sizes are increased until the system reaches an equilibrium state after some calculation cycles. Due the particle radii expansion, particles overlap and thus repulsive forces are generated at the contacts. Cycling is required to achieve equilibrium between the unbalanced forces and the forces generated by the isotropic boundary stresses from consolidation.

The study of the effects of rolling resistance on shear band formation in granular materials was performed for biaxial and simple shear loading conditions. The models consisted of assemblies of poly-dispersed disk-like particles with radii varying from 0.3 to 0.6 mm. The sample sizes were made large enough to simulate a representative element volume of granular materials, but not too large to require extensive calculation times. The model parameters values used in the study are the same as those used by Ng (2006) in the DEM simulation of the stress-strain behavior of granular materials. Some of the parameters were kept constant while some were given a range of values to test the sensitivity of model response to changes in parameter values.

3 Modeling Results and Discussion

Figure 1a summarizes the effects of the rolling resistance parameter α on the peak friction angle ϕ and the peak dilation angle ψ of the simulated granular material under biaxial loading conditions. As can be seen, the friction angle ϕ continuously increased from about 27° for $\alpha = 0$ to about 39.5° for $\alpha = 1$. The most significant increase in friction angle occurs for $\alpha < 0.6$, and for $\alpha > 0.6$ the changes in the friction angle are very small. The dilation angle ψ increases from about 14° for $\alpha = 0$ to a maximum of about 18.5° for $\alpha = 0.2$, then decreases to 15.5° for $\alpha = 1$.

Figure 1b shows the variation of the measured shear band orientation θ_m from the models against the rolling resistance parameter α. It is observed that θ_m first increased with increasing value of α. The measured shear band inclination angle for free rotation is about $\theta_m = 54.5°$, then it reached its highest value at about 60° for $\alpha = 0.25$ and decreased thereafter to $\theta_m = 58°$ for $\alpha = 1.0$. Figure 1b also shows the Mohr-Coulomb, Roscoe (1970), and Arthur-Vardoulakis (Arthur et al. 1977; Vardoulakis 1980) shear band orientations using the measured peak friction angle ϕ and dilation angle ψ from the DEM models. The measured shear band orientations are closest to the Arthur-Vardoulakis predictions. The Mohr-Coulomb orientation over predicts and the Roscoe orientation under predicts the measured shear band inclinations.

Due to the rolling resistance, columns of particles formed during loading of the granular material. The roles of column bucking and particle rotation in promoting shear band formation in granular materials are clearly shown in Fig. 2 for the simple

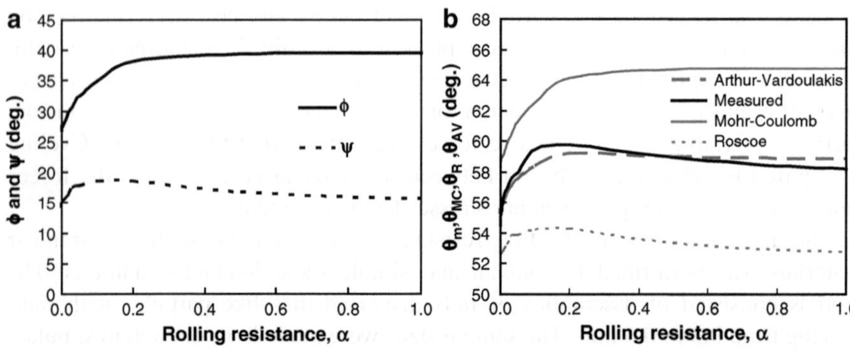

Fig. 1 Effects of the rolling resistance parameter on (**a**) the peak friction and dilation angles, and (**b**) on the shear band orientation

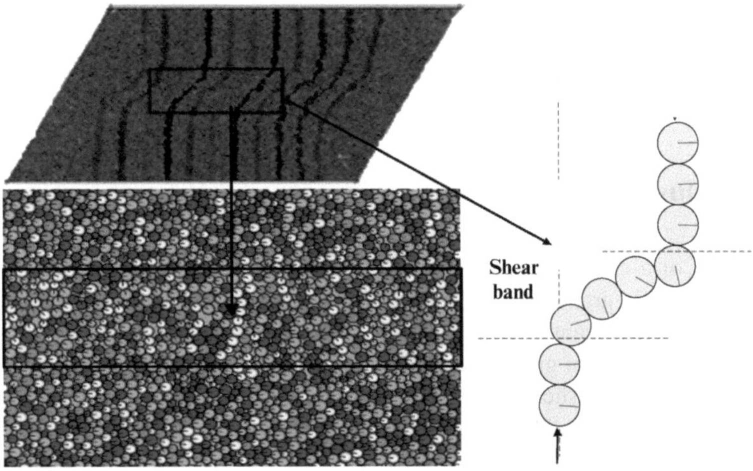

Fig. 2 Effects of column buckling and particle rotation in shear band formation

shear model with a rolling resistance of $\alpha = 0.2$. As the shear displacement is increased, the colored columns used to track the movement of particles in the model, which were initially vertical, started to buckle locally inside the shear band. The buckling shown in Fig. 2 occurred despite the fact that uniform shear displacements are applied at the lateral boundaries of the simple shear model. It was observed that column buckling was more pronounced as the rolling resistance parameter α was increased. These columns increased the shear strength but enhanced instability due to the tendency of the columns to buckle during loading. In contrast, in the case of low or no rolling resistance, particles were free to move and were unable to form columns. As a result, instability in the case of high rolling resistance was accompanied by larger reduction in load carrying capacity than in the case of low or no rolling resistance.

In Fig. 2, particle rotations are shown by the radial line drawn from the center of each particle again for simple shear loading and $\alpha = 0.2$. These radial lines were originally horizontal before shearing. Outside the shear band, it is observed that particles do not rotate at all stages of shear deformation. Once the deformation have localized, particle rotations occur extensively within shear band. As a result, shear band formation is accompanied by gradients in the magnitude of particle rotations. The gradients in the magnitude of particle rotation increases with increasing α. For low values of α, particles rotate randomly all over the volume of the sample and clear gradients in particle rolling were not observed.

Another important observation is that particle rotation in the shear band is associated with the formation of non-uniform void distribution within the shear band. As can be seen in Fig. 2, large voids are formed inside the shear band. In comparison, for the case of no rolling resistance, the void distribution is nearly uniform within the granular assembly. The occurrence of areas of large voids in the case where particles have rolling resistance is attributed to the fact that particles, in the presence of rotational moments, tend to push each other apart. This causes the granular assembly to expand and increase the overall volume locally.

4 Conclusions

The results of a comprehensive study of the effects of rolling resistance in shear band formation in granular materials were presented. It was concluded that rolling resistance has significant effects on the stress-strain, volumetric response, and strain localization behavior of granular materials. In general, the peak friction angle, peak dilation angle and the shear band inclination (measured from the σ_3-axis) increases with increasing value of the rolling resistance parameter α. The most significant effects of rolling resistance occurs for $\alpha < 0.25$ to 0.6. The micromechanical observations from the DEM modeling reinforce prior conclusions by Oda et al. (1982) on the importance of rolling resistance in promoting shear band formation in granular materials. Rolling resistance promotes the buckling of vertical columns of particles in narrow zones. The buckling of the columns, which is accompanied by large gradients in particle rotation, then leads to the development of the shear band and large voids within the band.

References

J.R.F. Arthur, T. Dunstan, Q.A.J.L.A. Al-Ani, Assadi Plastic deformation and failure in granular media. Geotechnique **27**, 53–74 (1977)

Itasca Consulting Group, Particle Flow Code in 2-D (PFC-2D): User's Manual, Version 4.0. (2008)

K. Iwashita, M. Oda, Rotational resistance at contacts in the simulation of shear band development by DEM. J. Eng. Mech. Div. ASCE **124**(3), 285–292 (1998)

A. Mohamed, M. Gutierrez, Comprehensive study of the effects of rolling resistance on the stress-strain and strain localization behavior of granular materials. Granul. Matter **12**(5), 527–541 (2010)

T.T. Ng, Input parameters of discrete element methods. J. Eng. Mech. ASCE **132**(7), 723–729 (2006)

M. Oda, H. Kazama, Microstructure of shear bands and its relation to the mechanisms of dilatancy and failure of dense granular soils. Geotechnique **48**(4), 465–481 (1988)

M. Oda, J. Konishi, S. Nemat-Nasser, Experimental micromechanical evaluation of strength of granular materials: effects of particle rolling. Mech. Mater. **1**, 269–283 (1982)

K.H. Roscoe, The influence of strains in soil mechanics – tenth rankine lecture. Geotechnique **20**(2), 129–170 (1970)

H. Sakaguchi, E. Ozaki, T. Igarashi, Plugging of the flow of granular materials during the discharge from a silo. Int. J. Mod. Phys. **7**(9–10), 1949–1963 (1993)

I. Vardoulakis, Shear band inclination and shear modulus of sand in biaxial test. Int. J. Num. Anal. Meth. in Geomechanics **4**, 103–119 (1980)

Modeling Porous Granular Aggregates

R. Affes, V. Topin, J.-Y. Delenne, Y. Monerie, and F. Radjaï

Abstract We rely on 3D simulations based on the Lattice Element Method (LEM) to analyze the failure of porous granular aggregates under tensile loading. We investigate crack growth by considering the number of broken bonds in the particle phase as a function of the matrix volume fraction and particle-matrix adhesion. Three regimes are evidenced, corresponding to no particle damage, particle abrasion and particle fragmentation, respectively. We also show that the probability density of strong stresses falls off exponentially at high particle volume fractions where a percolating network of jammed particles occurs. Decreasing the matrix volume fraction leads to increasingly broader stress distribution and hence a higher stress concentration. Our findings are in agreement with 2D results previously reported in the literature.

Keywords Damage • Stress transmission • Rheology • Cemented granular materials • Lattice element method

1 Introduction

Dense granular materials are characterized either in terms of the network of solid particles or by the properties of the pore space which can be fully or partially filled by a solid binding matrix or a liquid. At high particle volume fractions ρ^p (typically, for $\rho^p > 0.57$), the stress transmission is basically guided by a

R. Affes (✉) · J.-Y. Delenne · F. Radjaï
LMGC, CNRS-Université Montpellier II, Place Eugène Bataillon, 34095 Montpellier Cedex, France and Laboratoire MIST, IRSN-CNRS-Université Montpellier 2, Montpellier, France
e-mail: jean-yves.delenne@univ-montp2.fr; radjai@lmgc.univ-montp2.fr

V. Topin · Y. Monerie
IRSN, DPAM, CE Cadarache, Bat. 702, 13115 St. Paul-lez-Durance Cedex, France and Laboratoire MIST, IRSN-CNRS-Université Montpellier 2, Montpellier, France

S. Bonelli et al. (eds.), *Advances in Bifurcation and Degradation in Geomaterials*, Springer Series in Geomechanics and Geoengineering 11,
DOI 10.1007/978-94-007-1421-2_32,

percolating network of inter-particle contacts (Satake and Jenkins 1988). This role of the contact network in force transmission and rheological behavior has been mainly investigated in granular materials in the absence of a binding matrix and under compressive confining stresses (Mueth et al. 1998; Radjaï et al. 1996).

The issue of stress concentration and the role of particles are much less evident in the presence of a binding matrix and under tensile loading. Such *porous granular aggregates* have been recently studied in some detail in a 2D geometry by means of numerical simulations (Van Mier et al. 2002; Topin et al. 2007). The overall stiffness and tensile strength of these materials are dependent on the matrix volume fraction ρ^m, particle volume fraction ρ^p and particle-matrix adhesion σ^{pm}.

In this paper, we introduce the lattice element method (LEM) in a 3D geometry for the simulation of porous granular aggregates of spherical particles with a solid matrix. Based on a lattice discretization of both phases and their interface as well as an efficient quasi-static time-stepping scheme, the LEM algorithm allows us to analyze the fracture of cohesive aggregates as a function of phase volume fractions and local binding strength.

2 Lattice Element Method

The *lattice element method* (LEM) has been recently employed as an alternative to the finite element method for the investigation of the fracture properties of granular materials mixed with a binding matrix (Van Mier et al. 2002; Topin et al. 2007). Such materials, to which we refer in this paper as porous granular aggregates or cemented granular materials can be found in very different forms in nature and industry. Well-known examples are conglomerates and concrete.

In LEM, the space is discretized as a regular or disordered grid of points (nodes) interconnected by one-dimensional elements (bonds). Each bond can transfer normal force, shear force and bending moment up to a threshold in force or energy. Various rheological behaviors can be carried by these material lattice bonds, in contrast to the finite element approach where the local behavior is carried by volume elements. When several phases are present as in a porous granular aggregates, each phase and its boundaries are materialized by lattice elements sharing the same properties and belonging to the same portion of space. We use linear elastic-brittle elements, each element characterized by a Hooke constant and a breaking force threshold. The bonds transmit only normal forces between the lattice nodes and thus the strength of the lattice in shear and distortion is ensured only by the high connectivity of the nodes. This simple kinematics allows to investigate high sampling statistical approach. A sample is defined by its contour and the configuration of the phases in space. The samples are deformed by imposing displacements or forces to nodes belonging to the contour. The initial state is the reference (unstressed) configuration. The total elastic energy of the system is a convex function of node displacements and thus finding the unique equilibrium configuration of the nodes amounts to a minimization problem (implemented here by means of the conjugate gradient method). Performing this minimization for stepwise loading corresponds to subjecting the system to a quasi-static deformation process. The overloaded elements (exceeding a threshold) are removed according to a breaking rule. This corresponds to irreversible micro-cracking of the lattice. The released elastic energy between two successive equilibrium states is thus fully dissipated by micro-cracking. In the fast implementation used in the present work, all overloaded elements occurring within the same step are removed, as well as those appearing recursively after energy minimization (within the same step). This corresponds physically to unstable growth of the micro-cracks compared to the imposed strain rate.

The 3D LEM has the advantage to be cheap in computational effort, making it possible to simulate systems with an large number of nodes for reasonable computing time. It should be remarked that due to the simple additivity of the potential energy, the computation time depends only linearly on the number of nodes. It is also obvious that the LEM is a convenient model of brittle fracture in which the generation and propagation of cracks are "naturally" taken into account.

3 Application to Granular Aggregates

In a granular aggregate, there are three bulk phases: particles, matrix and voids. There are also two interface phases: particle–particle and particle-matrix; see Fig. 1. To construct the samples, we first generate a large dense packing of rigid spherical particles compressed isotropically by means of the contact dynamics method. A cubic portion of this three-dimensional packing is overlaid on a disordered tetrahedral lattice. The particle properties are attributed to the bonds falling in the bulk of the particle phase. The binding matrix is then added in the form of bridges of variable width connecting neighboring particles within a prescribed gap between

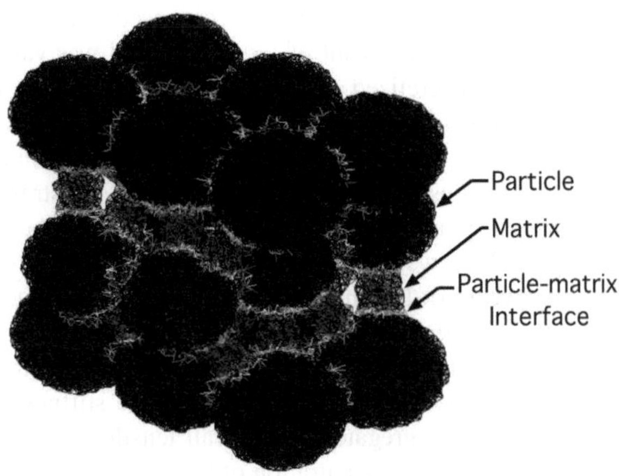

Fig. 1 Example of a discretized aggregate involving particles, matrix, voids and particle–particle and particle-matrix interface zones

particles. The bonds belonging to these bridges are given the properties of the matrix. In the same way, the bonds falling between a particle and the matrix or between two particles are given the properties of the corresponding interface. The width of solid bridges between particles is proportional to the total volume of the binding material. At higher levels of the matrix volume fraction, the bridges overlap and the porosity declines to zero. The particles are polydisperse with diameters varying nearly uniformly is size in a range $[0.8d, d]$. The total particle volume fraction is about 0.6 corresponding to a dense close packing. The samples consist of the bulk phases: (1) particles, denoted p; (2) matrix, denoted m; and (3) void space or pores, denoted v, as well as the interface phases: (1) particle–particle interface, denoted pp, and (2) particle-matrix interface, denoted pm. The elements belonging to each phase ϕ (bulk or interface) are given a Hooke constant $k^{\phi}$and a breaking force f^{ϕ}. We have $f^{v} = 0$ and the choice of the value of k^{v} is immaterial. The interface phases pm and pp are transition zones of finite width. But for large systems, the volume fractions of these transition zones are negligible compared to those of the particles and matrix. The interface phases affect the global behavior through their specific surface and their strengths represented by the Hooke constants k^{pp} and k^{pm} and the corresponding tensile force thresholds f^{pp} and f^{pm}. In our simulations, we model the interface phases by a one bond-thick layer linking two particles or a particle to the matrix. The volume fractions of the interface phases are thus assumed to be zero ($\rho^{pp} = \rho^{pm} = 0$) and the volume fractions ρ^{p}, ρ^{m} and ρ^{v} are attributed only to the three bulk phases, with $\rho^{p} + \rho^{m} + \rho^{v} = 1$. It is dimensionally convenient to express the bond characteristics in stress units. We thus define the bond breaking (or debonding) stresses $\sigma^{\phi} = \frac{f^{\phi}}{a^2}$ and the moduli $E^{\phi} = \frac{k^{\phi}}{a}$ where a is the length of the lattice vector. These bond moduli E^{ϕ} of the lattice should be carefully distinguished from the equivalent phase moduli which depend both on the bond moduli and the geometry of the lattice. We will use below square brackets to represent the phase moduli: $E^{[p]}$, $E^{[m]}$, $E^{[pp]}$ and $E^{[pm]}$. It can be shown that the overall Young modulus and Poisson ratio of an disordered isotropic tetrahedral lattice are $E^{eff} = \frac{5}{4\sqrt{2}} E^{\phi}$ and $\nu^{eff} = 0.25$. We performed a serie of simple tension tests over samples composed of 516 particles. The particle volume fraction was kept constant $\rho^{p} = 0.6$, and ρ^{m} was varied from $\rho^{p}/10$ to $4\rho^{p}$. Each sample was discretized over a lattice containing about 1.5×10^{6} elements. The results presented below were obtained for hard particules $E^{p} = 3E^{m}$, $\sigma^{p} = \sigma^{m}$ and $\sigma^{pp} = 0$. The cubic samples were subjected to uniaxial tension with free lateral sides. The nodes belonging to the base were constrained to be immobile. Upward step-wise displacements were applied to the nodes belonging to the upper surface.

Figure 2 shows the stress–strain plot under for $\rho^{m} = 28\%$ and $\rho^{m} = 13\%$. We observe a brittle behavior with a well-defined initial stiffness E_{eff} and a tensile strength σ_{eff} at the stress peak. The post-peak behavior is characterized by nonlinear propagation of the main crack (initiated at the stress peak) in the form of a sequence of loading–unloading events. The stiffness declines due to progressive damage of the aggregate. The overall tensile strength is higher at larger ρ^{m} as a result of a weaker concentration of stresses. The probability distribution functions

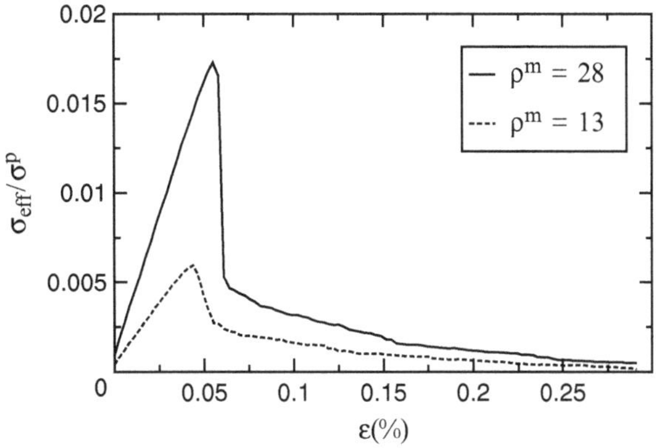

Fig. 2 Normalized vertical stress as a function of vertical strain in tension for two values of the matrix volume fraction ρ^m (in %)

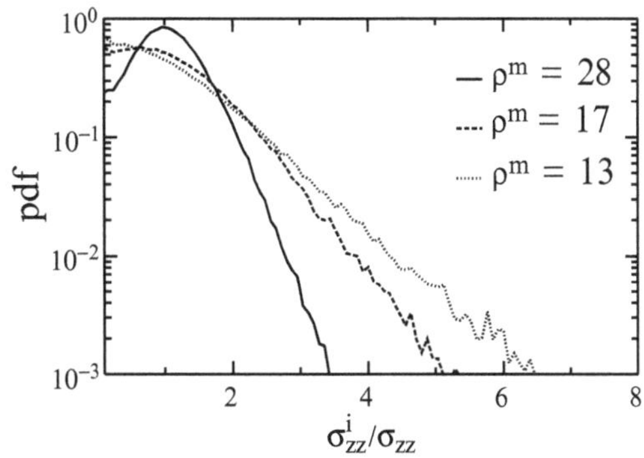

Fig. 3 Probability densities of normalized vertical node stresses σ_{zz}^i for three values of the matrix volume fraction ρ^m (in %)

of vertical node stresses σ_{zz} are shown in Fig. 3. From the shapes of the pdf's, we distinguish large stresses falling off exponentially as observed for large contact forces in granular media (Mueth et al. 1998; Radjaï et al. 1996). The weak stresses have nonzero probability (increasing as $\sigma_{zz} \rightarrow 0$) reflecting the arching effect whereas intermediate stresses are centered on the mean and define a nearly Gaussian distribution. The large stresses mostly concentrate at the contact zones and they form well-defined chains that cross the particles.

The tensile strength and crack propagation are controlled by both ρ^m and σ^{pm}. For a quantitative evaluation of this effect, we consider here the proportion n_b of broken bonds inside the particles with respect to the total number of broken bonds. Figure 4 shows a map of n_b in the parameter space (ρ^m, σ^{pm}) following failure. We see that below a well-defined frontier, no particle damage occurs ($n_b \simeq 0$). For this range of parameter values, the cracks propagate either in the matrix or at the particle-matrix interface. Above this "particle-damage" limit, the isovalue lines become nearly parallel to the limit line with an increasing level of n_b. This suggests three distinct regimes of crack propagation: (1) below the particle-damage limit, the cracks bypass the particles and propagate through the matrix, the pores or along the particle-matrix interface; (2) above this limit and for $\rho^m < 20$, the cracks penetrate

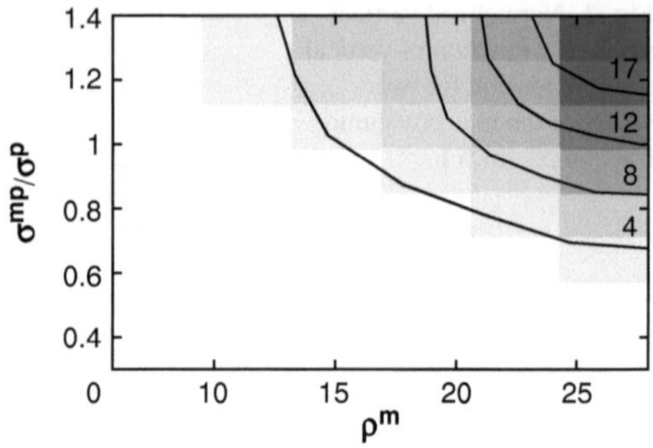

Fig. 4 Grey level map of the fraction of broken bonds in the particle phase for different values of matrix volume fraction and particle-matrix adhesion

into the particles from solid bridges that strongly concentrate stresses; (3) Above this limit, the cracks propagate inside the matrix as well as across the particles, causing the *fragmentation* of the particles. These results are qualitatively similar in 2D cohesive granular aggregates (Topin et al. 2007).

4 Conclusion

In this paper, a lattice-based discretization approach (lattice element method) was introduced and illustrated by application to the brittle failure of porous granular aggregates. In contrast to dilute particle-reinforced composites, such materials involve a high level of particle volume fraction and thus a jammed skeleton of solid particles interconnected via a binding matrix. The overall behavior depends on the bulk phase volume fractions and the properties of the particle–particle and particle-matrix interface zones. We found that the presence of the particle skeleton controls stress concentration and thus the strength properties of these materials. It was also shown that for a range of the values of the particle-matrix adhesion and matrix volume fraction, no particle damage occurs. The trends are very similar to those previously established for 2D aggregates by the same model.

References

D.M. Mueth, H.M. Jaeger, S.R. Nagel, Force distribution in a granular medium. Phys. Rev. E. **57**, 3164 (1998)

F. Radjaï, M. Jean, J.-J. Moreau, S. Roux, Force distribution in dense two-dimensional granular systems. Phys. Rev. Lett. **77**(2), 274 (1996)

M. Satake, J.T. Jenkins, *Micromechanics of Granular Materials* (Elsevier, Amsterdam, 1988)

V. Topin, J.-Y. Delenne, F. Radjaï, L. Brendel, F. Mabille, Strength and fracture of cemented granular matter. Eur. Phys. J. E. Soft Matter. **23**, 413–429 (2007)

J.G.M. Van Mier, M.R.A. Van Vliet, T.K. Wang, Fracture mechanisms in particle composites: statistical aspects in lattice type analysis. Mech. Mater. **34**(11), 705–724 (2002)

Comparative Modelling of Shear Zone Patterns in Granular Bodies with Finite and Discrete Element Model

Jacek Tejchman

Abstract The evolution of patterns of shear zones in cohesionless sand for quasi-static earth pressure problems of a retaining wall under conditions of plane strain was analyzed with a discrete element method (DEM). The passive and active failure of a retaining wall was discussed. The numerical calculations were carried out with a rigid and very rough retaining wall undergoing horizontal translation, rotation about the top and rotation about the toe. The geometry of calculated shear zones was qualitatively compared with experimental results of laboratory model tests using X-rays and a Digital Image Correlation (DIC) technique, and quantitatively with corresponding finite element results obtained with a micro-polar hypoplastic constitutive model.

Keywords DEM • FEM • Earth pressure • Hear localization • Granular body • Grain rotation

1 Introduction

Earth pressure on retaining walls is one of the soil mechanics classical problems. In spite of an intense theoretical and experimental research over more than 200 years, there are still large discrepancies between theoretical solutions and experimental results due to the complexity of the deformation field in granular bodies near the wall caused by localization of shear deformation (which is a fundamental phenomenon of granular material behavior (Tejchman 2008)). It was experimentally observed that localization can appear as single, multiple or pattern of shear zones, depending upon both initial and boundary conditions. It can be plane or curved. Within shear zones,

J. Tejchman (✉)
Gdańsk University of Technology, Faculty for Civil and Environmental Engineering,
80–233 Gdańsk-Wrzeszcz, Narutowicza 11/12, Poland
e-mail: tejchmk@pg.gda.pl

S. Bonelli et al. (eds.), *Advances in Bifurcation and Degradation in Geomaterials*, Springer Series in Geomechanics and Geoengineering 11,
DOI 10.1007/978-94-007-1421-2_33,

pronounced grain rotations and curvatures connected to couple stresses, large strain gradients, and high void ratios together with material softening (negative second-order work) are expected. The thickness of shear zones depends on many various factors, as: the mean grain diameter, pressure level, initial void ratio, direction of deformation, grain roughness and grain size distribution. The knowledge of both the distribution of shear zones and shear and volumetric strains within shear zones is important to explain the mechanism of granular deformation. The multiple patterns of shear zones are not usually taken into account in engineering calculations.

The intention of the paper is to check the capability of DEM (Kozicki and Donze 2008) to simulate a pattern of quasi-static shear zones in initially dense sand and to describe their formation mechanism at the micro-level. The plane strain DEM calculations were carried out with sand behind a rigid and very rough retaining wall, undergoing passive and active movements: horizontal translation, rotation about the top and rotation about the toe. The layout of calculated shear zones was qualitatively compared with corresponding experimental results of laboratory model tests performed at University of Cambridge employing X-rays (Leśniewska 2000) and also with some tests performed by Niedostatkiewicz et al. (Niedostatkiewicz et al. 2010) using a Digital Image Correlation technique (DIC). The experiments with X-rays and DIC were carried out with different sands, granular specimen sizes and initial void ratios. The results of discrete simulations were also quantitatively compared with FE results obtained by modeling a sand behaviour with a micro-polar hypoplastic constitutive model (Tejchman 2008; Tejchman and Górski 2008) for the same sand, its initial void ratio, specimen size and boundary conditions. The capability of DEM to simulate a single shear zone during plane strain compression, direct and simple shearing was several times confirmed in the literature. However, its capability to simulate complex patterns of shear zones in the interior of granulates has not been comprehensively checked yet. This paper is focused on a direct comparison between finite and discrete results at the global and local level.

2 DEM Results

A three-dimensional spherical discrete model YADE was developed at University of Grenoble (Kozicki and Donze 2008) by taking advantage of the so-called soft-particle approach (i.e. the model allows for particle deformation which is modeled as an overlap of particles). Spherical elements were used only. To simulate grain roughness, additional contact moments were introduced into a 3D model, which were transferred through contacts and resisted particle rotations (Iwashita and Oda 1998).

The following five main local material parameters are needed for discrete simulations: E_c, (modulus of elasticity of grain contact) υ_c (Poisson's ratio of grain contact), μ (the inter-particle friction angle), β (rolling stiffness coefficient) and η (moment limit coefficient) which were calibrated with corresponding triaxial

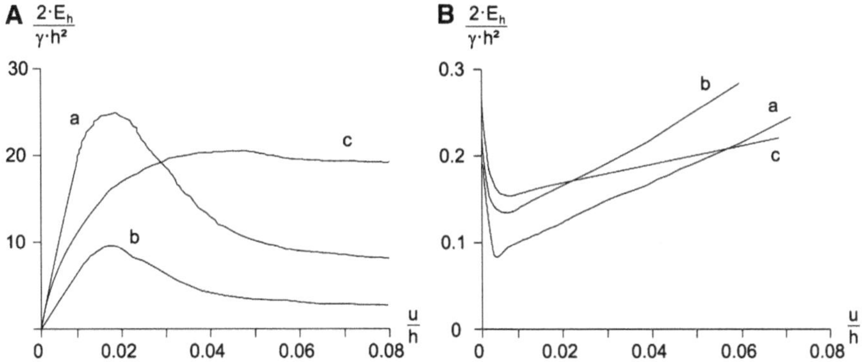

Fig. 1 Resultant normalized earth pressure force $2E_h/(\gamma h^2)$ versus normalized wall displacement u/h from DEM for: (**A**) passive case, (**B**) active case, (**a**) translating wall, (**b**) wall rotating around top, (**c**) wall rotating around toe (E_h horizontal earth pressure force, γ initial sand density, h wall height, u horizontal wall displacement; $e_o = 0.63$, $d_{50} = 1.0$ mm, $E_c = 30$ GPa, $\upsilon_c = 0.3$, $\mu = 30°$, $\eta = 1.0$, $\beta = 0.15$)

laboratory test results with Karlsruhe sand (Wu 1992) ($E_c = 30$ GPa, $\upsilon_c = 0.3$, $\mu = 30°$, $\eta = 1.0$, $\beta = 0.15$ (Widuliński et al. 2009)). In addition, the particle radius R, particle density ρ and damping parameters α are required.

The plane strain discrete calculations were performed with a sand body of a height of $H = 200$ mm and length of $L = 400$ mm (Widuliński et al. 2011). The height of the retaining wall located on the left-hand side of the granular body was assumed to be $h = 200$ mm. The vertical retaining wall and the bottom of the granular specimen were assumed to be stiff and very rough. The granular specimen depth was equal to the grain size.

Some discrete simulations results at the global level for a passive and an active earth pressure problem are shown for initially dense sand (initial void ratio $e_o = 0.63$, mean grain diameter $d_{50} = 1.0$ mm) in Figs. 1 and 2 (Widuliński et al. 2011).

The evolution of the passive and active horizontal earth pressure force $2E_h/(\gamma h^2)$ versus the normalized horizontal wall displacement u/h is similar in three different wall movements (Fig. 1). For a passive case, the horizontal force increases, reaches a maximum for about $u/h = 2$–5%, next shows softening and tends to an asymptotic value. For the wall rotation around the bottom, a decrease of the curve after the peak is smaller (in the considered range of u/h). In the case of the active earth pressure, the horizontal normalized forces drop sharply at the beginning of the wall movement, reach the minimum at $u/h = 0.1$–0.2% and next increase continuously. The maximum normalized passive horizontal forces $2E_h/\gamma h^2$ are between 9 and 25 (in FEM: $2E_h/\gamma h^2 = 12$–31, respectively), and the minimum normalized active earth pressure forces $2E_h/\gamma h^2$ lie between 0.08 and 0.16 (in FEM: $2E_h/\gamma h^2 = 0.10$–0.16, respectively).

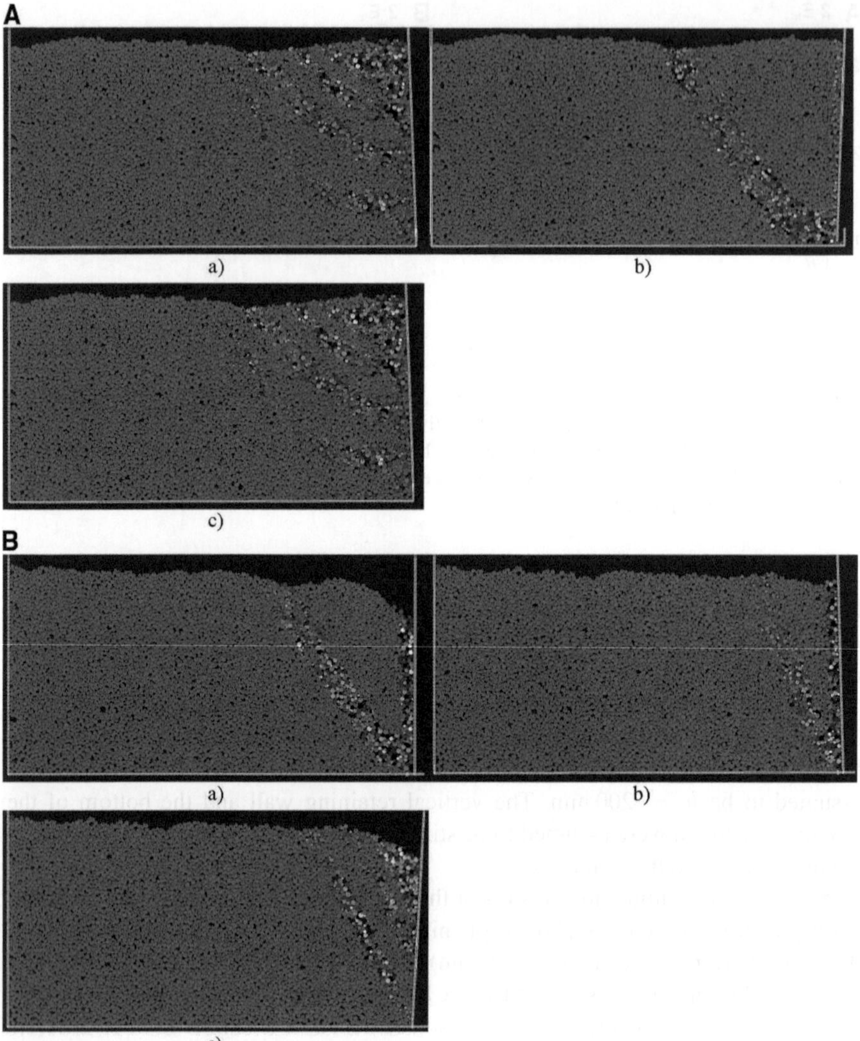

Fig. 2 Deformed granular body with distribution of rotations for initially dense sand from DEM: (**A**) passive case, (**B**) active case, (**a**) translating wall, (**b**) wall rotating around top, (**c**) wall rotating around toe ($e_o = 0.63,\ d_{50} = 1.0\,\mathrm{mm}$)

The geometry and thickness of shear zones on the basis of grain rotations in DEM simulations (Fig. 2) are similar as in FE calculations with a micro-polar constitutive model and in experiments with X-rays and DIC (Widuliński et al. 2011).

With respect to local discrete outcomes across shear zones, the results of a micro-polar rotation (on the basis of grain rotations), void ratio, stresses (on the basis of normal and tangential forces and couple stresses (on the basis of contact moments) were similar as in FE calculations. The maximum grain rotation in the shear zones from DEM was about ±(15–35°) at the residual state during passive wall translation. In turn, the resultant grain rotation in the middle of a radial shear zone (Fig. 2Aa) from the area of $5d_{50} \times 5d_{50}$ was about 10°. The largest internal work at peak was performed by contact normal forces (45% of the total work) and contact tangential forces (40% of the total work) and the smallest one by contact moments (15% of the total work). In the residual state, the work performed by normal and tangential contact forces was similar and the work performed by contact moments was three times smaller than the remaining ones. The dissipation in the granular specimen during deformation was about 25% at peak and 10% at the residual state, respectively (as compared to the total external work).

3 Conclusions

The numerical simulations show that a discrete element method is capable to reproduce the most important macroscopic properties of cohesionless granular materials without being necessary to describe the granular structure perfectly. Comparing discrete simulations with experimental tests and continuum calculations shows that a discrete model is able to realistically predict the experimental results of a pattern of shear zones in sand. It can be also used study and describe the mechanism of the initiation, growth and formation of a pattern of shear zones at the micro-level.

The experimental patterns of shear zones are realistically reproduced in discrete calculations. They depend strongly on the direction and type of the wall movement (passive or active, translation or rotation). They are similar as in FE computations.

The largest passive earth pressures occur with the horizontal translation of the wall, they are smaller with the wall rotation around the bottom and again smaller with the wall rotation around the top. The smallest active earth pressures are created during wall translation, and the largest during wall rotation around the top.

A mean grain size has a significant effect on a load-displacement diagram in DEM when shear localization is taken into account. The largest internal work in DEM simulations is performed by contact normal forces and the smallest one by contact moments.

The granular material tends to a critical state inside of shear zones. The grain rotations are noticeable only in shear zones.

The maximum horizontal passive force on the wall grows with increasing micro-mechanical parameters μ, β and η. In turn, the residual horizontal passive force depends on β only.

Conventional earth pressure mechanisms with slip surfaces are roughly reproduced. Realistic earth pressure coefficients can be obtained with actual values of internal friction angles only.

References

K. Iwashita, M. Oda, Rolling resistance at contacts in simulation of shear band development by DEM. ASCE. J. Eng. Mech. **124**(3), 285–292 (1998)

J. Kozicki, F.V. Donze, A new open-source software developed for numerical simulations using discrete modelling methods. Comput. Methods Appl. Mech. Eng. **197**, 4429–4443 (2008)

D. Leśniewska, *Shear Band Pattern Formation in Soil, Habilitation Monograph* (Institute of Hydroengineering of the Polish Academy of Science, Gdańsk, 2000)

M. Niedostatkiewicz, D. Leśniewska, J. Tejchman, Experimental analysis of shear zone patterns in sand for earth pressure problems using Particle Image Velocimetry. Strain (2010). doi: 10.1111/j.1475-1305.2010.00761.x2010

J. Tejchman, *FE Modeling of Shear Localization in Granular Bodies with Micro-Polar Hypoplasticity, Series in Geomechanics and Geoengineering*, ed. by W. Wu, R. Borja, Springer (Springer, Berlin-Heidelberg, 2008)

J. Tejchman, J. Górski, Computations of size effects in granular bodies within micro-polar hypoplasticity during plane strain compression. Int. J. Solids Struct. **45**(6), 1546–1569 (2008)

L. Widuliński, J. Kozicki, J. Tejchman, Numerical simulation of a triaxial test with sand using DEM. Arch. Hydro-Eng. Environ. Mech. **56**(3–4), 3–26 (2009)

L. Widuliński, J. Tejchman, J. Kozicki, D. Leśniewska, Discrete simulations of shear zone patterning in sand in earth pressure problems of a retaining wall. Int. J. Solids Struct. **48**(7–8), 1191–1209 (2011)

W. Wu, Hypoplastizität als mathematisches Modell zum mechanischen Verhalten granularer Stoffe, Ph.D. thesis, University of Karlsruhe, Karlsruhe (1992)

Influence of Weak Layer Heterogeneity on Slab Avalanche Release Using a Finite Element Method

J. Gaume, G. Chambon, M. Naaim, and N. Eckert

Abstract Snow avalanches generally result from the collapse of a weak layer underlaying a cohesive slab. We use the finite element code Cast3m to build a complete mechanical model of the {weak layer-slab} system including inertial effects. We model the weak layer as a strain-softening interface whose properties are spatially heterogeneous. The softening accounts for the breaking of ice bridges. The overlying slab is represented by a Drucker-Prager elasto-plastic model, with post-peak softening to model the crack opening. The two key ingredients for the mechanical description of avalanches releases are the heterogeneity of the weak layer and the redistribution of stresses by elasticity of the slab. The heterogeneity is modeled through a Gaussian stochastic distribution of the friction angle with spatial correlations. We first study the effect of the weak layer's heterogeneity and the slab depth on the release on a simple uniform slope geometry. We observe two releases types, full slope releases corresponding to a crown rupture and partial slope releases for which the traction rupture occurs inside the slope and thus only a part of the slope is released. The influence of slab depth on the relative proportion of these two rupture types, as well as on the avalanche angle distributions is also studied.

Keywords Avalanche release • Shear softening • Weak layer • Heterogeneity • Drucker-Prager

1 Introduction

Recently, several models have been developed to simulate the flow of the various types of snow (Naaim et al. 2003; Naaim and Gurer 2004) involved in snow avalanches. Nevertheless, the systematic use of these models in operational

J. Gaume (✉) · G. Chambon · M. Naaim · N. Eckert
Cemagref, BP 76, 38402 St Martin d'Heres, France
e-mail: johan.gaume@cemagref.fr

S. Bonelli et al. (eds.), *Advances in Bifurcation and Degradation in Geomaterials*, Springer Series in Geomechanics and Geoengineering 11,
DOI 10.1007/978-94-007-1421-2_34,

applications still faces a number of difficulties including the evaluation of the avalanche release volume which is an input parameter of these models which very strongly influences the results. However, there is currently no clear and well-defined methodology which would enable, for a given avalanche path, to predetermine the initial volume of potential avalanches for different return periods. Slab avalanches result from the collapse of a thin weak layer underlaying a cohesive slab. Recently, some cellular automaton models have been developed providing important insight into the mechanisms of avalanche release (Failletaz et al. 2004; Fyffe and Zaiser 2004, 2006). In particular, they pointed the two basic ingredients that are essential for the mechanical description of avalanche release, namely the heterogeneity of the weak-layer and the stress redistribution effect conveyed by the elasticity of the overlaying slab. However, the treatment of the mechanical behavior of the different layers is very simplified in these models. In addition these models are unable to include a detailed description of topography and geomorphology of release zones. The objective of this study is to develop a complete mechanical model of the {weak layer-slab} system. In this paper, we focus in particular on the role of the weak layer heterogeneity on the avalanche release mode.

2 Formulation of the Model

We use the Finite Element code Cast3m to build a complete mechanical model for slab avalanche release including inertial effects. The code solves the mass and momentum balance equations under the small-strain assumption. The used procedure enables to perform non-linear incremental computations with an implicit integration scheme.

2.1 Geometry, Boundary Conditions and Loading

We study a simple 2D uniform slope geometry (Fig. 1). The system is composed of a weak layer modeled as an interface and an upperlaying slab. The gravity is the only

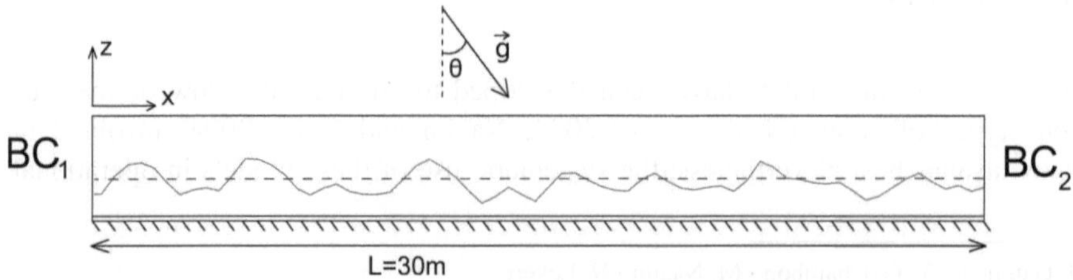

Fig. 1 Geometry of the system. The *curve* represents a realization of the heterogeneity (friction angle) of the weak layer

external force and we load the system by increasing the slope angle θ. The length of the slope is $L = 30$ m and we perform simulations for different slab depths. The weak layer's base is fixed to the ground. The boundary condition at the base of the slope (BC_2) consists in imposing a nil displacement in the x direction. At the crown of the slope (BC_1) we apply a shear stress varying with the depth in order to avoid bending due to limit size effects.

2.2 *Mechanical Behavior of Snow: Constitutive Laws*

Snow is a very complex material whose mechanical behaviour remains largely unknown. Nevertheless, it has been shown that under certain conditions (McClung 1979) snow behaves as an elasto-plastic material. Thus, we model the weak layer as an elasto-plastic interface with shear softening (McClung 1979; Fyffe and Zaiser 2004, 2006; Kronholm and Birkeland 2005). This shear softening is justified by the breaking of ice bonds at the micro-scale. In a first step, we have considered a Mohr-Coulomb rupture criterion without cohesion. Once rupture is reached, shear softening occurs over a characteristic tangential displacement $u_c = 2$ mm (McClung 1979) after which the shear stress reaches a residual value equal to the half of the peak shear stress.

The slab layer is modeled using a Drucker-Prager constitutive law with a softening post-peak behavior and a very low residual stress in order to represent the opening of the traction crack. The value of the mechanical parameters were chosen according to (Schweizer 1999). We have taken a slab density of $\rho = 250\,\text{kg/m}^3$, a Poisson's ratio $\nu = 0.3$, a Young's modulus $E = 10^6$ Pa and a traction and compression elastic limit $\sigma_t = 1$ kPa, $\sigma_c = 10$ kPa. Hence, the parameters of the plastic flow law are not important since the plastic limit only account for a slab rupture criterion.

2.3 *Spatial Heterogeneity of the Weak Layer*

The spatial variability of the week layer mechanical properties are modeled through an heterogeneous stochastic friction angle ϕ. Several field studies (Jamieson and Johnston 2001; Schweizer et al. 2008; Kronholm and Birkeland 2005) show that the spatial variability of mechanical properties of snow can be approximated by a Gaussian distribution. Thus, in agreement with the literature, ϕ is modeled as a Gaussian stochastic field with spatial correlations. The associated covariance matrix is expressed as follow: $C_{ij} = s^2 exp(-0.5\left(d_{ij}/\epsilon\right)^2)$, where s is the standard deviation, d_{ij} is the distance between P_i and P_j and ϵ is the correlation length. Note that no nugget effect is considered. According to Conway and Abrahamson (1984), the typical correlation length of strength variation should be between 0.2 and 1.3 m.

In this study, we took $\epsilon = 1$ m. According to Schweizer et al. (2008), we have taken a coefficient of variation $CV = \tan s / \tan <\phi> = 15\%$ and the mean friction angle $<\phi> = 30°$. One realization of this heterogeneity is represented on Fig. 1.

3 First Results and Discussion

For different slab depths ($h \in \{0.3, 0.6, 0.9, 1.2, 1.5, 2.1\,\text{m}\}$) we performed 100 simulations with different realizations of the heterogeneity. This gives us some distributions of release angle. For each slab depths, we then studied these distributions and the release type. Knowing the evolution of the mean and the standard deviation of release angle distributions in function of the slab depth, we will be able to determine the distributions of release depths for different slope angles.

3.1 Release Types and Criteria

By looking at the plastic deformation in the slab for each simulation, we can distinguish two kinds of releases: Full slope releases (the entire slope is released without traction rupture) and partial slope releases (only a part of the slope is released with a traction rupture). Both of them are induced by a shear rupture in the weak layer. Indeed, the slab rupture is a secondary process and always results from an instability inside the weak layer ($u > u_c$). Partial releases are generally associated to an important local heterogeneity (difference of shear strength between adjacent elements) around the weakest zone. In the case of a global release, the shear rupture propagates in the whole weak layer since the local heterogeneity is not sufficient to make the slab rupture ($\epsilon_{pl}^{slab} = 0$). According to these rupture modes, we defined two release criteria. The first one is based on the plastic deformation of the slab and is only relevant for partial releases. The second one is based on the displacement of the base of the slab and can be used for both full and partial releases.

Figure 2a shows the evolution of the probability of traction rupture with the slab depth h calculated from the finite element method. The traction rupture probability is very low for $h < 0.5$ m and then increases almost linearly until $h \approx 1.2$ m where it starts to level off to 1 more slowly. This means that for very thick slabs ($h > 1.2$ m), we will a significant proportion of partial slope releases. On the contrary, full slope releases will be more frequent for thin slabs ($h < 0.5$ m). This evolution can be interpreted with a simple small model. We considered the occurrence of traction rupture is directly related to the shear-stress difference $\Delta\tau = |\tau_2 - \tau_1|$ between two neighboring elements 1 and 2 and occurs when $\Delta\tau$ is greater than the slab traction rupture limit σ_T (neglecting spatial correlations). We calculated this probability $P(\Delta\tau > \sigma_T)$ by computing many independent realizations of ϕ_1 and ϕ_2. The results of this model are represented on Fig. 2a (dotted line) and show very good agreement with the results from the finite element method.

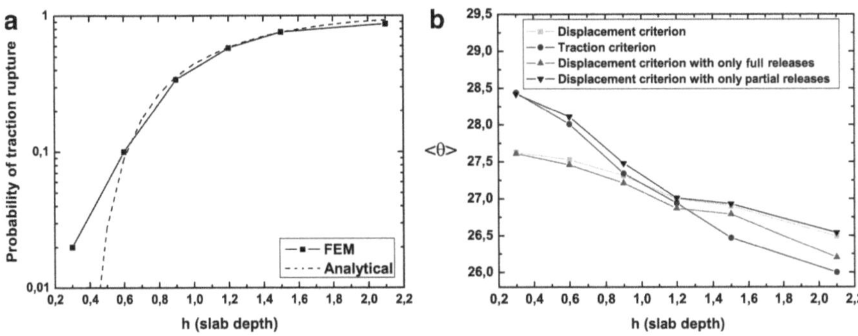

Fig. 2 Probability of traction rupture (**a**) and mean release angle (**b**) as a function of the slab depth

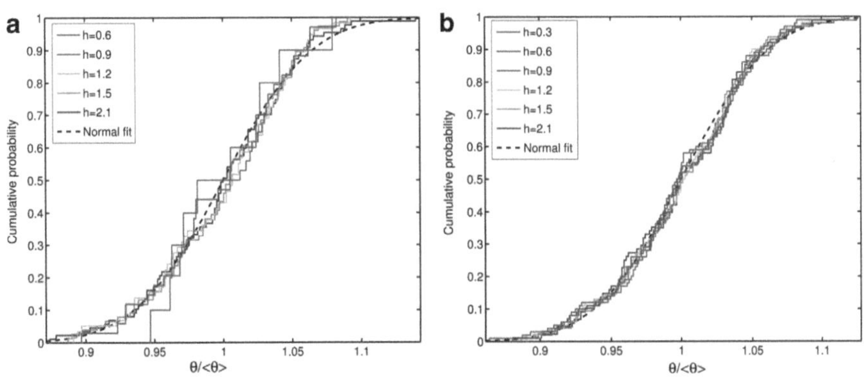

Fig. 3 Distributions of release angle: traction criterion (only partial releases) (**a**) and displacement criterion (all releases) (**b**)

3.2 *Release Angle Distributions*

Figure 3 shows the distributions of release angles obtained using the traction and the displacement criterion. We first note that for both rupture criteria, the distributions are normally distributed with a variance independent of h. This particular shape is presumably a direct consequence of the Gaussian heterogeneity.

Figure 2b reports the average of these distribution and we also separate full and partial slope releases from the displacement criterion. We can firstly note that the slab depth does not strongly influence the slope stability since the release angle variations remain low. This is due to the frictional rupture criterion used for the weak-layer. This criterion also explains why we observe that the slab traction rupture is more sensitive to the slab depth h. Moreover, the mean release angle is always lower than the mean friction angle $<\phi>$ since the rupture occurs around weakest spots. Finally, the rupture type (full slope or partial slope) has a greater influence on the release angle for thin slabs ($h < 1.2$ m) than for thick ones ($h > 1.2$ m).

4 Conclusions and Perspectives

Slab snow avalanches are triggered by the rupture of a weak layer underlaying an elasto-plastic slab. In this paper, we study the influence of the heterogeneity of the weak layer mechanical properties on avalanche release using a finite element method. We have shown that the slab rupture always results from an instability inside the weak layer. We have seen that the slab depth has an important influence on the release type but almost no influence on the release angle. Finally, the distributions of release angles have the same shape as the distribution of the heterogeneity (Gaussian). As perspectives, we will first repeat this study in the cohesive case. Indeed, the slab depth has a little influence on the release angle because of the frictional rupture criterion. Then we will analyse the influence of the correlation length ϵ on the release type.

References

H. Conway, J. Abrahamson, Snow slope stability – a probabilistic approach. J. Glaciol. (1984)

J. Failletaz, F. Louchet, J.-R. Grasso, Two-threshold model for scaling laws of noninteracting snow avalanches. Phys. Rev. Lett. **93**, 208001 (2004)

B. Fyffe, M. Zaiser, The effects of snow variability on slab avalanche release. Cold Reg. Sci. Technol. **40**, 229–242 (2004)

B. Fyffe, M. Zaiser, Interplay of basal shear fracture and slab rupture in slab avalanche release. Cold Reg. Sci. Technol. **49**, 26–38 (2006)

J.B. Jamieson, C.D. Johnston, Evaluation of the shear frame test for weak snowpack layers. Ann. Glaciol. **32**, 59–69 (2001)

K. Kronholm, K. Birkeland, Integrating spatial patterns into a snow avalanche cellular automata model. Geophys. Res. Lett. **32**, L19504, doi:10.1029/2005GL024373 (2005)

D. McClung, Shear fracture precipitated by strain softening as a mechanism of dry slab avalanches. J. Geophys. Res. **84**(B7), 3519–3526 (1979)

M. Naaim, I. Gurer, Two-phase numerical model of powder avalanche theory an application. Nat. Hazard. **24**, 569–585 (2004)

M. Naaim, T. Faug, F. Naaim-Bouvet, Dry granular flow modelling including erosion and deposition. Surv. Geophys. **17**, 129–145 (2003)

J. Schweizer, Review of dry snow slab avalanche release. Cold Reg. Sci. Technol. **30**, 43–57 (1999)

J. Schweizer, K. Kronholm, B. Jamieson, K. Birkeland, Review of spatial variability of snowpack properties and its importance for avalanche formation. Cold Reg. Sci. Technol. **51**, 253–272 (2008)

Part IV
Failure, Localization and Multi-physical Couplings

Coupled Mechanical-Hydrological-Chemical Problems in Elasto-plastic Saturated Soils and Soft Rocks Using *escript*

Arash Mohajeri, Hans Muhlhaus, Lutz Gross, and Thomas Baumgartl

Abstract The understanding of chemical effects on mechanical behavior of porous media is a key element in the solution of problems in geology, mining, soil, rock and environmental engineering. In order to develop this understanding, the current work employs a new set of equations to numerically investigate the coupled mechanical-hydrological-chemical (MHC) problem for soils and soft rocks. The objectives of this research are to observe the soil and soft rock behavior under mechanical loading and chemical erosion and also to validate the application of our solver algorithm written in a finite element programming environment "escript".

Keywords Coupled problems • Chemical reactions • Mechanical behaviour • Elasto plasticity • Finite element • Escript

1 Introduction

To predict and analyze engineering problems like consolidation of chemically polluted soil, in situ leaching, dam stability due to acidic upstream flow and mineralization; understanding soil and rock real mechanical and chemical behavior is very important. Such problems, which are in fact the presence and occurrence of two or more different phenomena simultaneously, lead to systems of differential equations. These systems of equations are generally called coupled problems. Problems like

A. Mohajeri (✉) · H. Muhlhaus · L. Gross
Earth Systems Science Computational Center, School of Earth Sciences, The University of Queensland, St. Lucia, QLD 4072, Australia
e-mail: arash.mohajeri@uqconnect.edu.au

T. Baumgartl
Center for Mined Land Rehabilitation, Sustainable Minerals Institute, The University of Queensland, St. Lucia, QLD 4072, Australia

S. Bonelli et al. (eds.), *Advances in Bifurcation and Degradation in Geomaterials*, Springer Series in Geomechanics and Geoengineering 11,
DOI 10.1007/978-94-007-1421-2_35,

the ones in this study are coupled mechanical-hydrological-chemical problems (MHC). The main challenges in coupled problems lie in defining the mathematical formulations, finding accurate and stable solution and numerical approximations. Previous studies have investigated the coupled hydrological-chemical-mechanical process in engineering problems. Experimental and numerical techniques have been employed to solve the problem of which a few also cover the elasto-plasticity behavior of soils and rocks (Hueckel 1992; Gajo et al. 2002). Simplifications made in many of these studies increase the uncertainty about the accuracy of the results (Nova 2006). Some recent studies, however, have suggested new models and employed more accurate numerical tools to simulate the porosity instability and rock alteration as a part of the mechanical-chemical coupled model (Muhlhaus et al. 2001a; Freij-Ayoub et al. 2002). In spite of the work done to investigate coupled hydrological-chemical-mechanical problems, due to complexity of the problems, they are yet to be very well understood (Lanru and Xiating 2003).

The focus of this work is on the 1D and 2D coupled mechanical-hydraulic-chemical (MHC) problems. This mathematical model of finite elasto-plastic deformation and pore water pressure along with the chemical transport equation for fully saturated porous media is expressed as an algorithm in the finite element program "escript". The solution of this new set of equations will lead to understanding the behavior of chemically polluted/eroded soils and soft rocks under mechanical loading. The mass transfer equation along with the pore pressure equation, itself, could be very challenging due to having two different length scales in the equations (Muhlhaus et al. 2001). Variable porosity, permeability and seepage velocity are always expected in a coupled MHC problem. Moreover, the mechanical response may also vary for different type and extent of a chemical reaction. Putting all these difficulties together, they create some problematic issues in constitutive and numerical models which we will face in this research. The paper starts with the outline of the mechanical-hydrological model and the governing equations. It is followed by the governing equations of chemical transport and its effect on mechanical properties. The performance of our code is analyzed using two examples of MHC coupling problems.

2 Governing Equation

2.1 Soil Skeleton Constitutive Equation and Strain-Stress Relationship

The incremental form of the mechanical equilibrium equations are written as:

$$-\left(C_{ijkl} d u_{kl}\right)_{,j} = \left(\bar{\sigma}_{ij} - dp\delta_{ij}\right)_{,j} + f_i \tag{1}$$

In the above equations C_{ijkl} is the elasto-plastic constitutive tensor, $\bar{\sigma}_{ij}$ is the stress value from the previous time step and f_i is the volume force due to gravity. The Drucker-Prager yield criterion is written as:

$$F = \tau^e - \alpha p^e - \tau_y \tag{2}$$

where τ_y and α are functions of the plastic shear stress and are called the current shear length and the friction parameter respectively. Also p is the first stress invariant and τ^e is the second deviatoric stress invariant (Gross et al. 2008).

2.2 Fluid Flow Equations

We use an updated Lagrangian scheme, the mass balance equation and the continuity equation. Neglecting the fluid phase acceleration, the pore fluid can be written as (Zienkiewicz and Shiomi 2005):

$$-u_{i,i} + (k/\gamma^f)p_{,ii} - (n/K_f)p_{,t} = 0 \tag{3}$$

where u_i, K_f, n, k, γ_f are the displacement vector, bulk modulus of the fluid phase, porosity, permeability coefficient and unit weight of fluid respectively.

2.3 Chemical Transport Governing Equations

Calcium carbonate is a common mineral and can be found in different type of rocks (Fernandez-Merodo et al. 2007):

$$\underbrace{CaCO_3}_{Solid} + 2\underbrace{H_3O^+}_{Aqueous} \Leftrightarrow \underbrace{Ca^{2+}}_{Aqueous} + 3\underbrace{H_2O}_{Aqueous} + \underbrace{CO_2}_{gas}$$

The general form of mass transport equation in porous media can be written in below form (Muhlhaus et al. 2001; Liu et al. 2004):

$$(nC_1)_{,t} = (-nC_1 w_i + nD_1 C_{1,i})_{,i} + (\rho_s/\rho_w)N\psi n_{,t} \tag{4}$$

$$n_{,t} = K(n_{final} - n_{initial})^{1/3}(n_{final} - n)^{2/3}C_1 \tag{5}$$

where in these equations, C_1 is the concentration of acid. w_i is the seepage velocity, n_{final} is the final porosity, ρ_s and ρ_w are the density of calcium carbonate and water. N presents the stoichiometric coefficient, ψ is the ore grade and K shows a rate constant. The total dynamic porosity n includes both chemical and mechanical changes of porosity (Muhlhaus et al. 2001; Liu et al. 2004):

$$n = n_{initial} + \Delta n_{chemical} + \Delta n_{mechanical} \tag{6}$$

In real soils and rocks, the permeability varies with porosity (Liu et al. 2004):

$$k = k_c\,(n/n_{initial})^3 \tag{7}$$

where k_c is the initial permeability. The influence of chemical rock alteration on the yield strength is considered by the following relationship (Muhlhaus et al. 2001a; Freij-Ayoub et al. 2002):

$$\tau_y = \tau_{residual} + (\tau_0 - \tau_{residual})e^{-\alpha C_1} \tag{8}$$

Here, $\tau_{residual}$ is the residual yield strength, τ_0 is the initial yield strength and $\alpha > 1$ is an arbitrary coefficient.

Equations 1, 3 and 4 are the governing equations of our MHC coupled problem which have to be solved for u_i, p, n and C_1.

3 Examples

In this section we illustrate our theory by means of 1D and 2D examples.

3.1 One Dimensional Elasto-plastic Consolidation for Chemically Polluted Soil

Here we consider the influence of an erosive chemical on consolidation process. To simulate this, we assume that besides the mechanical load, a chemical source distributes erosive chemicals on the top surface. The problem domain is shown in Fig. 1a with R representing the mechanical load. Figure 2a,b show the variation of normalized displacement and stress with time in a point right underneath the top

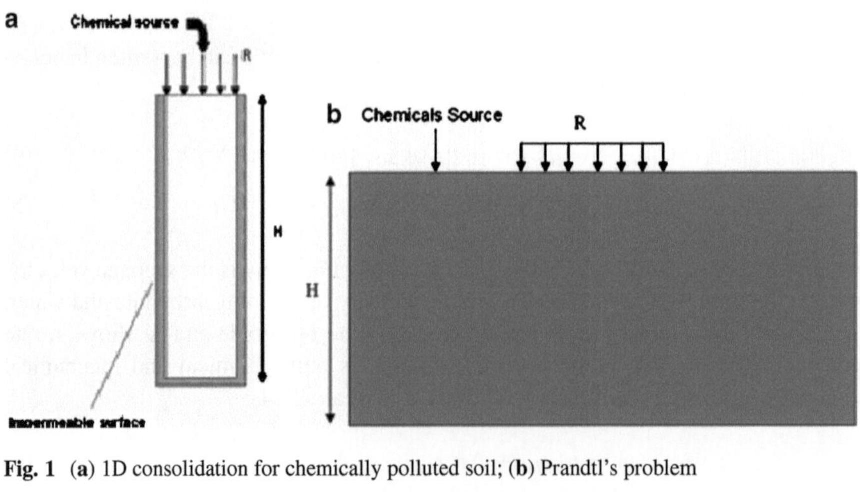

Fig. 1 (**a**) 1D consolidation for chemically polluted soil; (**b**) Prandtl's problem

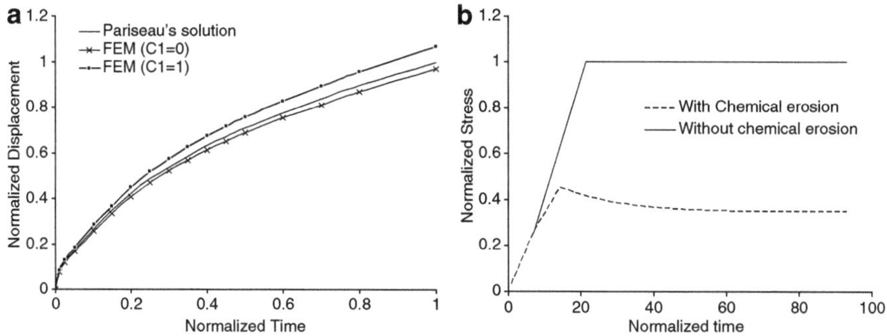

Fig. 2 (**a**) The effect of erosion on displacement; (**b**) variation of yield value with chemical erosion

surface. The displacement has been compared to Pariseau's analytical solution for a porous medium undergoing only mechanical forces (Pariseau 1999). As it is shown, our code was able to produce close results to the analytical solution when $C = 0$. Also, the chemical erosion is seen to increase the displacement in the domain. Looking at stress variation, chemical erosive helps the domain reaches its new yield value quicker which is smaller compare to the case when $C = 0$.

3.2 *Elasto-plastic Semi-infinite Medium with a Plane Surface*

As can be seen in Fig. 1b this problem includes a semi infinite layer of a soil or soft rock with strip loading exerting on the top surface. We prescribed a constant chemical concentration ($C_1 = 1$) right beside the distributed mechanical load (outside of the foundation). From Fig. 3a,b the deformation of the domain is seen to influence the direction of chemical transport. Figure 3c presents the symmetric displacement contours when the concentration of chemicals is zero whereas Fig. 3d shows the case with $C_1 = 1$. The comparison of Fig. 3c,d shows that change in concentration results in having different value for displacement which in turn results in non symmetric displacement contours in the domain. In Fig. 3e the evolution of porosity with time is demonstrated for different chemical concentrations in a point underneath the foundation. When the concentration is zero the porosity does not remain constant and it decreases with time because of the change in porosity due to mechanical loading. Figure 3f shows that increase in concentration leads to increase in erosion and porosity. In a medium with higher porosity, the pore water can be expelled more quickly.

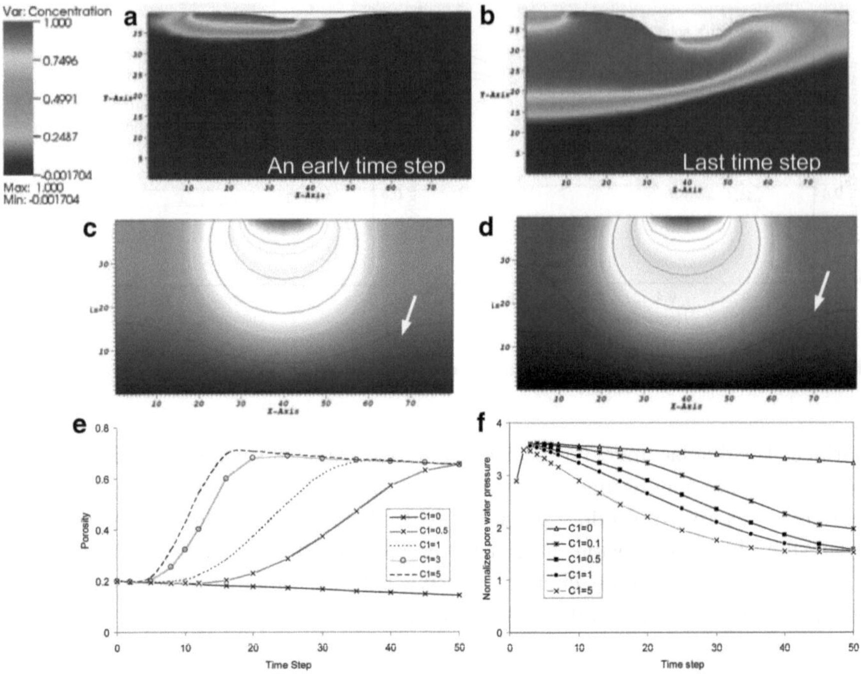

Fig. 3 (**a**) chemical transport in an early time step; (**b**) chemical transport in last time step; (**c**) symmetric velocity contours in the absence of chemical erosion; (**d**) non symmetric velocity contours in the presence of chemical erosion; (**e**) porosity-time graph for different concentrations; (**f**) variation of pore pressure with time for different concentrations

4 Conclusions

In this research, a new set of equations was coded in "escript" with the aim of investigating the coupled hydrological–mechanical-chemical problems in elasto-plastic soil and soft rocks. The model was applied to three different examples. The chemical reaction in all these examples was assumed to be the erosion of carbonate rock due to acidic rain. In order to evaluate the solver algorithm, a coupled MHC problem was solved for a 1D domain. The domain was studied under simultaneous mechanical and chemical loading and it was observed that the presence of the erosive element causes the increase in the displacement. Moreover, chemical erosion reduces the yield value. The second example was a 2D foundation problem. The examples demonstrated that the settlement of the foundation, the porosity and the pore water pressure can change significantly due to chemical erosion in the domain. The reduction in domain strength can vary depending on the nature and magnitude of the chemical erosion. The two last examples show the capability of the new algorithm to simulate the coupled problems. It also highlighted the importance of involving the chemical transport in engineering calculations.

References

J.A. Fernandez-Merodo, R. Castellanza, M. Mabssout, M. Pastor, R. Nova, M. Parma, Coupling transport of chemical species and damage of bonded geomaterials. Comput. Geotech. **34**(4), 200–215 (2007)

R. Freij-Ayoub, H.-B. Muhlhaus, L. Probst, in *Multicomponent Reactive Transport Modelling: Applications to Ore Body Genesis and Environmental Hazards Porous Media: Theory, Experiments, and Numerical Applications*, ed. by W. Ehlers, J. Bluhm (Springer, Berlin/Heidelberg/New York, 2002), pp. 415–434

A. Gajo, B. Loret, T. Hueckel, Electro-chemo-mechanical couplings in saturated porous media: elastic-plastic behaviour of heteroionic expansive clays. Int. J. Solids Struct. **39**(16), 4327–4362 (2002)

L. Gross, H. Mühlhaus, E. Thorne, K. Steube, A new design of scientific software using Python and XML. Pure Appl. Geophys. **165**(3–4), 653–670 (2008)

T. Hueckel, On effective stress concepts and deformation in clays subjected to environmental loads: discussion. Can. Geotech. J. **29**, 1071–1086 (1992)

J. Lanru, F. Xiating, Numerical analysis of thermo-hydro-mechanical and chemical processes (THMC) of geological media – International and chinese experinces. China J. Rock Mech. Eng. **22**(10), 1704–1715 (2003)

J. Liu, C. Mallett, A. Beath, D. Elsworth, B. Brady, A coupled flow-transport-deformation model for underground coal gasification, in *Elsevier Geo-Engineering Book Series (Coupled Thermo-Hydro-Mechanical-Chemical Processes in Geo-Systems, Fundamentals, Modelling, Experiments and Applications)*, vol. 2 (Elsevier, London, 2004), pp. 611–616

H.B. Muhlhaus, B.E. Hobbs, R. Freij-Ayoub, J.L. Walshe, A. Ord, Multi scales in rock alteration and ore deposit genesis, in *Proceedings of the 5th International Workshop of Bifurcation and Localisation Theory in Geomechanics*, Perth, 2001a

H.B. Muhlhaus, J. Liu, B.E. Hobbs, Modeling of In-situ Solution Mining Processes, in *Solid Mechanics and Its Applications*, vol. 87 (Springer, Netherlands, 2001), pp. 301–306

R. Nova, Modelling of bonded soils with unstable structure, in *Springer Proceedings in Physics, Modern Trends in Geomechanics* (Springer, Berlin/Heidelberg, 2006), pp. 129–141

W.G. Pariseau, Poroelastic-Plastic consolidation – analytical solution. Int. J. Numer. Anal. Methods Geomech. **23**(7), 577–594 (1999)

C. Zienkiewicz, T. Shiomi, Dynamic behaviour of saturated porous media; the generalized Biot formulation and its numerical solution. Int. J. Numer. Anal. Methods Geomech. **8**(1), 71–96 (2005)

Analytical Evidence of Shear Band Bifurcations for Softening Materials

D. Caillerie, R. Chambon, and Frédéric Collin

Abstract Considering a boundary value problem for a softening material on a 2D annulus with axisymmetrical boundary conditions, we prove the uniqueness of the radial displacement solution even in the softening regime and we analytically show the existence of shear band bifurcation by giving exact solutions with orthoradial displacement.

Keywords Shear band • Analytical solution • Bifurcation • Softening • Deformation theory of plasticity

1 Introduction

Strain localization is a phenomenon occurring very often in anelastic media close to rupture and it is very important to take it into account in most of applications, particularly in geotechnics. Strain localization has been widely studied experimentally, see for instance Desrues (2004) or Desrues and Viggiani (2004). It has also been put in evidence in finite element simulations even if it presents mesh dependencies, see Chambon et al. (2001) for instance. From a theoretical point of view, the Rice condition (Rice 1976; Rudnicki and Rice 1975), with some assumptions, enables to determine the onset of strain localization but does not necessarily prove the existence of shear bands.

D. Caillerie (✉) · R. Chambon
Laboratoire Sols Solides Structures – Risques, Grenoble Université & CNRS, BP 53, 38041 Grenoble Cedex, France
e-mail: Denis.Caillerie@inpg.fr; Rene.Chambon@inpg.fr

F. Collin
Université de Liège, ArGEnCo – *Geo*3, Bat B52/3 chemin des 132 chevreuils 1, 4000 Liège, Belgium
e-mail: f.collin@ulg.ac.be

S. Bonelli et al. (eds.), *Advances in Bifurcation and Degradation in Geomaterials*, Springer Series in Geomechanics and Geoengineering 11,
DOI 10.1007/978-94-007-1421-2_36,

The purpose of this paper is to establish the existence of shear bands by exhibiting analytical solutions. The studied problem is the deformation of a 2D thick walled cylinder (an annulus) made of an isotropic hardening/softening "elastic" material and submitted to internal and external pressures. The considered constitutive equation is a 2D generalization of that one considered in Chambon et al. (1998) for 1D problems. More precisely, the hardening/softening behaviour considered in Chambon et al. (1998) holds true for the deviatoric part of the 2D constitutive relation in any fixed direction of the deviatoric strain (see Sect. 2). Unlike that of Chambon et al. (1998) which is affine by parts, the strain stress relation considered in this paper is truly non linear. It is said to be "elastic" because the datum of the strain and not that of its history yields the stress. In fact, this "elastic" behaviour fits into the framework of deformation theory of plasticity and can be thought of as the response of some elasto plastic constitutive equation to a monotonous loading (see Budiansky 1959). In the present paper elastic unloading can only occur on the initial elastic curve.

2 Constitutive Equation

Let $\mathbf{u}$ be the displacement, $\nabla\mathbf{u}$ be its gradient and the symmetrical part $\varepsilon = \frac{1}{2}\left(\nabla\mathbf{u} + \nabla\mathbf{u}^t\right)$ of it be the strain.

$\mathbb{I}$ being the identity second order tensor, the strain ε is split in its isotropic and deviatoric part: $\varepsilon = \frac{\mathrm{tr}\varepsilon}{2}\mathbb{I} + \varepsilon^d$ with $\mathrm{tr}\varepsilon^d = 0$ and the constitutive equation reads:

$$\sigma = (\lambda + \mu_d)\,\mathrm{tr}\varepsilon\mathbb{I} + 2\left(\mu_m + (\mu_d - \mu_m)\frac{M}{\left\|\varepsilon^d\right\|}\right)\varepsilon^d \tag{1}$$

where $\|A\|$ is the norm of the second order tensor A (in components $\|A\|^2 = A_{ij}A_{ij}$) and $M = \min\left[\left\|\varepsilon^d\right\|, \frac{C}{\sqrt{2}}\right]$, C being a positive constant.

For $\left\|\varepsilon^d\right\| \leq \frac{C}{\sqrt{2}}$, $M = \left\|\varepsilon^d\right\|$ and the constitutive equation is the usual constitutive equation of an isotropic elastic material, λ and μ_d being its Lamé constants.

The constitutive equation (1) can be written:

$$\sigma = (\lambda + \mu_d)\,\mathrm{tr}\varepsilon\mathbb{I} + \Sigma\frac{\varepsilon^d}{\left\|\varepsilon^d\right\|}$$

with $\Sigma = 2\left(\mu_m\left\|\varepsilon^d\right\| + (\mu_d - \mu_m)\,M\right)$. The coefficient μ_m being assumed to be negative, the plot of $\frac{\Sigma}{2}$ versus $\left\|\varepsilon^d\right\|$ is:

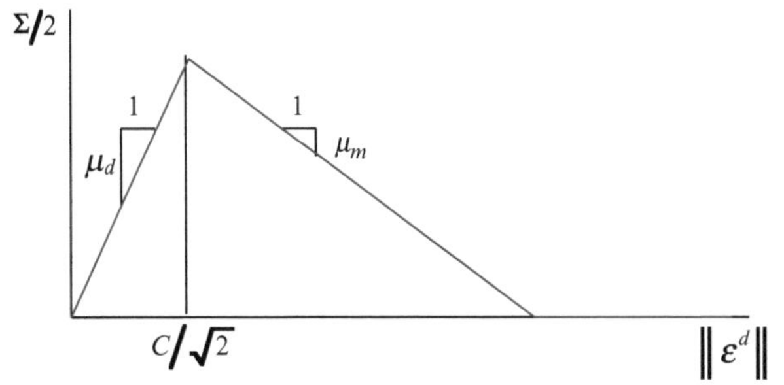

The relation $\|\varepsilon^d\| \to \Sigma$ which is the deviatoric strain stress relation for a fixed deviatoric strain direction $\frac{\varepsilon^d}{\|\varepsilon^d\|}$ is the same as that considered in Chambon et al. (1998) for 1D problems. Σ is clearly a continuous function of $\|\varepsilon^d\|$, so σ is a continuous function of ε.

In the following Λ denotes $\lambda + \mu_d$, so the constitutive equation (1) reads:

$$\sigma = \Lambda \mathrm{tr}\varepsilon \mathbb{I} + 2\left(\mu_m + (\mu_d - \mu_m)\frac{M}{\|\varepsilon^d\|}\right)\varepsilon^d \tag{2}$$

and it is assumed that $0 < \Lambda + \mu_m$.

3 Axisymmetric Problem and Elementary Solutions

We consider the deformation at equilibrium of a circular annulus of inner aud outer radii R_i and R_e made of a material the constitutive law of which is (2). The inner and outer circles of the annulus are respectively subject to the pressures p_i and p_e, no surface force density is applied to the annulus. As no orthoradial forces are applied on the boundaries, the orthoradial displacement v is defined up to a constant, to fix this constant, v is set to zero on the inner circle. The solution of the problem is looked for in the form:

$$\mathbf{u} = u\mathbf{e}_r + v\mathbf{e}_\theta \text{ with } \partial_\theta u = \partial_\theta v = 0 \tag{3}$$

r and θ are the polar coordinates of the space variable $\mathbf{x}$ and $\mathbf{e}_r$ and $\mathbf{e}_\theta$ are the associated basis vectors. From (3), we successively deduce that:

$$\varepsilon = \partial_r u \mathbf{e}_r \otimes \mathbf{e}_r + \frac{1}{2}\left(\partial_r v - \frac{v}{r}\right)(\mathbf{e}_\theta \otimes \mathbf{e}_r + \mathbf{e}_r \otimes \mathbf{e}_\theta)$$
$$+\frac{u}{r}\mathbf{e}_\theta \otimes \mathbf{e}_\theta \text{ and } \mathrm{tr}\varepsilon = \partial_r u + \frac{u}{r}$$

$$\varepsilon^d = \frac{1}{2}\left(\left(\partial_r u - \frac{u}{r}\right)(\mathbf{e}_r \otimes \mathbf{e}_r - \mathbf{e}_\theta \otimes \mathbf{e}_\theta) + \left(\partial_r v - \frac{v}{r}\right)(\mathbf{e}_r \otimes \mathbf{e}_\theta + \mathbf{e}_\theta \otimes \mathbf{e}_r)\right)$$

$$\left\|\varepsilon^d\right\|^2 = \frac{1}{2}\left(\left(\partial_r u - \frac{u}{r}\right)^2 + \left(\partial_r v - \frac{v}{r}\right)^2\right)$$

The carrying of those relations in the constitutive equation (2) yields the stress field σ in the form:

$$\sigma = \sigma_{rr}\mathbf{e}_r \otimes \mathbf{e}_r + \sigma_{r\theta}\left(\mathbf{e}_\theta \otimes \mathbf{e}_r + \mathbf{e}_r \otimes \mathbf{e}_\theta\right) + \sigma_{\theta\theta}\mathbf{e}_\theta \otimes \mathbf{e}_\theta$$

with:

$$\sigma_{rr} = \Lambda\left(\partial_r u + \frac{u}{r}\right) + \left(\mu_m + (\mu_d - \mu_m)\frac{M}{\left\|\varepsilon^d\right\|}\right)\left(\partial_r u - \frac{u}{r}\right)$$

$$\sigma_{\theta\theta} = \Lambda\left(\partial_r u + \frac{u}{r}\right) - \left(\mu_m + (\mu_d - \mu_m)\frac{M}{\left\|\varepsilon^d\right\|}\right)\left(\partial_r u - \frac{u}{r}\right)$$

$$\sigma_{r\theta} = \left(\mu_m + (\mu_d - \mu_m)\frac{M}{\left\|\varepsilon^d\right\|}\right)\left(\partial_r v - \frac{v}{r}\right)$$

Due to the fact that the components of the stress do not depend on θ, the equilibrium equation $\mathrm{div}\sigma = 0$ reads:

$$\left(\partial_r \sigma_{rr} + \frac{1}{r}\left(\sigma_{rr} - \sigma_{\theta\theta}\right)\right)\mathbf{e}_r + \left(\partial_r \sigma_{r\theta} + \frac{2}{r}\sigma_{r\theta}\right)\mathbf{e}_\theta = 0$$

The equation $\partial_r \sigma_{r\theta} + \frac{2}{r}\sigma_{r\theta} = 0$ for $\sigma_{r\theta}$ integrates into $\sigma_{r\theta} = \frac{K}{r^2}$. That proves, together with the possible interface conditions on circles between softening and hardening parts of the material, that $\sigma_{r\theta}$ is continuous. Therefore, due to the boundary conditions ($\sigma_{r\theta} = 0$ for $r = R_i$ and $r = R_e$):

$$\sigma_{r\theta} = 0$$

That entails that either $\mu_m + (\mu_d - \mu_m)\frac{M}{\left\|\varepsilon^d\right\|} = 0$, which can hold true only for $\frac{C}{\sqrt{2}} \leq \left\|\varepsilon^d\right\|$ and yields $\left\|\varepsilon^d\right\| = \frac{\mu_d - \mu_m}{-\mu_m}\frac{C}{\sqrt{2}}$, or $\partial_r v - \frac{v}{r} = 0$ and $v = Dr$.

Let's call elementary solutions the displacement fields that are solutions of the equation $\partial_r \sigma_{rr} + \frac{1}{r}\left(\sigma_{rr} - \sigma_{\theta\theta}\right) = 0$ in some internal annulus of the studied domain. Three cases are to be considered, $\left\|\varepsilon^d\right\| \leq \frac{C}{\sqrt{2}}$ (and necessarily $v = Dr$), $\frac{C}{\sqrt{2}} \leq \left\|\varepsilon^d\right\|$ with $v = Dr$ and lastly $\left\|\varepsilon^d\right\| = \frac{\mu_d - \mu_m}{-\mu_m}\frac{C}{\sqrt{2}}$.

Type 1 elementary solution – case $\left\|\varepsilon^d\right\| \leq \frac{C}{\sqrt{2}}$.

This is the usual case of linear elasticity, we have $u = A_1 r + \frac{B_1}{r}$, $v = D_1 r$ and $\sigma_{rr} = 2\Lambda A_1 - 2\mu_d \frac{B_1}{r^2}$, A_1, B_1 and D_1 being integration constants to be determined by interface or boundary conditions. In that case $\left\|\varepsilon^d\right\| = \frac{1}{\sqrt{2}}\left|\partial_r u - \frac{u}{r}\right| = \sqrt{2}\frac{|B_1|}{r^2}$ and this solution is valid in a annulus $R_b \leq r \leq R_e$ if $2\frac{|B_1|}{R_b^2} \leq C$.

Type 2 elementary solution – case $\frac{C}{\sqrt{2}} \leq \left\|\varepsilon^d\right\|$ and $v = Dr$.

In that case we have $u = -\hat{C} s\, r \ln\frac{r}{A_2} + \frac{B_2}{r}$, $v = D_2 r$ and $\sigma_{rr} = -\Lambda \hat{C} s \ln\frac{r^2}{A_2^2} - 2\mu_m \frac{B_2}{r^2}$ A_2, B_2 and D_2 being integration constants and where $s = \operatorname{sgn}\left(\partial_r u - \frac{u}{r}\right)$ (sgn denotes the sign function) and $\hat{C} = \frac{\mu_d - \mu_m}{\Lambda + \mu_m} C$ (due to the hypothesis $0 < \Lambda + \mu_m$, $\hat{C}$ is positive). We too have $\left\|\varepsilon^d\right\| = \frac{1}{\sqrt{2}}\left|\partial_r u - \frac{u}{r}\right| = \sqrt{2}\left(\frac{|B_2|}{r^2} - \hat{C}\right)$ and this solution is valid in a annulus $R_b \leq r \leq R_e$ if $C\frac{\Lambda + \mu_d}{\Lambda + \mu_m} \leq \frac{2|B_2|}{R_e^2}$.

Type 3 elementary solution – case $\left\|\varepsilon^d\right\| = \frac{\mu_d - \mu_m}{-\mu_m}\frac{C}{\sqrt{2}}$.

We have $u = A_3 r + \frac{B_3}{r}$, $v = \pm \int \frac{1}{r^3}\left(\frac{(\mu_d - \mu_m)^2}{\mu_m^2} C^2 r^4 - 4B_3^2\right)^{1/2} \mathrm{d}r + D_3 r$ and $\sigma_{rr} = 2\Lambda A_3$. A_3, B_3 and D_3 being integration constants. In that case $\left\|\varepsilon^d\right\| = \frac{\mu_d - \mu_m}{-\mu_m}\frac{C}{\sqrt{2}}$ and the validity condition of this solution comes down to the positivity of $\frac{(\mu_d - \mu_m)^2}{\mu_m^2} C^2 r^4 - 4B_3^2$, in a annulus $R_b \leq r \leq R_e$, it reads $2\frac{|B_3|}{R_b^2} \leq \frac{\mu_d - \mu_m}{-\mu_m} C$.

4 Solutions of the Boundary Value Problem

The elementary solutions of Sect. 3 satisfy the bulk equations (equilibrium and constitutive equation) but they still comprise integration constants. To form solutions of the boundary value problem, let's consider a succession of the elementary solutions defined in a succession of concentric annuli and apply the boundary conditions as well as continuity conditions for the displacement and the radial stress σ_{rr} at the interfaces between the concentric annuli. Let's first consider the classical type 1 solution in the whole annulus ($R_i \leq r \leq R_e$). It is obviously purely radial ($D_1 = 0$) and the other integration constants A_1 and B_1 are determined using the boundary conditions . According to Sect. 3 this solution is valid for $|p_i - p_e| \leq C\mu_d \frac{R_e^2 - R_i^2}{R_e^2}$.

For $C\mu_d \frac{R_e^2 - R_i^2}{R_e^2} < |p_i - p_e|$, the previous classical solution is no more valid since in points near the inner circle, the norm $\|\varepsilon^d\|$ of the deviatoric part of the strain exceeds the threshold $\frac{C}{\sqrt{2}}$. We then consider a solution made of two elementary solutions, one of type 2, the other of type 1 in two concentric annuli. The six corresponding integration constants are determined using the boundary and the interface continuity conditions. It turns out that solution is purely radial ($v = 0$) and that the validity conditions for the two elementary solutions entail that, in the inner annulus ($R_i \leq r \leq R_1$), the solution is of type 2 and of type 1 in the outer annulus ($R_1 \leq r \leq R_e$), the radius of the interface between the two annuli being solution of the transcendental equation:

$$\begin{aligned}&\left|(\Lambda + \mu_m)\,\mu_d\,R_i^2\left(R_e^2 - R_1^2\right) + (\Lambda + \mu_d)\,\mu_m\,R_e^2\left(R_1^2 - R_i^2\right)\right| C \\ &= (\Lambda + \mu_m)\,R_i^2 R_e^2 \left|p_e - p_i - 2\Lambda\hat{C}s\ln\frac{R_1}{R_i}\right|\end{aligned}$$

the solving of which can only be performed numerically.

The previous solution is the only purely radial solution in the case $C\mu_d \frac{R_e^2 - R_i^2}{R_e^2} < |p_i - p_e|$. That does not mean that the solution is unique, indeed bifurcation occurs in the form of non radial solutions, therefore showing shear bands. To prove that, we consider a solution made of a succession of a type 2, a type 3 and a type 2 elementary solutions in three concentric annuli. Due to the same validity conditions as for the purely radial solution, the type 2 solution has to take place in a more inner annulus than that of the type 1 solution. Here we consider a solution which is of type 2 in the inner annulus ($R_i \leq r \leq R_1$), of type 3 in the intermediate annulus ($R_1 \leq r \leq R_2$) and of type 2 in the outer annulus ($R_2 \leq r \leq R_e$).

Due to the condition $v(R_i) = 0$ on the orthoradial displacement, the type 2 solution in the inner annulus is purely radial but the orthoradial displacement v is necessarily different from 0 in the type 3 intermediate annulus and, by continuity condition, so it is in the type 1 outer annulus. The intermediate annulus can be seen as a shear band. Contrary to the case of the purely radial solution, the validity conditions of the three elementary solutions do not sum up in an equality but in a set of inequalities for the integration constant B_2 of the type 2 solution:

$$CR_1^2 \leq 2\,|B_2| \leq \min\left(\frac{\Lambda}{\Lambda + \mu_d}\,\frac{\mu_d - \mu_m}{-\mu_m}\,CR_1^2, CR_2^2\right)$$

That shows that, most probably, the values of the two radii R_1 and R_2 are not completely determined but are bounded by some inequalities. This is consistent with the expected fact that the width of the shear band (the intermediate annulus ($R_1 \leq r \leq R_2$) is not determined as there is no internal length in the modeling.

5 Conclusion

Considering the deformation of a hardening/softening "elastic" material in a 2D annulus submitted to inner and outer pressures, which enables the determination of analytical solutions, it has been proved that, past the softening threshold, the purely radial solution remains unique but that shear banding may occur. The analytical solutions with shear banding have been determined bringing possible benchmark for numerical simulation. It has also been confirmed that the shear bands have no determined length, as expected owing to the lack of any intrinsic length in the modeling.

References

B. Budiansky, A reassment of deformation theories of plasticity. J. Appl. Mech. **26**, 259–264 (1959)

R. Chambon, D. Caillerie, N. El Hassan, One dimensional localization studied with a second grade model. Eur. J. Mech. A/Solids **17**(4), 637–656 (1998)

R. Chambon, S. Crochepeyre, R. Charlier, An algorithm and a method to search bifurcation points in non-linear problems. Int. J. Numer. Methods Eng. **51**, 315–332 (2001)

J. Desrues, Tracking strain localization in geomaterials using computerized tomography. in *Proceedings of the International Workshop on X-ray CT for Geomaterials Kumamoto, Japan, November 6–7 2003* (Balkema, Lisse, 2004), pp. 15–41

J. Desrues, G. Viggiani, Strain localization in sand : an overview of the experimental results obtained in Grenoble using stereophotogrammetry. Int. J. Numer. Anal. Methods Geomech. **28**(4), 279–321 (2004)

J. Rice, The localization of plastic deformation, in *International Congress of Theoretical and Applied Mechanics*, ed. by W.D. Koiter (North Holland, Amsterdam, 1976)

J.W. Rudnicki, J.R. Rice, Conditions for the localization of deformation in pressure-sensitive dilatant materials, J. Mech. Phys. Solids **23**, 371–394 (1975)

5 Conclusion

Shear Banding in Cross-Anisotropic Sand Tests with Stress Rotation

Poul V. Lade

Abstract Shear banding in Santa Monica Beach sand deposited by dry pluviation in hollow cylinder specimens is studied in drained torsion shear tests with rotation of principal stress directions. Each test was conducted with the same, constant inside and outside confining pressure, σ_r, thus tying the value of $b = (\sigma_2 - \sigma_3)/(\sigma_1 - \sigma_3)$ to the inclination, β, of the major principal stress. Shear bands can develop freely without significant restraint for soft rubber membranes. Strain localization and shear banding were observed in the hollow cylinder specimens, and this created failure conditions in plane strain and in tests with higher b-values. The results indicate the influence of the cross-anisotropic fabric on the stress-strain behavior, on the shear band inclination and on the shape of the failure surface.

Keywords Cross-anisotropy • Principal stress rotation • Sand • Shear band • Torsion shear tests

1 Introduction

Development of shear bands under three-dimensional conditions have been investigated under conditions of no stress rotation, and their occurrence have been compared with the type of instability that may lead to liquefaction (Lade 2002). Further, the experimentally observed shear banding was successfully predicted for these three-dimensional conditions without stress rotation (Lade 2003). Both experiments and predictions showed that shear banding has an important influence on the three-dimensional shape of the failure surface. Thus, peak failure is caused by

P.V. Lade (✉)
Department of Civil Engineering, The Catholic University of America,
Washington, DC 20064, USA
e-mail: Lade@cua.edu

S. Bonelli et al. (eds.), *Advances in Bifurcation and Degradation in Geomaterials*, Springer Series in Geomechanics and Geoengineering 11,
DOI 10.1007/978-94-007-1421-2_37,

shear banding in the hardening regime in the approximate range of $b = (\sigma_2 - \sigma_3)/(\sigma_1 - \sigma_3)$ from 0.18 to 0.85, while it occurs in the softening regime outside this range of b-values. Therefore, a smooth, continuous 3D failure surface is not generally obtained for granular materials.

The true triaxial experiments were performed in the first sector of the octahedral plane, and they were predicted using an isotropic elasto-plastic constitutive model. Thus, neither the experiments nor the predictions involved any serious or strong effects of cross-anisotropy, which usually emerge in the second and are strongest in the third sector of the octahedral plane (Abelev and Lade 2004).

Presented here are the results of a study of the behavior of pluviated sand in hollow cylinder specimens exposed to torsion shear tests in which the principal stress directions are rotated during shearing to failure. Shear banding was observed in the hollow cylinder specimens, and this created failure conditions in plane strain and in tests with higher b-values. The results clearly indicate the influence of the cross-anisotropic fabric on the shape of the failure surface.

A series of true triaxial tests was also performed on the sand in the first sector of the octahedral plane in which the effect of cross-anisotropy is not pronounced. The friction angles obtained from tests with and without stress rotations are compared to show the effects of cross-anisotropy on failure.

2 Torsion Shear Apparatus

The experiments on hollow cylinder specimens presented here were performed in the torsion shear apparatus presented by Lade (1981). The torsion shear apparatus was designed to provide individual control of confining pressure, vertical deviator stress and shear stress applied to the hollow cylinder specimen. Both the vertical deviator stress and the torsion shear stress can be either stress- or strain-controlled. The specimen has inside and outside diameters of 18 and 22 cm (wall thickness = 2 cm), and experiments were performed on specimens with heights of 40 and 25 cm. The inside and outside pressures were maintained at the same value in all tests presented here. For this condition the value of $b = (\sigma_2 - \sigma_3)/(\sigma_1 - \sigma_3)$ is related to the angle, β, between vertical and the direction of the major principal stress, σ_1:

$$b = \sin^2 \beta \tag{1}$$

Thus, it may be difficult to distinguish the effect of the intermediate principal stress from the effect of stress rotation. However, an attempt will be made to separate the effects on the basis of results from true triaxial tests performed in a cubical triaxial apparatus.

3 Sand Tested

All tests were performed on the portion of Santa Monica Beach sand passing the No. 40 sieve (sieve opening = 0.425 mm). The specific gravity was 2.66, the coefficient of uniformity was 1.58, and the maximum and minimum void ratios were 0.91 and 0.60. The particle shapes were angular to subangular. Specimens were prepared by dry pluviation with a void ratio of 0.68 corresponding to a relative density of 70%.

4 Torsion Shear Tests

A total of thirty four drained torsion shear tests were performed. Twenty six tests were performed on the tall specimens with height of 40 cm. Fourteen of the twenty six tests were performed to investigate the effect of stress rotation on the soil behavior during primary loading or small stress reversals and eight tests were performed to investigate the soil behavior during large stress reversals. Eight tests on short specimens with height of 25 cm were performed to investigate the effect of the specimen-height on the soil behavior in hollow cylinder specimens.

Additional details regarding the testing program and the performance of the torsion shear tests as well as the true triaxial tests have been presented by Lade et al. (2008, 2009).

5 Failure Criterion

In order to investigate the strength characteristics observed in the torsion shear tests, the strengths have been compared internally and with those from the true triaxial compression tests. The observed strengths were also compared with those predicted by an isotropic three-dimensional failure criterion for soils.

For this purpose an isotropic three-dimensional failure criterion expressed in terms of stress invariants has previously been developed for frictional material, such as clay, sand, concrete and rock (Lade 1977). This isotropic failure criterion is expressed in terms of the first and the third stress invariants of the stress tensor as follows:

$$\left(\frac{I_1^3}{I_3} - 27\right)\left(\frac{I_1}{p_a}\right)^m = \eta_1 \tag{2}$$

in which

$$I_1 = \sigma_z + \sigma_r + \sigma_\theta = \sigma_1 + \sigma_2 + \sigma_3 \tag{3}$$

$$I_3 = \sigma_z \sigma_r \sigma_\theta - \sigma_r \tau_{z\theta} \tau_{\theta z} \tag{4}$$

and p_a is atmospheric pressure expressed in the same units as the stresses. The parameters η_1 and m are constants. Results of conventional triaxial compression tests on Santa Monica Beach sand were used to determine the parameters η_1 and m.

6 Comparison of Experimental Data and Failure Criterion

The variation of the measured effective stress friction angle φ with the value of b is shown in Fig. 1 for the true triaxial tests and the torsion shear tests. The friction angles from torsion shear tests were obtained from tall and short hollow cylinder specimens. The friction angles obtained from the true triaxial tests all correspond to the first sector in the octahedral plane, thus indicating essential isotropic behavior.

The thick solid line in Fig. 1 represents the isotropic failure surface given by Eq. 2. Considering the variety of test equipment employed for testing, the isotropic failure surface fits well with the experimental points from the true triaxial tests, except in the midrange of b-values, where the shear banding influences the failure condition (Lade 2002, 2003).

Figure 1 also shows that the variation of the friction angles with b-value obtained from torsion shear and true triaxial tests are almost the same from triaxial compression ($b = 0$) to plane strain conditions (near $b = 0.3$). The thin solid line is a trend line is a trend line for the failure conditions in the torsion shear tests. Here the friction angles decreased from plane strain to triaxial extension conditions. This line, although drawn through the upper friction angles, is substantially different from

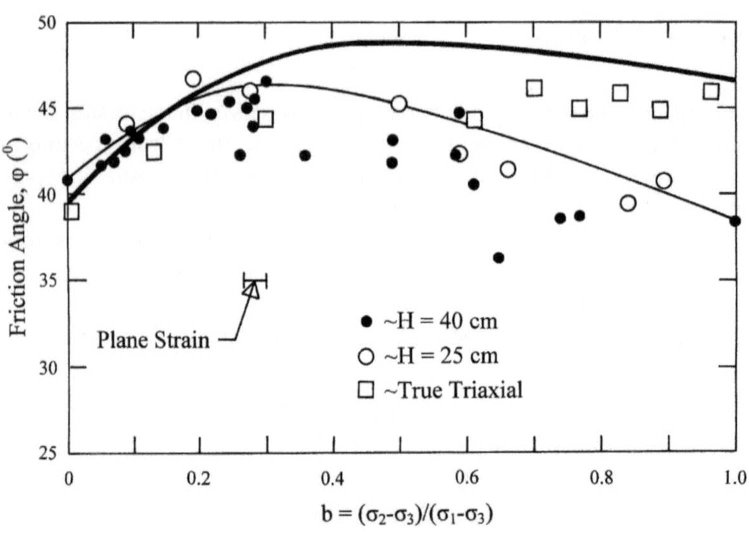

Fig. 1 Variation of friction angles obtained from torsion shear tests on Santa Monica Beach sand with void ratio of 0.68 corresponding to a relative density of 70%

the thick solid line at higher b-values. This is due to the cross-anisotropic behavior of the pluviated sand deposit and the shear banding that occurs in this range of b-values. Effects of shear banding in the hardening regime for b-values beyond the plane strain condition result in lower friction angles than indicated by the thin trend line. In comparison, the friction angles in true triaxial tests increase slightly until $b = 0.8$ followed by a small decrease close to $b = 1.0$. The largest difference between measured friction angles of approximately 7° occurs in triaxial extension where $b = 1.0$.

7 Occurrence of Shear Banding

As mentioned in the Introduction, shear banding occurs in the hardening regime in the mid-range of b-values in true triaxial tests, resulting in abrupt decrease in load-carrying ability and clearly visible effects on the stress-strain and volume change relations. In triaxial compression tests with $b = 0.0$ and in tests with b-values up to just before plane strain, shear banding occurs in the softening regime and has no influence on the measured friction angle. The friction angle steadily increases in this range of b-values and relatively little scatter is observed in this range. This is also true for the torsion shear tests, as seen in Fig. 1.

Judging from the variation in friction angle in Fig. 1 and the knowledge of occurrence of shear bands in true triaxial tests as confirmed by analysis (Lade 2003), the effect of shear banding on the measured strength begin near the plane strain condition (where b is approximately 0.3) and is present in the mid-range of b-values. Similar effects appear to be present in the torsions shear tests. Thus, the friction angles in the mid-range of b-values may be affected by shear banding. But the difference in friction angle of 7° observed at $b = 1.0$ is likely the effect of cross-anisotropy. However, this may be further studied by analysis similar to that performed for the true triaxial tests (Lade 2003). Note that cross-anisotropy causes the friction angle in triaxial extension, $\varphi_e = 38.4°$ at $\theta = 180^0$, to be several degrees lower than the friction angle in triaxial compression, $\varphi_c = 40.8°$. Thus, there is no reason to believe that the friction angle in extension should be equal to the friction angle in compression, as is occasionally asserted.

8 Shear Band Inclination

The shear band inclinations observed in the torsion shear tests are compared with theoretical values. Based on force equilibrium, Coulomb's theory states that failure occurs at the point of maximum obliquity, and the inclination of shear bands therefore coincides with the inclination of planes on which the ratio of shear to normal stress reaches its maximum value. The angle between the shear band and the direction of minor principal stress is then calculated from:

$$\theta_C = 45^0 + \varphi/2 \tag{5}$$

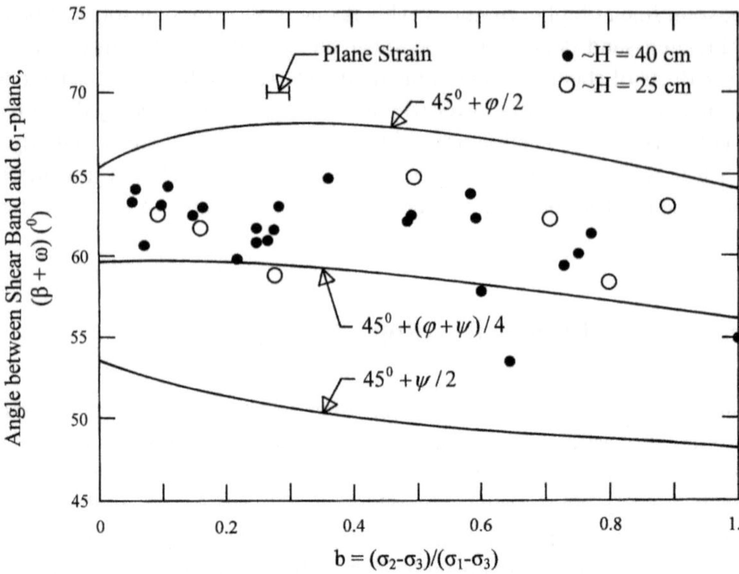

Fig. 2 Comparison of experimental shear band inclinations with three theoretical values

Roscoe (1970) emphasized the importance of strains and suggested that shear bands correspond to lines of zero extension. This resulted in shear band inclinations being expressed in terms of the angle of dilation, ψ:

$$\theta_R = 45^0 + \psi/2 \tag{6}$$

Arthur and Dunstan (1982) found that shear bands were inclined between the Coulomb and Roscoe directions:

$$\theta_A = 45^0 + (\varphi + \psi)/4 \tag{7}$$

In the latter two expressions, the angle of dilation is defined as the ratio of the plastic volumetric strain rate to the maximum plastic shear strain rate (Roscoe 1970):

$$\sin \psi = -\frac{\dot{\varepsilon}_1 + \dot{\varepsilon}_3}{\dot{\varepsilon}_1 - \dot{\varepsilon}_3} = -\frac{\dot{\varepsilon}_v/\dot{\varepsilon}_1}{2 - \dot{\varepsilon}_v/\dot{\varepsilon}_1} \tag{8}$$

The angles of dilation were determined from the torsion shear tests, and together with the friction angles in Fig. 1 used to calculate the theoretical inclinations according to the three expressions above. Figure 2 shows that the experimental shear band inclinations decrease with increasing *b*-values. This is likely because the horizontal bedding planes in the sand deposit act as attractor to the shear bands. Thus, the cross-anisotropy has an influence on the direction of the shear bands in the torsion shear tests.

In comparison, the experimental results in Fig. 2 shows the shear band inclinations determined from the three expressions above. It is seen that the experimental results are located between the Coulomb and the Arthur expressions.

9 Conclusions

A series of drained torsion shear tests on hollow cylinder specimens of Santa Monica Beach Sand have been performed along various stress paths for the purpose of investigating the strength during rotation of principal stress axes. The strength observed in torsion shear tests was compared with the measured strength in true triaxial tests performed in the first sector of the octahedral plane in which the sand tends to behave as an isotropic material.

In addition to the influence of the intermediate principal stress, the experimental results from the torsions shear tests showed that the strength was influenced by cross-anisotropy and by shear banding.

The effect of cross-anisotropy was clearly evident with increasing inclinations of the σ_1-direction. Cross-anisotropy was most strongly indicated by comparison of the friction angles in extension. In the first sector of the octahedral plane at $\theta = 60°$ ($b = 1.0$) the friction angle was 7° higher than in the third sector at $\theta = 180°$ ($b = 1.0$).

Superimposed on the effect of cross-anisotropy was the effect of shear banding, which also appeared in the same range of high b-values beyond plane strain conditions where $b = 0.3$. Shear banding initiated before smooth peak failure had been reached and resulted in further decreases in the friction angles. In comparison, the friction angles measured in true triaxial test remained almost constant beyond plane strain conditions.

References

A.V. Abelev, P.V. Lade, Characterization of Failure in Cross-Anisotropic Soils. J. Eng. Mech. ASCE **130**(5), 599–606 (2004)

J.R.F. Arthur, T. Dunstan, Rupture layers in granular materials, in *IUTAM Symposium Deformation and Failure of Granular Materials* (Balkema, Delft, 1982), pp. 453–459

P.V. Lade, Elasto-plastic stress-strain theory for cohesionless soil with curved yield surfaces. Int. J. Solids Struct. (Pergamon, New York, 1977) **13**, 1019–1035 (1977)

P.V. Lade, Torsion shear apparatus for soil testing, in *Laboratory Shear Strength of Soil*, ASTM STP 740 (American Society for Testing and Materials, Philadelphia, 1981), pp. 145–163

P.V. Lade, Instability, shear banding, and failure of granular materials. Int. J. Solids Struct. **39**(13–14), 3337–3357 (2002)

P.V. Lade, Analysis and prediction of shear banding under 3D conditions in granular materials. Soils Found. **43**(4), 161–172 (2003)

P.V. Lade, J. Nam, W.P. Hong, Shear banding and cross-anisotropic behavior observed in laboratory sand tests with stress rotation. Can. Geotech. J. **45**(1), 74–84 (2008)

P.V. Lade, J. Nam, W.P. Hong, Interpretation of strains in torsion shear tests. Comput. Geotech. **36**(1–2), 211–225 (2009)

K.H. Roscoe, The influence of strains in soil mechanics. Geotechnique **20**(2), 129–170 (1970)

Instability of Unsaturated Soil During the Water Infiltration

F. Oka, E. Garcia, and S. Kimoto

Abstract It is known that the unstable behavior of unsaturated soil includes collape behavior which is due to the decrease of suction as well as the shear failure. In addition, we often meet the numerical instability for the simulation of unsaturated soil during the wetting process. In the present study, we have performed a one-dimensional linear instability analysis for the air-water-soil coupled three-phase viscoplastic material. From the instability analysis of the viscoplastic unsaturated soil subjected to the water infiltration, we have found that the specific moisture capacity, the matric suction and the hardening parameter affect the instability. Then, we have performed numerical simulations of the behavior of unsaturated soil during the water infiltration using a multi-phase coupled elasto-viscoplastic finite element analysis method. The conditions under which the numerical instability occurs during the simulation agree well with the results by linear instability analysis.

Keywords Instability • Unsaturated soil • Numerical analysis

1 Introduction

The failure of soil structures can be triggered by a wetting process from an unsaturated stage resulting from an increase in moisture content and a decrease in suction. Jennings and Burland (1962) conducted a series of consolidation tests and showed that partly saturated soil upon wetting undergoes additional settlement

F. Oka (✉) · S. Kimoto
Department of Civil and Earth Resources Engineering, Kyoto University, Kyoto, Japan
e-mail: oka.fusao.2s@kyoto-u.ac.jp; kimoto.sayuri.6u@kyoto-u.ac.jp

E. Garcia
Department of Civil and Earth Resources Engineering, Kyoto University, Kyoto, Japan

On leave from the Department of Civil Engineering, University of Antioquia, Colombia

S. Bonelli et al. (eds.), *Advances in Bifurcation and Degradation in Geomaterials*, Springer Series in Geomechanics and Geoengineering 11,
DOI 10.1007/978-94-007-1421-2_38,

or "collapses"; this phenomenon is commonly referred to as collapse behavior. To study this behavior in unsaturated soil, several constitutive models have been developed (e.g., Alonso et al. 1990; Oka et al. 2006; Wheeler and Sivakumar 1995).

The instability of saturated porous media has been widely studied by many researchers from both experimental and analytical points of view. However, studies on the instability of unsaturated porous materials have not been completed. Many experimental and numerical researches have been conducted on the deformation behavior of unsaturated soil (e.g., Kimoto et al. 2010; Oka et al. 2010). Nevertheless, the deformation of unsaturated soil is an interesting topic that has not yet been fully investigated. And theoretical analyses, such as instability analyses, have not yet been performed.

In the present study, we have theoretically and numerically analyzed the effect of parameters and state variables on the unstable behavior of unsaturated materials when they are subjected to water infiltration. For the numerical analysis a seepage-deformation coupled method and an elasto-viscoplastic model (Oka et al. 2006) were employed. The consistency between the results obtained by the theoretical analysis and the results obtained by the numerical analysis is shown.

2 Governing Equations and One-Dimensional Instability Analysis

2.1 Governing Equations

In the followings, the governing equations for the one-dimensional infiltration problem are described. Stress variables are defined as follows:

$$\sum_{\alpha} \sigma^{\alpha} = \sigma, \quad (\alpha = S, W, G) \quad \mathrm{S} = \mathrm{Soil}, \quad \mathrm{W} = \mathrm{Water}, \quad \mathrm{G} = \mathrm{Gas}. \tag{1}$$

$$\sigma^S = \sigma' - n^S P^F; \quad \sigma^W = -n^W P^W; \quad \sigma^G = -n^G P^G \tag{2}$$

where n^{α} is a volume fraction of α phase, P^F is the average pressure of the fluid surrounding the soil skeleton, and it is given as a function of the saturation as

$$P^F = sP^W + (1 - s)\, P^G \tag{3}$$

A simplified constitutive model is used in this analysis. The stress-strain relation can be written as

$$\sigma' = H\varepsilon + \mu\dot{\varepsilon} \tag{4}$$

where ε is the strain, the superimposed dot denotes the differentiation with respect to time, H is the strain hardening-softening parameter, which is a function of suction P^C $(P^G - P^W)$, and μ is the viscoplastic parameter.

The equilibrium and continuity equations are given, respectively, by

$$\frac{\partial \sigma}{\partial x} = \frac{\partial \sigma'}{\partial x} - \frac{\partial P^F}{\partial x} = 0 \tag{5}$$

$$s\dot{\varepsilon} + n\dot{s} = -\frac{\partial V^W}{\partial x}; \quad (1-s)\,\dot{\varepsilon} - n\dot{s} + n\,(1-s)\,\frac{\dot{\rho}_G}{\rho_G} = -\frac{\partial V^G}{\partial x} \tag{6}$$

where the fluid flows V^W and V^G are described by the Darcy's law. Details of the governing equations are described in Garcia et al. (2011).

2.2 Linear Instability Analysis

A linear instability analysis is conducted on an unsaturated material in order to examine which constitutive parameters and conditions lead to the onset of a growing instability during a wetting process. The perturbation of pore water pressure, pore air pressure, and strain in periodic form for the one-dimensional infiltration problem is considered for the governing equations of the seepage-deformation coupled method. From the theoretical analysis we obtain the matrix form of the perturbed equilibrium and the continuity equations as

$$\begin{bmatrix} -\left(A\varepsilon + BP^C + s\right) & \left(A\varepsilon + BP^C + s - 1\right) & (H + \mu\omega) \\ -\left(B\dot{\varepsilon} + Bn\omega - q^2\frac{k^W}{\gamma_W}\right) & (B\dot{\varepsilon} + Bn\omega) & (s\omega) \\ (B\dot{\varepsilon} + Bn\omega) & -\left(B\dot{\varepsilon} + Bn\omega - q^2\frac{k^G}{\gamma_G}\right) & (1-s)\,\omega \end{bmatrix} \begin{Bmatrix} P^{W*} \\ P^{G*} \\ \varepsilon^* \end{Bmatrix}$$

$$= [A]\{y\} = \{0\} \tag{7}$$

For nonzero values of P^{W*}, P^{G*}, and ε^*, the determinant of matrix [A] has to be equal to zero. From $\det[A] = 0$, we have the factors for a polynomial function of the growth rate of the fluctuation ($\omega^2 + \alpha_1\omega + \alpha_2 = 0$) as

$$\begin{aligned} \alpha_1 = \Big[q^2 \big\{&\gamma_W k^G s^2 + \left(A\varepsilon + BP^C\right)\left(\gamma_W k^G s - \gamma_G k^W (1-s)\right) \\ &+ \gamma_G k^W \left(s^2 - 2s + 1\right) - B\left(\gamma_W k^G + \gamma_G k^W\right)(nH + \dot{\varepsilon}\mu) + k^W k^G \mu q^2\big\} \\ &- B\dot{\varepsilon}\gamma_W\gamma_G\Big] > 0 \end{aligned} \tag{8}$$

$$\alpha_2 = \left\{-B\dot{\varepsilon}\left(\gamma_W k^G + \gamma_G k^W\right) + k^G k^W q^2\right\} q^2 H > 0 \tag{9}$$

Adopting the Routh-Hurwitz criteria, we can say that the material system is stable when the factors α_1 and α_2 are larger than zero. On the other hand, the material system is unstable when these factors are negative. The possibility for α_1 and α_2 to be negative in Eqs. 8 and 9 depends on:

1. Large $B(\partial s/\partial P^C)$. B is negative and corresponds to the slope of the soil water characteristic curve (moisture capacity). When B increases, term BP^C in α_1 becomes negative more easily, namely, the potential for instability is higher. In addition, if strain rate $\dot{\varepsilon}$ is negative, terms $-B\dot{\varepsilon}$ included in α_1 and α_2 become negative.
2. Large Suction (P^C). If P^C increases, term BP^C becomes more negative; consequently, α_1 becomes negative more easily.
3. Large A $(\partial H/\partial P^C)$ and negative strain $\varepsilon < 0$. A is positive and corresponds to the slope of the relation between the hardening-softening parameter and the suction. When A increases while the strain is negative, term $A\varepsilon$ is more negative. Consequently, there is a possibility for α_1 to become more negative.

From the theoretical analysis, it can be said that the onset of the instability of a material in a viscoplastic state mainly depends on the level of suction, the moisture capacity, the rate of variation of the hardening-softening parameter with respect to the suction and the strain. In the next section, results of various simulations of the one-dimensional infiltration problem will be presented in order to study the material instability by a seepage-deformation coupled method.

3 Numerical Analysis

Weak forms of the continuity equations for water and air and the rate type of conservation of momentum are discretized in space and solved by the finite element method. In the finite element formulation, the independent variables are the pore water pressure, the pore air pressure, and the nodal velocity. An eight-node quadrilateral element with a reduced Gaussian integration is used for the displacement, and four nodes are used for the pore water pressure and the pore air pressure. The backward finite difference method is used for the time discretization. An elasto-viscoplastic model is used for the soil skeleton (Oka et al. 2006).

The finite element mesh and the boundary conditions for the simulations are shown in Fig. 1. A homogeneous soil column with a depth of 1 m is employed in the simulations. The top of the column is subjected to a pore water pressure equal to 4.9 kPa. The simulations of the infiltration process start from an unsaturated condition where initial suction ${P^C}_i$ is the same along the column. At $t = 0$, water starts to infiltrate due to the pore water pressure applied at the top until a hydrostatic condition is attained in the column. Different values for the van Genuchten parameters (van Genuchten 1980); parameter α, from $\alpha = 0.1$ to $\alpha = 10$ (1/kPa), and parameter n', from $n' = 1.01$ to $n' = 9.0$, as well as two different levels of initial suction, ${P^C}_i = 25.5$ (Case 1) and 100 kPa (Case 2), were considered to study the instability of the unsaturated material system.

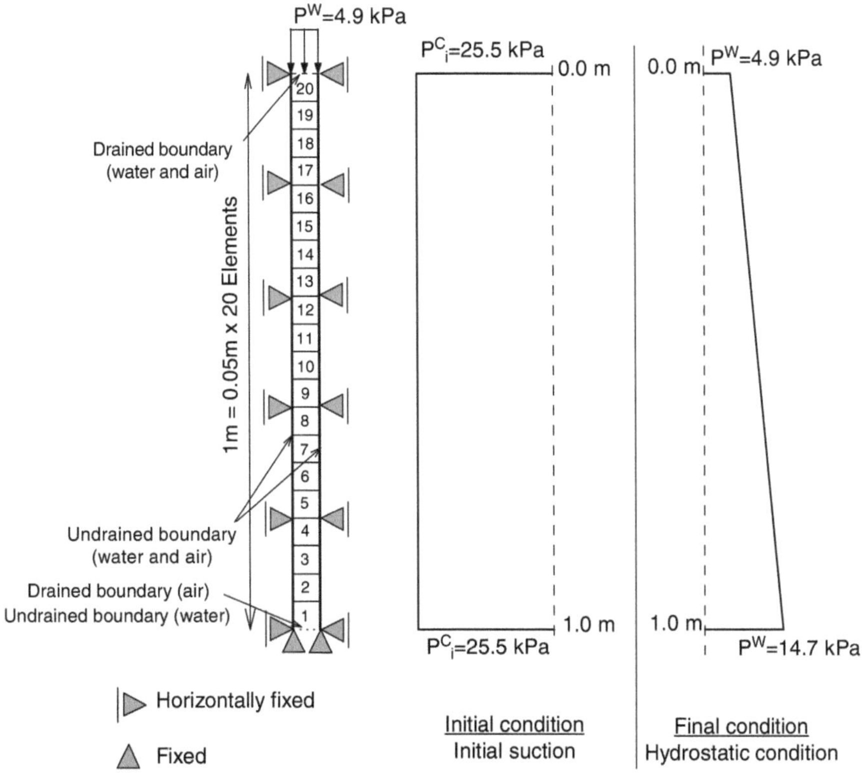

Fig. 1 Finite element mesh and boundary conditions

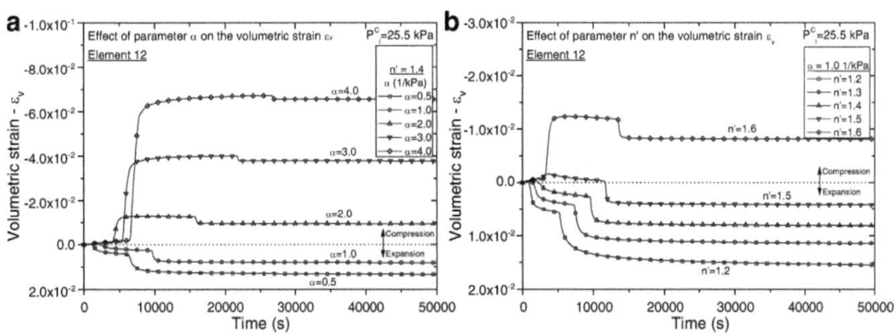

Fig. 2 Influence of parameter α on B and the development of volumetric strain

3.1 Numerical Results

Figures 2a, b show the influence of parameters α and n' on the development of volumetric strain ε_v for Element 12 (in Fig. 1) when the material is subjected to

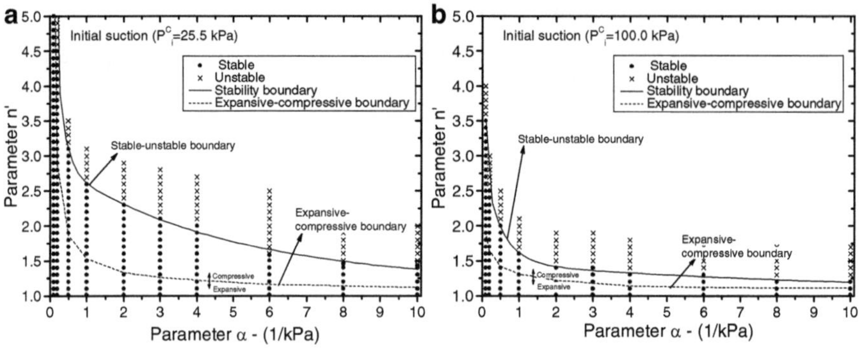

Fig. 3 Stable and unstable region in $\alpha - n'$ space

infiltration from an initial suction, $P^C{}_i = 25.5\,\text{kPa}$. From these figures, it is seen that the volumetric strain changes from expansive to compressive when parameters α and n' increases. The increase in parameters α and n' means an increase in B (slope of the soil water characteristic curve) and an increase in the compressive behavior of the soil, and therefore, the potential for instability.

The simulation results are presented in Fig. 3a, b for the $\alpha - n'$ space and for two different levels of initial suction, namely, $P^C{}_i = 25.5$ and $100\,\text{kPa}$, respectively. In these figures, an stable result is shown by a solid circle, whereas x indicates an unstable result. The boundaries between expansive-compressive behavior and stable-unstable results are also shown.

The following characteristic can be drawn from the figures:

1. The expansive behavior is presented for material with smaller parameters α and n', but the compressive behavior is presented when these parameters increases.
2. The potential for instability increases when the parameters α and n' are large. Larger parameters α and n' leads to the increases in B and the increase of the compressive behavior of the soil.
3. The expansive-compression and the stable-unstable boundaries shrink when suction increases, namely, when suction is increased from 25.5 to 100 kPa.
4. The instability obtained by the numerical results is consistent with the onset of instability obtained by the theoretical analysis.

4 Conclusions

We have conducted a theoretical and a numerical one-dimensional infiltration analysis in order to analyze the material parameters and the conditions that lead to the onset of instability in an unsaturated material. A seepage-deformation method based on the theory of porous media and an elasto-viscoplastic model were used. The results show that the instabilities obtained by the numerical analysis are consistent with the onset of instability obtained by the theoretical analysis.

References

E.E. Alonso, A. Gens, A. Josa, A constitutive model for partially saturated soils. Geotechnique **40**, 405–430 (1990)

E. Garcia, F. Oka, S. Kimoto, Numerical analysis of a one-dimensional infiltration problem in unsaturated soil by a seepage-deformation coupled method. Int. J. Numer. Anal. Methods Geomech. **35**, 544–568 (2011). doi:10.1002/nag908

J.E. Jennings, J.B. Burland, Limitations to the use of effective stresses in partly saturated soils. Geotechnique, **12**, 125–144 (1962)

S. Kimoto, F. Oka, T. Fushita, A chemo–thermo–mechanically coupled analysis of ground deformation induced by gas hydrate dissociation. Int. J. Mech. Sci. **52**, 365–376 (2010)

F. Oka, T. Kodaka, S. Kimoto, Y.-S. Kim, N. Yamasaki, An elasto-viscoplastic model and multiphase coupled FE analysis for unsaturated soil, in *Proceedings of the 4th International Conference on Unsaturated Soils*, Geotechnical Special Publication, No. 147, vol. 2 (ASCE, Reston, 2006), pp. 2039–2050

F. Oka, T. Kodaka, H. Suzuki, Y.S. Kim, N. Nishimatsu, S. Kimoto, Experimental study on the behavior of unsaturated compacted silt under triaxial compression. Soils Found. **50**, 27–44 (2010)

M.T. van Genuchten, A closed form equation for predicting the hydraulic conductivity of unsaturated soils. Soil Sci. Soc. Am. J. **44**, 892–898 (1980)

S.J. Wheeler, V. Sivakumar, An elasto-plastic critical state framework for unsaturated soil. Geotechnique **45**, 35–53 (1995)

A Revisiting of Undrained Shear Banding in Water-Saturated Sand

Peijun Guo

Abstract The paper revisits the analysis of undrained shear banding in water saturated sand. By taking into account the coupling between potential volume change and excess pore water pressure as well as the continuity of stresses and pore pressure across the shear band, it is theoretically shown that incipient localisation may take place in loose sand but is precluded for dense sand under undrained conditions. The undrained shear band orientation primarily depends on the Poisson's ratio and the dilatancy characteristics of the material. Numerical examples are given to demonstrate the variation of shear band inclination with void ratio and the initial consolidation pressure.

Keywords Shear band • Incipient localization • Saturated sand • Isochoric constraint • Dilatancy

1 Introduction

Numerous researches, both analytical and experimental, have been carried out to investigate the emergence and development of shear band in granular materials under drained or undrained conditions (i.e., isochoric constraint). The theoretical work is mainly orientated to the analysis of deformation instability, which may evolve into either diffuse or localized deformation modes (e.g., Rice and Rudnicki 1980; Vardoulakis 1985; Vardoulakis and Sulem 1995). While generally good agreements are reached among researchers regarding shear band under drained conditions, inconsistence for undrained shear band still largely exists.

P. Guo (✉)
Department of Civil Engineering, McMaster University, Hamilton, ON, Canada, L8S 4L7
e-mail: guop@mcmaster.ca

S. Bonelli et al. (eds.), *Advances in Bifurcation and Degradation in Geomaterials*, Springer Series in Geomechanics and Geoengineering 11,
DOI 10.1007/978-94-007-1421-2_39,

In their pioneering work, Han and Vardoulakis (1991) observed shear banding in saturated dense and medium dense sand, but no localization in loose sand in undrained strain controlled tests. More recent experimental study by Khoa et al. (2006) confirmed that bifurcation did not give any localized failure for loose sand in undrained tests. On the other hand, Finno et al. (1997) and Mooney et al. (1997) observed shear band in loose specimens using stereophotogrammetry, which helped in identification of the bands that were not apparent by visual observation. Mokni and Desrues (1998) reviewed the strain localization in undrained plane-strain biaxial tests on Hostun RF sand. They observed shear banding in both contractive and dilative specimens, but for the latter the onset of localization was delayed until cavitation took place in the pore-fluid. They concluded that the isochoric constraint imposed to deformation was too restrictive in the case of dense dilative sand for shear banding to become the dominant deformation mode. Even though they did not disagree with the theoretical prediction by Vardoulakis (1996a) that localization could not take place because it would imply shock waves developing in pore water, they still proposed that the isochoric constraint did not preclude localization of loose contractive sand.

Inconsistence also exists in the literature in terms of the theoretical analysis for undrained shear band. Vardoulakis (1996a) suggested that undrained shear banding was physically not possible since the prerequisite for this phenomenon was the formation of pore-water pressure shocks across the shear band boundaries; or otherwise localization could only occur under local drained conditions (Vardoulakis 1985, 1996b). Pietruszczak (1995) showed that the bifurcation analysis under undrained constraint predicted a homogeneous deformation with no localization when incorporating undrained compliance, unless the inhomogeneous deformation mode (i.e., discontinuous localization) was introduced. Zhang and Schrefler (2001) showed that undrained shear band under plane strain conditions occurred with inclination angle of $\pm 45°$ regardless of the adopted constitutive model (associative or non-associative).

Motivated by the above review, this paper re-visits the strain localization analysis for saturated granular materials subjected to isochoric constraint under biaxial stress conditions. By taking into account the coupling between potential volume change and excess pore water pressure as well as the continuity of stresses and pore pressure across the shear band, it is identified theoretically that incipient localization may take place in loose sand but is precluded for dense sand under undrained conditions. Numerical examples show that the shear band inclination angle is affected by void ratio and the initial consolidation pressure.

2 Incipient Shear Band Under Isochoric Constraint

The following definitions are recalled for biaxial stress conditions for later use and negative stresses refer to compression. Let $\sigma = \sigma_{kk}/2$, $p = -\sigma$, $s_{ij} = \sigma_{ij} - \sigma\delta_{ij}$, $\tau = \sqrt{s_{ij}s_{ij}/2}$, $\varepsilon_v = \varepsilon_{kk}$, $\gamma^p = \sqrt{2e^p_{ij}e^p_{ij}}$, $\mu = \sin\varphi_m = \tau/p$ and

$\beta = \sin\psi_m = d\varepsilon_v^p/d\gamma^p$. Here s_{ij} and e_{ij}^p represent deviator stress and deviator plastic strain tensors, respectively. Also repeated indices which vary from 1 to 2 refer to summation. The Mohr-Coulomb model is selected as the constitutive model, for which the yield function and plastic potential are expressed as

$$F = \tau + \mu\sigma,\ Q = \tau + \beta\sigma \tag{1}$$

The mobilized friction angle φ_m and the angle of dilatancy ψ (and hence μ and β) are both assumed to be functions of appropriate hardening variables. In order to carry out a stability analysis on the selected constitutive model, one should first establish the incremental stress-strain equations $\dot{\sigma}_{ij} = D_{ijkl}\,\dot{\varepsilon}_{kl}$ with the fourth order constitutive tensor $D_{ijkl} = D_{ijkl}^e + D_{ijkl}^p$. Following the standard exercise in the theory of elasto-plasticity (Vardoulakis and Sulem 1995), the plastic constitutive tensor is give by $D_{ijkl}^p = -(\langle 1\rangle/H)(\mathbf{D} : \partial Q/\partial\boldsymbol{\sigma}) \otimes (\mathbf{D} : \partial F/\partial\boldsymbol{\sigma})$, where H is the plastic modulus $H = H_t + (\partial F/\partial\boldsymbol{\sigma}) : \mathbf{D} : (\partial Q/\partial\boldsymbol{\sigma})$, and the switch function $<1> = 1$ if loading of the yield function $F = 0$ is taking place, and $<1> = 0$ otherwise. The hardening modulus H_t is defined as $H_t = -\partial F/\partial\zeta$ with ζ being appropriate hardening parameter(s). For the yield and plastic functions given in Eq. 1, one has

$$D_{ijkl}^p = -\frac{\langle 1\rangle}{H} G\left(s_{ij}/\tau + \alpha\beta\delta_{ij}\right)\left(s_{ij}/\tau + \alpha\mu\delta_{kl}\right) \tag{2}$$

where $H = 1 + h + \alpha\beta\mu$, $h = ph_t/G$, $h_t = d\mu/d\gamma^p$, $\alpha = 1/(1-2\upsilon)$ with υ being Poisson's ratio of the material.

The bifurcation condition for shear banding is expressed in terms of jump conditions for the stress increments as $[\Delta\pi_{ij}]n_j = 0$ with $\Delta\pi_{ij}$ being the first P-K stress increment, with $[\Delta\varepsilon_{ij}] = (g_i n_j + g_j n_i)/2$ and $[\Delta\omega] = (g_2 n_1 - g_1 n_2)/2$ across the boundary of the shear band. The static compatibility condition for total stresses is then expressed as

$$\begin{bmatrix} D_{1111}n_1^2 + (1-\xi)n_2^2 & (D_{1122} + 1 + \xi)n_1 n_2 \\ (D_{1122} + 1 - \xi)n_1 n_2 & (1+\xi)n_1^2 + D_{2222}n_2^2 \end{bmatrix} \begin{Bmatrix} g_1 \\ g_2 \end{Bmatrix} - [\Delta p_w] \begin{Bmatrix} n_1 \\ n_2 \end{Bmatrix} = 0 \tag{3}$$

where $\xi = (\sigma_1 - \sigma_2)/(2G)$ is the normalized stress difference. By applying the isochoric constraint $\Delta\varepsilon_v = 0$ or $g_1 n_1 + g_2 n_2 = 0$, Eq. 3 becomes:

$$\begin{aligned} &G\left\{[(D_{1111} - D_{1122}) - (1+\xi)]\,n_1^2 + (1-\xi)n_2^2\right\} g_1 + [\Delta p_w]\,n_1 = 0 \\ &G\left[-(1+\xi)n_1^2 + (D_{2211} - D_{2222} + 1 - \xi)n_2^2\right] g_1 \quad + [\Delta p_w]\,n_2 = 0 \end{aligned} \tag{4}$$

The existence of non-trivial solution $g_1 \neq 0$ with both the continuity conditions for stress and pore pressure being satisfied requires

$$\frac{n_1^2}{n_2^2} = \frac{1-\xi}{1 + \xi + D_{1122} - D_{1111}} = \frac{D_{2211} - D_{2222} + 1 - \xi}{1+\xi} \tag{5}$$

which yields the following expression (Vardoulakis, 1996a, b):

$$(1+\xi)\frac{n_1^2}{n_2^2} + (1-\xi)\frac{n_2^2}{n_1^2} = 2\left(1 - \frac{2G^*}{G}\right) \tag{6}$$

with G^* being the tangent modulus. The localization condition can be readily determined from Eq. 5 as

$$h_B = -\alpha\beta_B \left(\mu_B + \frac{\alpha\beta_B + \xi}{1 + \alpha\beta_B \xi} \right) \tag{7}$$

Herein the subscript B refers to quantities at the onset of localization. The shear band inclination and the tangent shear modulus at the onset of localization are

$$\tan^2 \theta_B = \left(-\frac{n_1}{n_2} \right)^2 = \frac{(1 - \alpha\beta_B)(1 - \xi)}{(1 + \alpha\beta_B)(1 + \xi)}; \ \frac{G_B^*}{G} = -\frac{\alpha\beta_B (\alpha\beta_B + \xi)}{1 - \alpha^2 \beta_B^2} < 0 \tag{8}$$

As a special case, if $h_B \rightarrow 0$ when localization occurs, one obtains from Eq. 7 $\mu_B \rightarrow -(\alpha\beta_B + \xi)/(1 + \alpha\beta_B\xi)$. It follows according to Eq. 8:

$$\tan^2 \theta_B = \frac{(1 + \mu_B)(1 + \xi)}{(1 - \mu_B)(1 - \xi)} \tag{9}$$

which recovers Coulomb-type shear band angle with $\theta_B = \pi/4 + \varphi_B/2$ when neglecting the effect of initial stresses (i.e. $\xi = 0$).

The following conclusions can be drawn according to the above analyses:

1. When $\alpha^2\beta^2 < 1$, under isochoric conditions, strain localization is possible only during deviator stress softening when $G_B^* \leq 0$. In other words, undrained strain localization can only take place in contractive materials with $\beta < 0$, which in turn implies $h_B = -\alpha\beta_B(\mu_B + \alpha\beta_B) > 0$ or friction angle hardening at the onset of localization.
2. Equation 6 has real roots the first time when $(1 - 2G^*/G)^2 = 1 - \xi^2$, which corresponds to $G_B^* = 0$, $h_B = -\alpha\beta\mu$ and $\theta_B = \pm 45°$ when $\xi = 0$. However, a non-zero pore pressure jump is obtained from Eq. 4 as $[\Delta p] = \sqrt{2}G\,(\alpha\beta + \xi)\,g_1$ (Vardoulakis 1996a), which implies that shear band cannot physically occur at $G_B^* = 0$. In other words, localization does not take place until Eq. 5 is satisfied even though Eq. 6 has real roots.
3. For dense dilative sand with $\beta > 0$, the requirement of $G_B^* < 0$ cannot be satisfied. Consequently, localization is excluded under isochoric constraint.
4. Localization at $G_B^* = 0$ is possible only when $\beta = 0$ with $h = 0$ when neglecting the geometric effect, which implies that localization may occur at or close to the state of phase transformation as defined by Ishihara (1993). The strain localization condition is also satisfied at the critical state, however, large strain required to mobilize the critical state may lead to diffuse deformation prior to localization mode.
5. For saturated loose sand, the orientation of shear band could be very close to Coulomb-type angle $\theta_B = \pi/4 + \varphi_B/2$.

3 Numerical Examples

Figure 1 presents the influence of density and initial consolidation pressure on shear band inclination under strict isochoric constraint. Figure 1a, b summarize the experimental results by Mokni and Desrues (1998) and Mooney et al. (1997), respectively, while Fig. 1c, d show the results of simulation using a simple stress-strain model for granular materials (Wan and Guo 1999). While no attempts are made to compare the simulation results with the experimental data, the simulations yield the same trend as the experimental data regarding the dependency of undrained shear band orientation on stress level and density.

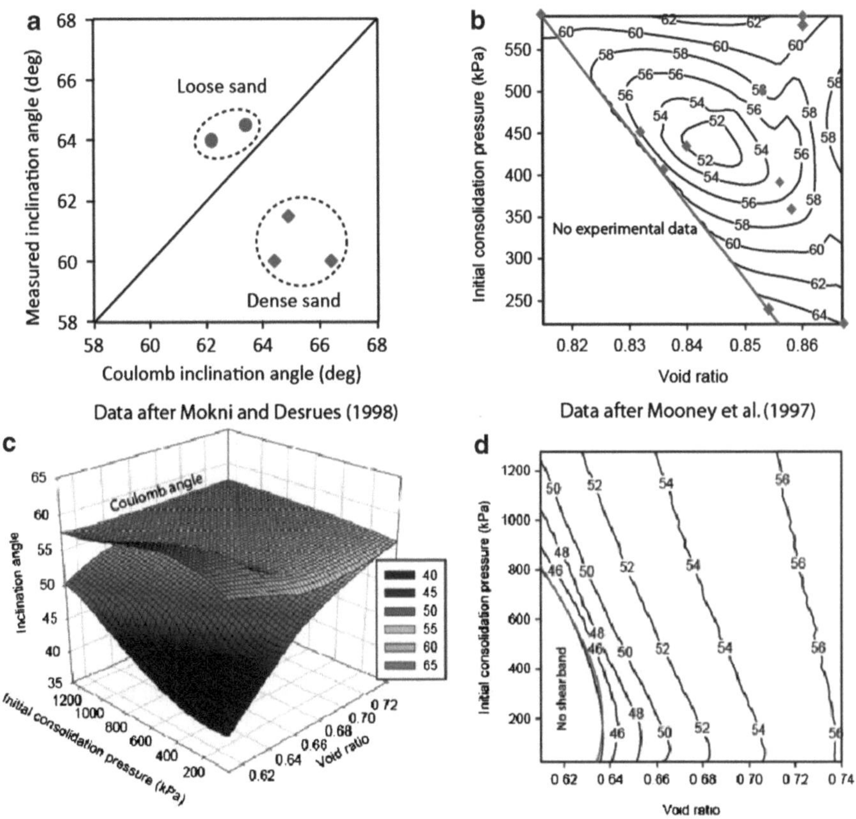

Fig. 1 Dependency of shear band orientation on density and stress level

4 Conclusions

The following conclusions are obtained from the revisiting of the analysis for incipient shear band in water saturated sand under undrained conditions:

(a) Undrained shear band is possible in loose, contractive sand with restrict isochoric constraint, and the onset of the shear band occurs in the regime of deviator stress softening;
(b) The inclination angle of undrained band in contractive sand depends on the Poisson's ratio and the angle of dilatancy of the material. For very loose sand, the inclination angle is close to the Mohr-Coulomb angle; and
(c) Shear banding in saturated dense, dilative sand is precluded by restrict isochoric constraint

The above conclusions are generally consistent with experimental observations by Finno et al. (1995), Mooney et al. (1997), and Mokni and Desrues (1998). It should be mentioned that, for dense dilative sand, localization may take place under local drained condition (Han and Vardoulakis 1991; Mokni and Desrues 1998). The re-examination of this case shows that the onset of shear band will be in the regime of friction angle softening if pore fluid cavitation is avoided.

Acknowledgments Funding provided by the Natural Sciences and Engineering Research Council of Canada is gratefully acknowledged.

References

R.J. Finno, W.W. Harris, M.A. Mooney, G. Viggiani, Shear bands in plane strain compression of loose sand. Geotechnique **47**(1), 149–165 (1997)

C. Han, I. Vardoulakis, Plane strain compression experiments on water-saturated fine-grained sand. Geotechnique **41**(1), 49–78 (1991)

K. Ishihara, Liquefaction and flow failure during earthquakes. Geotechnique **43**(3), 351–415 (1993)

H.D.V. Khoa, F. Darve, I.O. Georgopoulos, Diffuse failure in geomaterials: experiments and modelling. Comput. Geotech. **33**(1), 1–14 (2006)

M. Mokni, J. Desrues, Strain localisation measurements in undrained plane-strain biaxial tests on Hostun RF sand. Mech. Cohesive Frictional Mater. **5**(2), 419–441 (1998)

M.A. Mooney, G. Viggiani, R. J. Finno, Undrained shear band deformation in granular media. J. Eng. Mech. ASCE **123**(6), 577–585 (1997)

S. Pietruszczak, Undrained response of granular soil involving localized deformation. J. Eng. Mech. **121**(12), 1292–1297 (1995)

J.R. Rice, J.W. Rudnicki, A note on some features of the theory of localization of deformation. Int. J. Solids Struct. **16**, 597–605 (1980)

I. Vardoulaksi, Stability and bifurcation of undrained, plane rectilinear deformations on water-saturated granular soils. Int. J. Numer. Anal. Methods Geomech. **9**, 399–414 (1985)

I. Vardoulakis, Deformation of water-saturated sand: I. uniform undrained deformation and shear banding. Geotechnique **46**(3), 441–456 (1996a)

I. Vardoulakis, Deformation of water saturated sand: II. Effect of pore water flow and shear banding. Geotechnique **46**(3), 457–472 (1996b)

I. Vardoulakis, J. Sulem, *Bifurcation Analysis in Geomechanic* (Chapman & Hall, London, 1985)

R.G. Wan, P.J. Guo, A pressure and density dependent dilatancy model for granular materials. Soils Found. **39**(6), 1–12 (1999)

H.W. Zhang, B.A. Schrefler, Uniqueness and localization analysis of elastic-plastic saturated porous media. Int. J. Numer. Anal. Methods Geomech. **25**(1), 29–48 (2001)

Strain Localization Conditions Under True Triaxial Stress States

Kathleen A. Issen, Mathew D. Ingraham, and Thomas A. Dewers

Abstract This work uses a bifurcation approach to develop theoretical predictions for deformation band formation for a suite of true triaxial tests on Castlegate sandstone. In particular, the influence of the intermediate principal stress on strain localization is examined. Using common simplifying assumptions (localization occurs at peak stress, and the failure surface is similar to the yield surface), theoretical predictions captured the overall trends observed experimentally. However, agreement between predicted and observed band orientations for individual specimens was varied. This highlights the importance of detailed data analyses to accurately determine key material parameter values at the inception of localization.

Keywords Sandstone • Bifurcation • True triaxial compression • Deformation band • Theoretical predictions

1 Introduction

Deformation band formation is a common failure mode for porous sandstones in field settings and in laboratory experiments. Rudnicki and Rice (1975) proposed a theoretical framework, based on bifurcation theory, to predict conditions required to form these bands. They found a relationship between the deformation band type (compaction, compactant shear, isochoric shear, dilatant shear, or dilation), and the

K.A. Issen (✉) · M.D. Ingraham
Clarkson University, 8 Clarkson Ave #5725, Potsdam, NY 13699, USA
e-mail: issenka@clarkson.edu; ingrahmd@clarkson.edu

T.A. Dewers
Sandia National Laboratories, PO Box 5800, Albuquerque, NM 87185-0751, USA
e-mail: tderwers@sandia.gov

S. Bonelli et al. (eds.), *Advances in Bifurcation and Degradation in Geomaterials*, Springer Series in Geomechanics and Geoengineering 11,
DOI 10.1007/978-94-007-1421-2_40,

stress conditions required to produce it. The band orientation and strain type inside the band is predicted to depend on the relationship of the intermediate principal stress to the maximum and minimum compressive stresses. In the companion paper (Ingraham et al. 2011), we discuss a suite of true triaxial tests designed to evaluate these theoretical predictions, using Castlegate sandstone, a porous reservoir analog. Tests covered a wide range of mean stresses, including brittle, ductile and transitional behavior. The present paper examines theoretical predictions using preliminary results from these experiments.

2 Theoretical Background

Rudnicki and Rice (1975) suggested that the inception of a band of localized strain could be modeled as a bifurcation from homogeneous deformation, due to a constitutive instability. Thus, these localization predictions depend on the assumed constitutive model. Since the current paper utilizes preliminary findings from recently completed experimental work, a simple single yield surface model, such as that utilized by Rudnicki and Rice (1975) will be employed. The constitutive equations can be derived via classical plasticity, assuming non-associated flow, where the plastic strain increment is perpendicular to a plastic potential (Holcomb and Rudnicki 2001). For now, the yield surface and plastic potential are both assumed to depend on only the first invariant of stress through the mean stress, $\sigma = -\sigma_{kk}/3$, and the second invariant of deviatoric stress through the Mises equivalent shear stress, $\tau = (s_{ij}s_{ij})^{1/2}$. The deviatoric stress is $s_{ij} = \sigma_{ij} - \sigma_{kk}/3$. More complex models, perhaps including dependence on the third invariant of deviatoric stress, will be considered after fully examining the constitutive response of the material.

Using this pressure dependent constitutive relation, Rudnicki and Rice (1975) determined the predicted band angle (angle between the band normal and the direction of maximum compression) to be (Rudnicki and Olsson 1998)

$$\theta = \frac{\pi}{4} + \frac{1}{2}\arcsin\left[\frac{\frac{2}{3}(1+\nu)(\beta+\mu) - N_{II}(1-2\nu)}{\sqrt{4-3N_{II}^2}}\right], \tag{1}$$

where the friction factor, μ, and the dilation coefficient, β, are the local slopes of the yield surface and plastic potential, respectively. Poisson's ratio is ν, and $N_{II} = (\sigma - \sigma_2)/\tau$ is a parameter representing the relationships between the three principal stresses, $\sigma_1 \geq \sigma_2 \geq \sigma_3$ (positive in compression). From the N_{II} term, the predicted band angle increases as σ_2 increases from σ_3 to σ_1, for fixed values of the material parameters (Issen and Challa 2008). Values for N_{II} for the five experimental stress states are shown in Table 1, along with the corresponding Lode angles.

Table 1 Relationships between principal stresses for five N_{II} values and the corresponding Lode angles

N_{II}	Lode angle	Relationship between principal stresses
$\frac{1}{\sqrt{3}}$	$+30°$	$\sigma_2 = \sigma_3$
$\frac{1}{2\sqrt{3}}$	$+14.5°$	$\sigma_2 = \frac{1}{2}\left[\sigma_1\left(1-\frac{1}{\sqrt{5}}\right)+\sigma_3\left(1+\frac{1}{\sqrt{5}}\right)\right]$
0	$0°$	$\sigma_2 = \frac{1}{2}(\sigma_1+\sigma_3)$
$-\frac{1}{2\sqrt{3}}$	$-14.5°$	$\sigma_2 = \frac{1}{2}\left[\sigma_1\left(1+\frac{1}{\sqrt{5}}\right)+\sigma_3\left(1-\frac{1}{\sqrt{5}}\right)\right]$
$-\frac{1}{\sqrt{3}}$	$-30°$	$\sigma_2 = \sigma_1$

The above equation for the predicted shear band angle is valid when

$$-\frac{3\left(N_I+\nu N_{II}\right)}{1+\nu} \leq \beta+\mu \leq -\frac{3\left(N_{III}+\nu N_{II}\right)}{1+\nu}. \tag{2}$$

Compaction bands, $\theta = 0°$, are predicted when the left side of the inequality is false, while dilation bands, $\theta = 90°$ are predicted when the right side is false. Localization theory also predicts the critical value of the hardening modulus, h_{cr}, at the inception of localization (h is the slope of the shear stress – plastic shear strain curve). The expression for h_{cr} for shear bands ($0° < \theta < 90°$) is given below, where G is the shear modulus for elastic unloading.

$$h_{cr} = G(1+\nu)\left\{\frac{(\beta-\mu)^2}{9(1-\nu)} - \frac{1}{2}\left[N_{II}+\frac{1}{3}(\beta+\mu)\right]^2\right\}. \tag{3}$$

3 Results

An initial assessment of localization predictions is now presented, using preliminary results from the experiments described in the companion paper (Ingraham et al. 2011). We focus on tests conducted at low mean stresses, $\sigma = 30$, 60 and 90 MPa, where strain localization was observed; at higher mean stresses, no localization occurred. Table 2 lists observed band angles, measured from the specimen surface. To compare observations with predictions from Eq. 1 requires values for ν, μ, and β. This requires a complete constitutive analysis, including partitioning of strains into elastic and plastic components, which is currently underway. In the interim, some useful insights can be obtained using common assumptions.

First, in their examination of Castlegate sandstone, Dewers et al. (2011), report that the slope of the yield surface, μ, at failure is similar to the slope of the failure surface. Therefore, using the failure surfaces from the companion paper (Ingraham et al. 2011), probable values for μ are provided in Table 2. Dewers et al. (2011) also

Table 2 Localization predictions for observed μ, $\nu = 0.2$, and $h = 0$

		Observed		Predicted			
Lode angle	σ (MPa)	θ	μ	β_1	θ_1	β_2	θ_2
+30°	30	59°	0.77	6.40	90°	−0.50	43°
	60	33°	0.35	4.53	90°	−0.59	36°
	90	[a]	−0.07	2.67	90°	−0.69	28°
+14.5°	30	[a]	0.81	5.09	90°	−0.15	50°
	60	59°	0.54	3.89	90°	−0.21	46°
	90	53°	0.27	2.69	90°	−0.27	42°
0°	30	73°	0.81	3.61	90°	0.18	57°
	60	64°	0.53	2.36	90°	0.12	53°
	90	58°	0.25	1.10	61°	0.06	49°
−14.5°	30	72°	0.69	1.57	90°	0.49	63°
	60	62°	0.53	0.85	66°	0.45	60°
	90	41°	0.36	0.42	57°	0.13	54°
−30°	30	65°	0.78	0.85	81°	0.50	71°
	60	70°	0.47	0.78	71°	−0.88	45°
	90	46°	0.16	0.71	63°	−2.27	20°

[a]No band observed on specimen surface

report values for Poisson's ratio; here we use an average of their values, $\nu = 0.20$. The dilation coefficient, β, is defined as the negative of the ratio of the increments of plastic volume and plastic shear strains; this requires the aforementioned strain partitioning, which is not yet complete. Instead, we note that a peak in the stress – strain curve is typically observed in specimens that form deformation bands (see, e.g., Fig. 2 of Ingraham et al. 2011). Because $h = 0$ at peak stress, Eq. 1 is set equal to zero and solved for β, in terms of ν, μ, and N_{II}. Substituting the appropriate N_{II} value with the assumed values of ν and μ provides two predicted values for β; these values are listed n Table 2, labeled β_1 and β_2. Using the two predicted β values, two predicted band angles, θ_1 and θ_2, are determined, also listed in Table 2.

As expected, both the observed band angle and the observed failure surface slope, μ, decrease with increasing mean stress, within a given Lode angle. The predicted β values decrease with increasing mean stress, corresponding to the commonly reported transition in material response from dilatant at lower mean stresses to compactant at higher mean stresses. The β_1 is more dilatant than normality ($\beta = \mu$), while β_2 is less dilatant/more compactant than normality. This leads to higher θ_1 band angles for β_1 values and lower θ_2 band angles for β_2. Overall, the trend in predicted band angle is similar to observed: band angles decrease with increasing mean stress. With few exceptions, the observed band angle falls between θ_1 and θ_2, with neither predicted angle within 5° of the observed, although for all tests at least one of the predicted angles is within 17° of observed. For most of the tests at Lode angles of +30° to −14.5°, θ_2 is closer to the observed angle, which corresponds to a β value that is less dilatant/more compactant than normality. Similar deviations from normality have been reported by others (e.g., Olsson 1999; Baud et al. 2006) for sandstone tests at $+30°$ Lode angle, where $\sigma_2 = \sigma_3$. However, for two of the tests at $-30°$ Lode angle, where $\sigma_2 = \sigma_1$, the θ_1 is closer to the observed angle, corresponding to a β value that is more dilatant than normality.

4 Discussion

Because the localization conditions determined above are based on preliminary results and some key assumptions, there are several possible reasons for the discrepancies between the predicted and observed band angles. The observed band angles were measured on the surface of the specimen after the test was concluded. It is possible that the band angle evolved during the test, such that the band angle at the inception of localization could be different. Fitting a plane through the localized acoustic emissions events (Dewers et al. 2011), would provide a more accurate observed band angle.

While strain localization is commonly reported to occur "at or near peak stress", corresponding to our assumption of $h = 0$, by locating acoustic emissions events, a more refined determination of the inception of strain localization is possible and an actual value of h could be determined. While the slope of the failure surface may be close to the slope of the yield surface (μ), localization predictions are sensitive to small changes in material parameter values. Completing the planned constitutive analysis, including partitioning of strains into elastic and plastic components, will enable determination of more accurate μ values, since the yield surface is defined as a contour of constant plastic shear strain. Similarly, actual values for the dilation coefficient, β, can be determined from the plastic volume and plastic shear strain increments at the onset of localization.

Actual values for the elastic shear modulus and Poisson's ratio can be determined by examining elastic unloading curves for each specimen. Using actual values for h, G, ν, μ and β, at the inception of localization will provide a more rigorous band angle prediction, for comparison to the observed band orientation determined by fitting a plane through the localized cloud of acoustic emissions events. It is anticipated that completion of these more thorough analyses of experimental results will lead to improved theoretical predictions.

5 Conclusions and Future Work

This work examined predictions for deformation band formation using preliminary experimental results from a suite of true triaxial tests on Castlegate sandstone. Predicted band orientations were determined using two common assumptions: (1) localization occurs at peak stress, and (2) slopes of the failure surface and yield surface are similar. Trends in both observed and predicted band angles were similar: angle decreases with increasing mean stress. However, predicted angles differed from observed by as much as 17°. Thus, predictions based on common simplifying assumptions capture experimentally observed trends, but do not match well with observations from individual specimens. This re-affirms the need for detailed data analyses to accurately determine material parameter values at the inception of localization. In future work, this will include locating acoustic emissions events to determine the onset and orientation of strain localization, and careful partitioning of strain into elastic and plastic components.

Acknowledgments The authors gratefully acknowledge funding from the National Science Foundation Award EAR-0711346 to Clarkson University.

Sandia is a multiprogram laboratory operated by Sandia Corporation, a Lockheed Martin Company, for the United States Department of Energy's National Nuclear Security Administration under contract DE-AC04–94AL85000.

References

P. Baud, V. Vajdova, T.-F. Wong, Shear-enhanced compaction and strain localization: inelastic deformation and constitutive modeling of four porous sandstones. J. Geophys. Res. **111**, B12401, (2006). doi:10.1029/2005JB004101

T.A. Dewers, K.A. Issen, D.J. Holcomb, W.A. Olsson, Strain localization, elastic-plastic coupling, and constitutive modeling of Castlegate sandstone experimental deformation. J. Geophys. Res. (2011) (in prep)

D.J. Holcomb, J.W. Rudnicki, Inelastic constitutive properties and shear localization in Tennessee marble. Int. J. Numer. Anal. Methods Geomech. **25**, 109–129 (2001)

M.D. Ingraham, K.A. Issen, D.J. Holcomb, Failure of Castlegate sandstone under true triaxial loading, S. Bonelli et al. (eds.), *Advances in Bifurcation and Degradation in Geomaterials*, Springer Series in Geomechanics and Geoengineering 11, DOI 10.1007/978-94-007-1421-2_40, ©Springer Science+Business Media B.V. (2011)

K.A. Issen, V. Challa, Influence of the intermediate principal stress on the strain localization mode in porous sandstone. J. Geophys. Res. **113**, B02103 (2008). doi:10.1029/2005JB004008

W.A. Olsson, Theoretical and experimental investigation of compaction bands in porous rock. J. Geophys. Res. **104**, 7219–7228 (1999)

J.W. Rudnicki, W.A. Olsson, Reexamination of fault angles predicted by shear localization theory. Int. J. Rock Mech. Min. Sci. **35**, 512–513 (1998)

J.W. Rudnicki, J.R. Rice, Conditions for the localization of deformation in pressure-sensitive dilatant materials. J. Mech. Phys. Solids **23**, 371–394 (1975)

Meso-Scale Evolution of Shear Localization Observed in Plane Strain Experiment on Kaolin Clay

Dunja Perić and Marte Gutierrez

Abstract Undrained plane strain compression experiments were conducted on slurry consolidated kaolin clay samples to elucidate the mechanism of shear localization in clays. The biaxial device was heavily internally instrumented, thus enabling the multiple boundary displacement measurements. These measurements were combined with boundary digital imaging to compute and verify displacements of the shear band. The orientation of shear band was determined from photographs thus enabling the computation of its dilatancy angle. Finally, the volumetric and shear strains of the shear band were computed for two limiting initial thicknesses of the shear band corresponding to 1 mm and 1.5 mm. The computed meso scale strains are several times larger than those reported in sand

Keywords Clay • Evolution • Meso-scale • Plane strain • Strain localization

1 Introduction

Strain localization is characterized by a sudden appearance of narrow zones, which are known as deformation bands. With continued loading large strains that develop inside deformation bands are accompanied with little or no strains outside. The onset and subsequent evolution of strain localization signifies a transition form a relatively homogeneously deforming amorphous soil mass to an increasingly structured medium. The internal structure comprises a number of nearly rigidly moving bodies

D. Perić (✉)
Kansas State University, Manhattan, KS 66506–5000, USA
e-mail: peric@ksu.edu

M. Gutierrez
Colorado School of Mines, Golden, CO 80401, USA
e-mail: mgutierr@mines.edu

S. Bonelli et al. (eds.), *Advances in Bifurcation and Degradation in Geomaterials*, Springer Series in Geomechanics and Geoengineering 11,
DOI 10.1007/978-94-007-1421-2_41, © Springer Science+Business Media B.V. 2011

that are bounded by deformation bands, thus ultimately resulting in a fragmented soil mass. It is because the soils around foundations, in slopes, excavations and embankments, earth dams and around tunnels experience such failure mechanisms under the conditions that are conducive to strain localization that this subject is of an enormous importance for the advancement of engineering practice. One of the simplest forms of strain localization takes place in a biaxial or plane strain apparatus, in which a homogeneous soil sample gradually becomes fragmented, usually comprising two nearly rigidly deforming bodies that are interconnected by an extensively deforming shear band. Although clayey soils are largely encountered in practice, a vast majority of experimental studies conducted in plane strain devices has been devoted to sands, whereby the grain scale processes have been successfully quantified to a large extent. The major goal of this study was to advance the understanding of strain localization in clays. The special emphasis herein is placed on quantifying the shear localization at the meso-scale.

2 Experimental Program

Undrained plane strain compression experiments were conducted on the one dimensionally, slurry consolidated kaolin clay. The liquid and plastic limits of the clay are 52% and 27% respectively while its specific gravity is 2.66. The 100% of particles by weight have an average diameter that is smaller than 0.0075 mm, with the mean particle size of about 0.00084 mm according to the results of the hydrometer test. The 140 mm × 40 mm × 80 mm prismatic clay samples were placed between lubricated and glass lined bottom and side platens, and subsequently subjected to a plane strain consolidation followed by an undrained shear to failure. The bottom platen rests on a linear sled bearing, which facilitates an uninhibited onset and evolution of strain localization. The device is heavily instrumented including seven internal linear variable differential transducers (LVDT), out of which one pair was used to measure the axial displacement, the additional two pairs to measure the horizontal displacements near the top and bottom of a clay sample, and the last LVDT to measure the sled displacement. More detailed description of the biaxial apparatus used in this study was provided in (Finno and Rhee 1993).

3 Evolution of Shear Localization at Meso Scale

Multiple boundary displacement measurements have been used to characterize the shear band development in sand (Han and Vardoulakis 1991) and soft rock (Labuz and Dai 2000). The essentially different particle sizes and shapes of clays in comparison to those of sands are bound to have significant effects on the grain scale processes, which are vitally important for strain localization due to the small thicknesses of shear bands. While the thickness of shear band in sands is about 10–20 times larger than the mean particle size (Mooney et al. 1998), the corresponding information for clays is scarce at best. According to (Hicher et al. 1994) the zone of large particle reorientation in kaolin and bentonite clays ranged from 2 to 7 mm. The same study suggested that the thickness of shear band was of the order of 1 mm, based on the density measurements by X-ray scanner.

Characterization of grain scale processes in clays presently remains elusive due to their extremely small particle sizes, which are three orders of magnitude smaller than sand particles. A shear band is a meso-scale structure because:

1. its thickness is at least one order of magnitude smaller than its width and length, the later being of the same order of magnitude as the entire biaxial sample, and
2. it interconnects the macro-scale, to the sub-meso scales.

Herein, the meso-scale characterization of strain localization is conducted based on the multiple boundary measurements by employing the following assumptions:

1. the shear band is inextensible, it deforms in a simple shear mode,
2. displacements of the shear band vary in a linear fashion along its thickness, but are constant along its length, thus rendering its deformation uniform,
3. after the shear band is fully formed the macro-scale axial and lateral displacements are nearly fully absorbed within the shear band, the outside material is either deforming as a rigid body or unloading elastically, and
4. the orientation of the shear band is fixed in time.

A slightly heavily overconsolidated plane strain sample is selected for further analysis. Lateral displacements of the portion of the biaxial sample that was not intersected by the shear band were extremely small resulting in about 0.15% of a possible decrease of the length of the shear band, thus confirming the assumption of the inextensible shear band. Several discrete measurements of the shear band orientation indicate the maximum change of 0.7° occurred with the progress of strain localization. The average orientation of 54.3° is used for the entire experiment to avoid making assumptions about the nature of the change of the shear band orientation.

The onset of strain localization was determined from multiple boundary measurements of lateral displacements as a point at which top and bottom widths of the sample started to differ. Figure 1 shows the deviator stress versus global axial strain response. A time lag of about 51 min was noted between the occurrence of the peak deviator stress and peak excess pore water pressure. This amount was determined based on the corresponding global vertical strain rate of 0.33% hour. Furthermore, clustering of various events around the peak deviator stress and peak excess pore water pressure was observed. The two events that are the closest to the peak deviator stress are the onset of strain localization and peak excess pore water pressure. The two events that are the closest to the peak excess pore water pressure are the peak deviator stress and point B. The later indicates a full formation of the shear band, which coincides with the commencement of the constant sled velocity. While both responses, the deviator stress and the excess pore water pressure appear to detect the strain localization, the later lags behind the former due to its remote location away from the shear band. Finally, point PPP indicates the attainment of the post peak plateau in deviator stress.

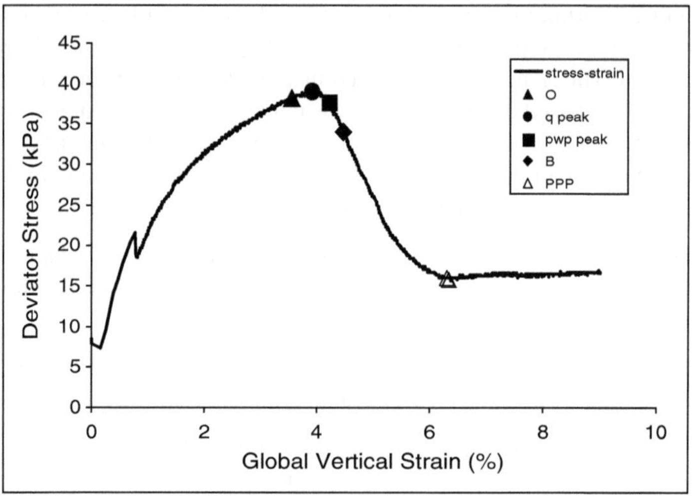

Fig. 1 Deviator stress vs. global vertical strain for lightly heavily overconsolidated kaolin clay sample (PS 55-F/S)

Displacements of the shear band are computed from multiple boundary displacement measurements. The horizontal displacement of the shear band is obtained directly from the sled displacement. The global vertical displacement is adjusted for an elastic unloading of the rigid blocks outside the shear band. Due to the one dimensional slurry consolidation that was followed by plane strain reconsolidation to much smaller mean effective stress, the sample is assumed to be transversely isotropic. Its elastic properties were determined from the series of conventional triaxial tests conducted on similarly prepared samples. Thus, the vertical displacements measured at the top boundary of the biaxial sample are decreased by the amount of an elastic unloading in an incremental fashion to get the vertical displacements of the shear band. Because of a degree of uncertainty in elastic parameters four different unloading scenarios are considered including no unloading, 0.67 of full unloading, 0.5 of full unloading and full unloading. Figure 2 shows the corresponding dilatancy angles, which are computed based on the multiple boundary displacement measurements and the average orientation of the shear band. A comparison of these dilatancy angles with the excess pore water pressure indicates that dilatancy angles obtained for 0.67 of full unloading are the most reasonable. In this particular case, dilatancy angle changes from positive (dilatancy) to negative (contractancy) at a global axial strain of 5.93%, while the minimum excess pore water pressure occurs at a global axial strain of 6.35%. This results in a time lag of about 64 min between the sign change in dilatancy angle and change in the trend of boundary excess pore water pressure.

A normal displacement of the shear band was also computed from photographs (boundary digital imaging), which clearly and directly showed the thickness of the shear band when proper magnification was used. The two methods used to compute

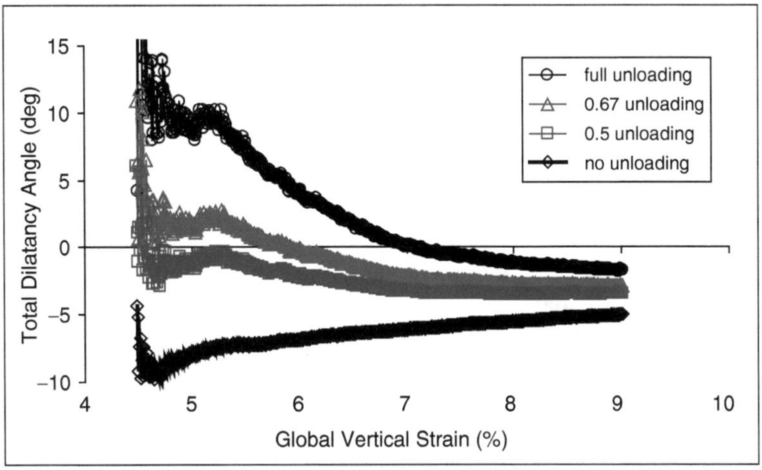

Fig. 2 Dilatancy angles for different amounts of unloading

the displacements of the shear band show a reasonable agreement up to about 8% of global axial strain, thus rendering the use of boundary displacement credible for calculation of meso scale strains.

Green-St. Venant finite volumetric and shear strains are computed next. Based on the previously stated assumptions the finite and infinitesimal in plane shear and volumetric strains turn out to be equal to each other. The shear strain is given by:

$$\varepsilon_{xy} = \frac{u}{2t_i}\cos\psi \tag{1}$$

where u is the total cumulative displacement of the shear band, t_i is the initial thickness of the shear band corresponding to the thickness at the full formation, and Ψ is dilatancy angle. The volumetric strain is given by:

$$\varepsilon_{vol} = \frac{u}{t_i}\sin\psi \tag{2}$$

The volumetric and shear strains are depicted in Fig. 3 for two different initial thicknesses of the shear band corresponding to 1 mm and 1.5 mm. These two thicknesses were determined to be the reasonable lower and upper bounds of the actual initial thickness, based on the direct measurement from the photographs combined with the boundary displacement measurements.

4 Conclusions

Meso scale displacements and strains show that the shear band, which formed in slightly heavily overconsolidated clay sample was initially dilating, followed by transition to a contraction. Although the strains within the shear band are

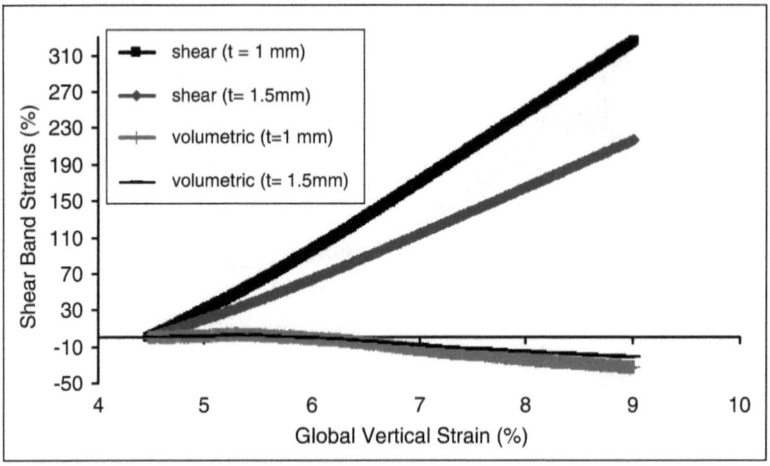

Fig. 3 Meso scale in plane shear and volumetric strains

highly dependent on its thickness, they appear to be larger than in sands, which is plausible due to the significantly different size and shape of clay particles. The actual sub-meso scale particle and particle aggregation mechanisms remain unknown at this time.

Acknowledgments The first author gratefully acknowledges the support provided by the National Science Foundation (CMS 9806052).

References

R.J. Finno, Y. Rhee, Consolidation, pre- and post-peak shearing responses from internally instrumented biaxial compression device. Geotech. Test. J. **16**, 496–509 (1993)

C. Han, I.G. Vardoulakis, Plane-strain compression experiments on water-saturated fine-grained sand. Géotechnique **41**, 49–78 (1991)

P.Y. Hicher, H. Wahyudi, D. Tessier, Microstructural analysis of strain localization in clay. Comput. Geotech. **16**, 205–222 (1994)

J.F. Labuz, S.T. Dai, Residual strength and fracture energy from plane-strain testing. J. Geotech. Geoenviron. **126**, 882–889 (2000)

M. Mooney, R.J. Finno, G. Viggiani, A unique critical state for sand? J. Geotech. Geoenviron. **124**, 1100–1108 (1998)

Failure of Castlegate Sandstone Under True Triaxial Loading

Mathew D. Ingraham, Kathleen A. Issen, and David J. Holcomb

Abstract A test series designed to investigate and quantify the effect of the intermediate principal stress on the failure of Castlegate sandstone was completed. Using parallelepiped specimens and a true triaxial testing system, constant mean stress tests were conducted. Stress states ranged from axisymmetric compression to axisymmetric extension. Results suggest a possible failure dependence on the third invariant of deviatoric stress at lower mean stresses.

Keywords Sandstone • True triaxial compression • Failure • Third stress invariant • Deformation band

1 Introduction

Sandstone is one of the most common rock types found on the surface of the earth. Therefore, deformation structures in sandstone can provide insight into the loading history of the earth at a particular location. Additionally, petroleum reservoirs are often found in high porosity sandstones. As such, understanding the failure modes of the petroleum host is important in drilling and extraction operations so that the sandstone host does not fail and close the well.

Failure behavior of high porosity sandstone has typically been determined using axisymmetric loading (e.g., Olsson and Holcomb 2000; Besuelle et al. 2000; Wong et al. 2001), where two of the principal stresses are equal (σ_1 is maximum compression, σ_2 is the intermediate principal stress, and σ_3 is minimum

M.D. Ingraham (✉) · K.A. Issen
Clarkson University, 8 Clarkson Ave #5725, Potsdam, NY 13699, USA
e-mail: ingrahmd@clarkson.edu; issenka@clarkson.edu

D.J. Holcomb
Sandia National Laboratories, PO Box 5800, Albuquerque, NM 87185–0751 USA
e-mail: djholco@comcast.net

S. Bonelli et al. (eds.), *Advances in Bifurcation and Degradation in Geomaterials*, Springer Series in Geomechanics and Geoengineering 11,
DOI 10.1007/978-94-007-1421-2_42,

compression; compression is positive). Haimson and colleagues (e.g., Haimson and Lee 2004) performed true triaxial drilling tests to investigate the formation and propagation of borehole breakouts. Christensen et al. (2004) performed true triaxial tests on Castlegate sandstone to establish a modified Mohr – Coulomb failure criterion. Otherwise, little work has been done to examine sandstone failure under non-axisymmetric stress states (σ_2 not equal to σ_1 or σ_3), even though the intermediate principal stress affects the failure behavior of other geomaterials. For example, Lade and Kim (1988) report that in sands and soils the failure surface is a rounded triangle in the pi-plane, indicating a dependence on the third invariant of deviatoric stress.

High porosity sandstone often fails by formation of deformation bands, e.g., shear bands or compaction bands. Rudnicki and Rice (1975) examined conditions for deformation band formation and predicted that the intermediate principal stress influences both the band orientation and the strain type inside the band (dilation, compaction, shear). However, these predictions have not been evaluated for high porosity sandstone under non-axisymmetric stress states. Therefore, the present work conducted a suite of true triaxial tests in which Castlegate sandstone was subjected to five deviatoric stress states. The intermediate principal stress varied from axisymmetric compression (ASC, $\sigma_2 = \sigma_3$) to axisymmetric extension (ASE, $\sigma_2 = \sigma_1$). This paper presents the resulting failure surfaces; the companion paper (Issen et al. 2011) compares observed deformation bands with preliminary theoretical predictions.

2 Materials and Methods

Parallelepiped specimens (25.40 × 57.15 × 57.15 mm) were cut from a Castlegate sandstone core purchased from TerraTek. Castlegate is a fluvial reservoir analog comprised primarily of ∼0.2 mm calcite cemented quartz grains. Specimens had a nominal unconfined compressive strength of 15 MPa and porosity of 26.4%. Specimens were jacketed in 0.127 mm thick copper to prevent infiltration of confining fluid. They were instrumented with LVDT's (Linear Variable Displacement Transformer) in the σ_1 and σ_3 directions and with a pair of strain gauges in the σ_2 direction. Figure 1 shows a sample with these sensors indicated. End caps were lubricated with a stearic acid and petroleum jelly mixture to reduce frictional effects (Labuz and Bridell 1993).

The localization theory of Rudnicki and Rice (1975) predicts a σ_2 dependence through the parameter N_{II}, where $N_{II} = (\sigma - \sigma_2)/\tau$, the mean stress is $\sigma = \sigma_{kk}/3$, the Mises equivalent shear stress is $\tau = (s_{ij}s_{ij})^{1/2}$, and the deviatoric stress is $s_{ij} = \sigma_{ij} - \sigma_{kk}/3$. Tests were conducted at five N_{II} values: $1/\sqrt{3},\ 1/(2\sqrt{3}),\ 0,\ -1/(2\sqrt{3})$ and $-1/\sqrt{3}$. These values correspond to Lode angles of $\alpha = 30°,\ 14.5°,\ 0°,\ -14.5°$ and $-30°$, respectively (0° is pure shear). Table 1 details the relationships between the Lode angle, N_{II}, and applied stresses. Tests were run at constant mean stress to assess dependence on the third invariant of deviatoric stress, $J_3 = \det(s_{ij})$.

Fig. 1 Image detailing the instrumentation of a test specimen

Table 1 Relation between Lode angles, N_{II}, and principal stresses for stress states presented

Lode Angle, α	N_{II}	Relationship between principal stresses
$+30°$	$\frac{1}{\sqrt{3}}$	$\sigma_2 = \sigma_3$
$+14.5°$	$\frac{1}{2\sqrt{3}}$	$\sigma_2 = \frac{1}{2}\left[\sigma_1\left(1 - \frac{1}{\sqrt{5}}\right) + \sigma_3\left(1 + \frac{1}{\sqrt{5}}\right)\right]$
$0°$	0	$\sigma_2 = \frac{1}{2}(\sigma_1 + \sigma_3)$
$-14.5°$	$-\frac{1}{2\sqrt{3}}$	$\sigma_2 = \frac{1}{2}\left[\sigma_1\left(1 + \frac{1}{\sqrt{5}}\right) + \sigma_3\left(1 - \frac{1}{\sqrt{5}}\right)\right]$
$-30°$	$-\frac{1}{\sqrt{3}}$	$\sigma_2 = \sigma_1$

To start each test, stresses were increased hydrostatically to just below the desired mean stress, then the system control mode was changed. The σ_1 direction (applied by the load frame) was placed in displacement control while a pair of self aligning hydraulic flat jacks (σ_2) and the confining pressure (σ_3) were in a calculated load control. The σ_2 and σ_3 channels acquired inputs from the system pressure transducers; the load cell measured σ_1. Using the relations in Table 1 and the current σ_1 value, the control software calculated the required levels for σ_2 and σ_3 to maintain the required mean stress. This system is similar to that of Lade and Duncan (1973), except it is capable of applying higher stresses (e.g., σ_2 of 200 MPa).

Applying the jacks to the specimen sides will cause some frictional load, particularly if the specimen were to undergo significant axial shortening. In testing sands and soils, this effect can be avoided by using fluid pressures to apply σ_2 and σ_3. However, to our knowledge, this approach cannot generate the high stresses required for rock testing. Additionally, because this research focuses on material

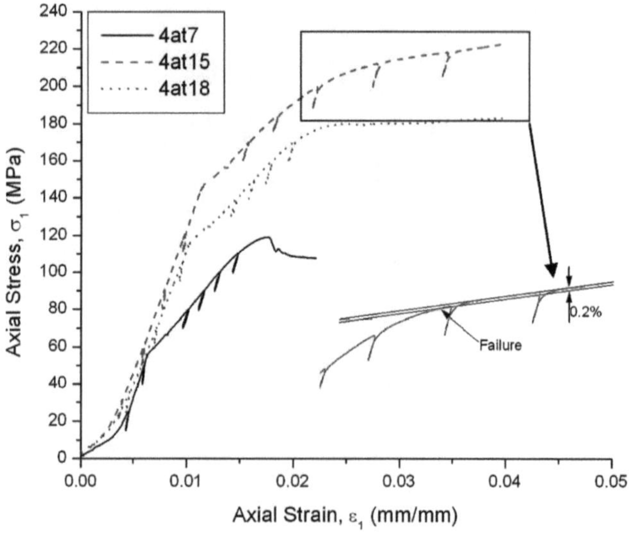

Fig. 2 Stress-strain curves for typical tests. Inset shows failure determination for tests without a stress peak. Test parameters: specimen 4at7 at 60 MPa mean stress and 14.5° Lode angle, 4at15 at 150 MPa mean stress and 14.5° Lode angle, 4at18 at 120 MPa mean stress and 0° Lode angle

Fig. 3 Failure surfaces for Castlegate sandstone using loading paths along different Lode angles

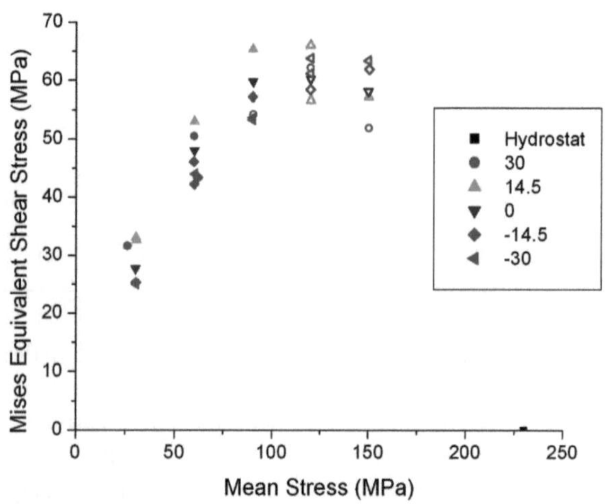

response prior to and including failure, relevant deformations are relatively small (axial strains of approximately 1–2%). Therefore, it is likely that the jack contact does not significantly influence strain localization. For example, in some specimens, the observed shear band had a band normal component in the σ_2 direction; thus the jacks applying σ_2 did not prevent the shear offset necessary to form these bands.

Failure was determined from the axial stress-strain response. For specimens that exhibited a stress peak (Fig. 2), the peak stress was defined as failure. These specimens are shown as closed symbols in Fig. 3. Most contained a deformation band; thus the peak likely corresponds to band formation. Specimens without a stress peak exhibited a knee in the axial stress – strain curve, after which the slope was relatively constant. This suggests a change in deformation mechanism (possibly the onset of grain crushing, although this has not yet been confirmed microstructurally). We elected to use the slope change to define "failure." A line was fit through the linear region after the knee; a 0.2% offset was used to determine failure stress (see inset of Fig. 2). These specimens (including the hydrostatic test), are indicated with an open symbol in Fig. 3.

3 Results and Discussion

Figure 3 shows the plot of the failure surfaces in the $\sigma - \tau$ plane. In this stress plane, for traditional J_3 dependence, the failure surface for ASE ($\sigma_2 = \sigma_1$, Lode angle $= -30°$) will be closest to the mean stress axis, the ASC failure surface ($\sigma_2 = \sigma_3$, Lode angle $= +30°$) will be farthest away, with the other stress states falling between. The pi-plane is perpendicular to the mean stress axis, and τ is related to the pi-plane radial direction. While there is scatter in the data, at low mean stresses (30–90 MPa) there appears to be some J_3 dependence. In general, as the Lode angle decreases, so does the Mises stress required to fail the material. However, there is a small increase in τ when the Lode angle changes from 30° to 14.5°. For tests at 120 and 150 MPa, it is difficult to discern a pattern that suggests J_3 dependence. This could be related to the brittle – ductile transition; however, additional analyses are required to validate this.

4 Conclusions and Future Work

For the failure surfaces determined here, as mean stress increases, so does the Mises equivalent shear stress, up to approximately 120 MPa, after which, as the mean stress increases the Mises shear stress decreases. This implies a cap on the failure surface, as is commonly observed in high porosity sandstones. The J_3 dependence of the material, while subtle, is more evident at lower mean stresses. At higher mean stresses any noticeable J_3 effect is lost in the scatter of the data. The decrease in J_3 effect with increasing mean stress seems to correspond to the brittle-ductile transition, but more work is required to confirm this.

Further analyses are underway, including analyses of the stress-strain curves, examination of acoustic emissions data, and construction of yield surfaces, including pi-plane plots. Determination of constitutive parameter values to evaluate predicted failure conditions from localization theory is also in progress.

Acknowledgments The authors gratefully acknowledge funding from the National Science Foundation Award EAR-0711346 to Clarkson University.

Sandia is a multiprogram laboratory operated by Sandia Corporation, a Lockheed Martin Company, for the United States Department of Energy's National Nuclear Security Administration under contract DE-AC04–94AL85000.

References

P. Bésuelle, J. Desrues, S. Raynaud, Experimental characterization of localization phenomenon inside a Vosges sandstone in a triaxial cell. Int. J. Rock Mech. Min. Sci. **37**, 1223–1237 (2000)

H.F. Christensen, M.G. Stage, N.K. Loe, B. Plischke, O. Havmoller, Impact of the intermediate principal stress on rock strength: polyaxial testing and numerical simulations, in *Proceedings North American Rock Mechanics Symposium. #04–469*, Houston, TX (2004)

B.C. Haimson, H. Lee, Borehole breakouts and compaction bands in two high-porosity sandstones. Int. J. Rock Mech. Min. Sci. **41**, 287–301 (2004)

K.A. Issen, M.D. Ingraham, T.A. Dewers, Strain localization conditions under true triaxial stress states, S. Bonelli et al. (eds.), *Advances in Bifurcation and Degradation in Geomaterials*, Springer Series in Geomechanics and Geoengineering 11, DOI 10.1007/978-94-007-1421-2_40, ©Springer Science+Business Media B.V. (2011)

J.F. Labuz, J.M. Bridell, Reducing frictional constraint in compression testing through lubrication. Int. J. Rock Mech. Min. Sci. **30**, 451–455 (1993)

P.V. Lade, J.M. Duncan, Cubical triaxial tests on cohesionless soils. J. Soil Mech. Found. Div. **99**, 793–812 (1973)

P.V. Lade, M.K. Kim, Single hardening constitutive model for frictional materials, III: comparisons with experimental data. Comput. Geotech. **6**, 31–47 (1988)

W.A. Olsson, D.J. Holcomb, Compaction localization in porous rock. Geophys. Res. Lett. **27**, 3527–3540 (2000)

J.W. Rudnicki, J.R. Rice, Conditions for the localization of deformation in pressure-sensitive dilatant materials. J. Mech. Phys. Solids **23**, 371–394 (1975)

T.-F. Wong, P. Baud, E. Klein, Localized failure modes in a compactant porous rock. Geophys. Res. Lett. **28**, 2521–2524 (2001)

Frequency Signature of Damage Localization

E. Pasternak and A.V. Dyskin

Abstract Spectral effect of temporal localization of acoustic events associated with their spatial localization is investigated. It is shown the temporal localization produces blue shift in the frequency spectrum as compared to non-localized (random) generation of acoustic events. This opens a possibility to detect localization on the basis of a single channel recording without the need of determining the pulse locations. This can be done by analyzing the difference between the spectra of registered acoustic pulses and their randomized series.

Keywords Temporal localisation • Arrival times • Energy spectrum • Frequency shift • Blue shift indicator

1 Introduction

Spatial localization of acoustic events is considered to be a part of fracture process and thus can cast light on the failure mechanisms and failure prediction. The detailed experimental investigation of the acoustic pulse localization involves spatial-temporal location of acoustic events which requires multichannel recording system with still limited accuracy achievable. It is therefore desirable to investigate the existence of indirect indicators of localization. One of such indicator is proposed in this paper.

E. Pasternak
School of Mechanical Engineering, University of Western Australia, Perth, WA, Australia
e-mail: elena@mech.uwa.edu.au

A.V. Dyskin (✉)
School of Civil and Resource Engineering, University of Western Australia, Perth, WA, Australia
e-mail: arcady@civil.uwa.edu.au

S. Bonelli et al. (eds.), *Advances in Bifurcation and Degradation in Geomaterials*, Springer Series in Geomechanics and Geoengineering 11,
DOI 10.1007/978-94-007-1421-2_43, © Springer Science+Business Media B.V. 2011

Fig. 1 Schematics of zone of localization of acoustic events

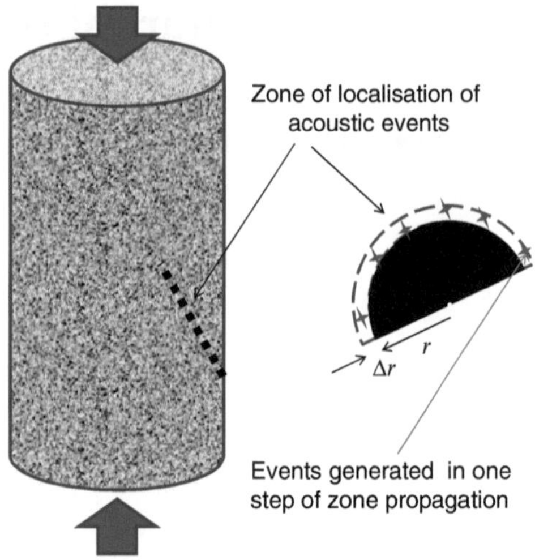

Results of experimental investigation of localization phenomenon in rocks under quasi-static loading (Lockner et al. 1991) suggest that the localization does not happen instantaneously. It is rather synchronized such that the zone of acoustic events propagates stepwise in a crack-like manner, increasing its size by dr each step, Fig. 1. While globally the localization of acoustic events can be perceived as a result of strain localization associated with bifurcation in the homogeneous solution of the equation of equilibrium (Rudnicki and Rice 1975), microscopically the acoustic event is produced by appearance of a defect whose position is dictated by the interaction with already existing defects. We assume that the influence of the defects on each other is transmitted via stress waves. The wave velocity of the order of 10^3 m/s suggests the interaction time to be in the order 10^{-4} s for a sample of 10 cm size. This is much smaller than the typical time of the load step. (Indeed, with the loading rate of 10^{-6} m/s (Lockner et al. 1992), during the defect interaction time the displacement is in the order 10^{-10} m, which is seven orders of magnitude smaller than the displacement achieved in the tests (Lockner et al. 1991, 1992). Therefore, one can assume that in the stepwise process of propagation of the zone of acoustic events, each step is characterized by almost instantaneous generation of a number of events proportional to the circumference of the zone, or to its radius r. On the other hand, if there is no localization, the acoustic events will be randomly distributed in time reflecting the random character of non-synchronized generation of defects. Thus we assume that the described temporal localization of acoustic events is indicative of the spatial localization. In the next section we will consider the difference in spectral characteristics of these two types of defect accumulation, localized vs. non-localized. The following sections will introduce a method of detecting the localization.

2 Blue Shift in Spectrum of Temporally Localized Acoustic Events

As discussed above, sequencing of acoustic events would indicate that the localization zone propagated rather than appeared instantaneously. For the sake of generality, we consider three geometries of the localization zone: a 1D crack-like zone in a plate (through its thickness), 2D crack-like zone in a bulk of material and a 3D 'cloud'. The rate of increase of the number of pulses as the localization zone propagates indicates the shape of the zone. Indeed, assuming that the pulses are generated from the front of the propagating zone, after k-th step of propagation of a 1D zone $O(k)$ pulses will be generated, 2D zone will generate $O(k^2)$ pulses and a 3D zone will generate $O(k^3)$ pulses. By assigning these geometries indexes 1, 2, 3 respectively one can write that number of acoustic events produced after K steps of propagation is $N_n = O(K^n)$.

We consider the spectral signatures of the outlined types of sequencing and compare them with the spectral signature of the random events. For the sake of simplicity we assume that all events emit square pulses of the same amplitude A_k at each step and the duration of 2. We also assume that the times required for each step of the propagation are independent random variables ξ_k. Given that the Fourier transform of a square pulse of unit half duration is $2\sin\omega/\omega$, the linearity of the Fourier transform and that a time shift of $(k+\xi_k)\Delta T$ attracts multiplication of the Fourier transform by $\exp(-i(k+\xi_k)\omega\Delta T)$, the power spectrum of the wave form resulting from the series of such signals is

$$P_n(\omega) = \left| 2\frac{\sin\omega}{\omega}\left[1 + \sum_{k=1}^{K} k^n A_k \exp\left(-i(k+\xi_k)\omega\Delta T\right)\right]\right|^2 \tag{1}$$

Figure 2 shows the results of simulations with $A_k = 1$, $\Delta T = 3$ and ξ uniformly distributed on $[-\varepsilon/2, \varepsilon/2]$, $\varepsilon = 0.5$ and regular time intervals ($\varepsilon = 0$). As an example the number of steps for the 3D case was assumed $K = 10$. The numbers of steps for other cases were adjusted to end up with the same total number of pulses (events) in each case. It is seen that both 2D crack-like propagating zone and the propagating 3D 'cloud' produce the 'blue shift': the power spectrum shifts to higher frequencies relative to the random event spectrum. Furthermore, random variations in the propagation rate have a minor effect on the spectrum. On the other hand, the 1D zone is almost indistinguishable from the random events.

To estimate the effect of attenuation, we have conducted further simulations with A_k being independent random variables uniformly distributed on [0, 1]. They show that such a strong variation in the amplitude does not change the above spectral signatures of the mechanisms of localization.

The above results suggest the following procedure for determination as to whether a localization of acoustic events has happened: record the acoustic pulses, compute their combined power spectrum, then randomize their arrival times and compute the power spectrum again. A blue shift in the original power spectrum as

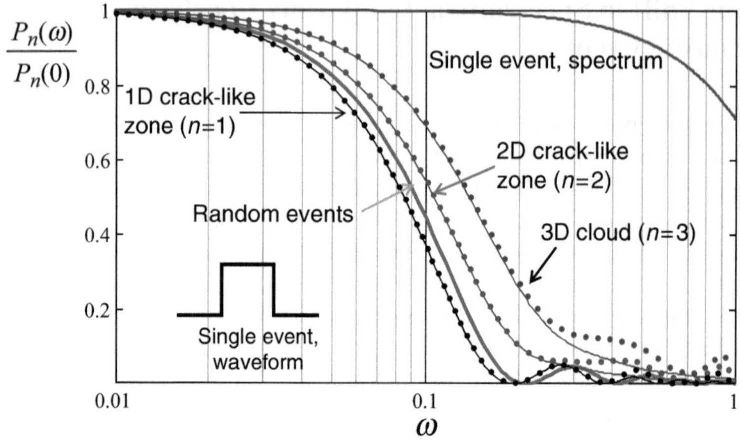

Fig. 2 Normalized power spectrum for all four modes of localization with events (*square pulses*, see insert) of the same magnitude and for a single event. The *solid lines* refer to constant propagation rate of localization zones, *filled circles* refer to randomly varying propagation rate ($\varepsilon = 0.5$)

compared to the randomized one would be an indication of localization. Of course, the visual inspection of the plots is not accurate, so a measure of blue shift is required. Such a measure will be proposed in the following section.

3 A Blue Shift Indicator: Ideal Case

We make an observation that in all cases of geometry of localization zone Eq. 1 has the same common factor, $|2\sin\omega/\omega|^2$ which is a square of the absolute value of the Fourier transform of the assumed rectangular pulse of duration 2. Thus, in the process of comparison between different geometries one can drop this factor and concentrate on only the pulse amplitudes and their generation or arrival times. (The differences between the generation and arrival times are absorbed in the subsequent computer simulations by introducing random variations in the arrival times.) We can now introduce the 'energy' of the arrival times in the frequency range $[0,\ \Omega]$. It is obtained by representing the spectrum of the time delays from (1) as product of the complex sum with its complex conjugate and integrating over $[0,\ \Omega]$ and using $t_k = k\Delta T$. To the accuracy of the constant factor, it reads

$$S_n(\Omega) = \int_0^{\Omega} \sum_{k,l=0}^{K} (k+1)^n\,(l+1)^n\,A_k A_l \exp\left(-i(t_k - t_l)\omega\right)d\omega \qquad (2)$$

For comparison between the random pulse sequence and a crack-like localized zone from Fig. 1 (subscripts 0 and 1) we introduce the relative energy shift

$$\Sigma(\Omega) = |S_1(\Omega) - S_0(\Omega)|/S_0(\Omega) \tag{3}$$

where the number of events in S_0 is taken to be equal to the total number of events in S_1, which is $(K+1)K/2$, where K is the number of steps in the propagation of the localized zone. In the case of constant amplitudes, $A_k = A$, it can be shown that for $\Omega^{\rightarrow}\infty$, $\Sigma(\Omega)$® $(K+2)(2K+1)/6 - 1$. This indicates the difference between the crack-like localization and random event generation. The larger the number of pulses involved in the analysis, the higher the limit $\Sigma(\Omega)$.

Thus ideally the estimate (3) in the limit of $\Omega^{\rightarrow}\infty$ can serve as a localization indicator. In reality however, the absolutely simultaneous pulse generation never happens. Rather they are generated in a (very small) interval $\Delta t = \Delta r/w$, where w is the rate of propagation of the localization zone. Then the limit $\Omega^{\rightarrow}\infty$ should be replaced with $\Omega^{\rightarrow}\Omega_{max}$, where $\Omega_{max} \sim 2\pi/\Delta t$. As Δt is not usually known *a priori*, it would be necessary to plot $\Sigma(\Omega)$ for different frequencies to have an indication of localization. The following section illustrates the approach based on computer simulations of random event generation and their crack-like localization.

4 A Blue Shift Indicator: Real Case Simulations

We now generate two time sequences: one to model localization, where at each step k, k pulses are randomly and uniformly generated in the segment $[k\Delta T,\ k\Delta T + \Delta t]$, and another, random uniformly distributed in the segment $[0,\ T]$, $T = (K+1)\Delta T$, both of length $M = (K+1)K/2$. Figure 3 shows the resulting plot of Σ vs. relative inverse frequency, $\tau/\Delta t$, where $\tau = 2\pi/\Omega$, for different values of M. It is seen that the curves have distinct peaks at $\tau \sim 10\Delta t$ which correspond to $\Omega_{peak} \sim 0.1\Omega_{max}$. Thus the curve could be used for an order of magnitude estimation of the time-step of the localization zone propagation.

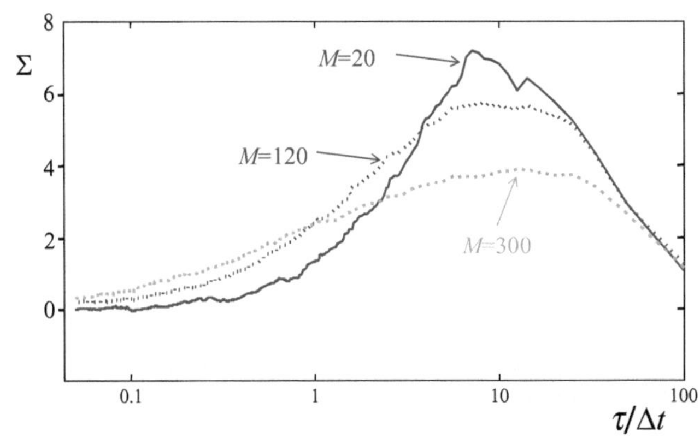

Fig. 3 Relative energy shift vs. relative inverse frequency for different numbers M of events

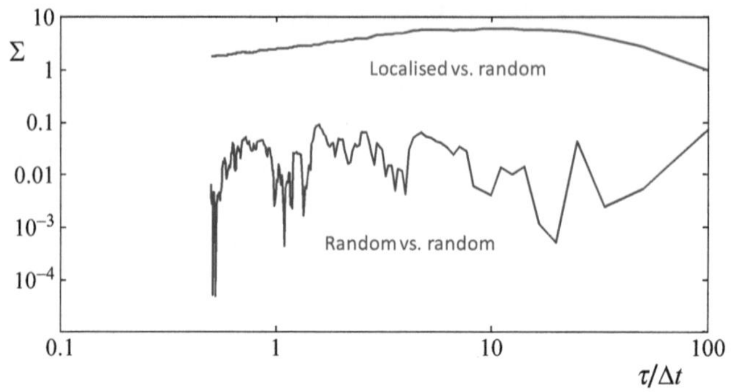

Fig. 4 The relative energy shifts between localised and random arrival times and between two realisation of random times for number of events $M = 120$

On the other hand, the relative energy shift for two realizations of random arrival times is almost negligible, Fig. 4. More importantly, it does not exhibit the peak, characteristic to the relative energy shift between the localized and random arrival times.

Further simulations showed that this feature is insensitive to both random and systematic variations in the pulse amplitudes.

5 Conclusions

The found effect of temporal localization of the acoustic events on the combined power spectrum of the acoustic emission allows the reconstruction of the localization phenomena and estimation of the time step of localization zone propagation without resorting to experimentally challenging determination of the locations of the acoustic events.

Acknowledgments This research is supported by ARC Discovery Grants DP0988449, DP0773839 and WA Geothermal Centre of Excellence. We grateful to Chris Hellsten for discussions and suggestions.

References

D.A. Lockner, J.D. Byerlee, V. Kuksenko, A. Ponomarev, A. Sidorin, Quasi-static fault growth and shear fracture energy in granite. Nature **350**, 39–42 (1991)

D.A. Lockner, J.D. Byerlee, V. Kuksenko, A. Ponomarev, A. Sidorin, Observations of quasistatic fault growth from acoustic emissions, in *Fault Mechanics and Transport Properties of Rocks*, ed. by B. Evans, T.-F. Wong (Elsevier, London, 1992), pp. 3–31

J.W. Rudnicki, J.R. Rice, Conditions for the localization of deformation in pressure-sensitive dilatant materials. J. Mech. Phys. Solids **23**, 371–394 (1975)

Modeling of Weathered and Moisture Sensitive Granular Materials

Erich Bauer and Zhongzhi Fu

Abstract The focus of this paper is on modeling the essential mechanical properties of weathered and moisture sensitive granular materials under dry and water saturated conditions using a state dependent solid hardness. Herein the solid hardness is related to the grain assembly in the sense of a continuum description and does not mean the hardness of an individual grain. Particular attention is paid to creep deformation initiated by water saturation of an initially dry material. First the constitutive model is outlined for isotropic loading paths. For general stress paths the influence of the stress deviator on creep deformation and stress relaxation is investigated with an extended hypoplastic model. The performance of the refined model is demonstrated for triaxial compression tests where the saturation of the material is investigated for different deviator stresses. The predictions of the model are compared with experiments.

Keywords Weathered granular materials • Solid hardness • Creep • Hypoplasticity

1 Degradation of the Solid Hardness

In this section the isotropic compression behavior of a granular material is modeled using a relation between the maximum void ratio e_i and the mean effective pressure $p = -(\sigma_{11} + \sigma_{22} + \sigma_{33})/3$ of the grain skeleton. In particular the following exponential function is considered (Bauer 2009):

E. Bauer (✉)
Institute of Applied Mechanics, Graz University of Technology, Graz, Austria
e-mail: erich.bauer@tugraz.at

Z. Fu
Institute of Hydraulic Engineering, Hohai University, Nanjing, China
e-mail: fu_zhongzhi@yahoo.com

S. Bonelli et al. (eds.), *Advances in Bifurcation and Degradation in Geomaterials*, Springer Series in Geomechanics and Geoengineering 11,
DOI 10.1007/978-94-007-1421-2_44,

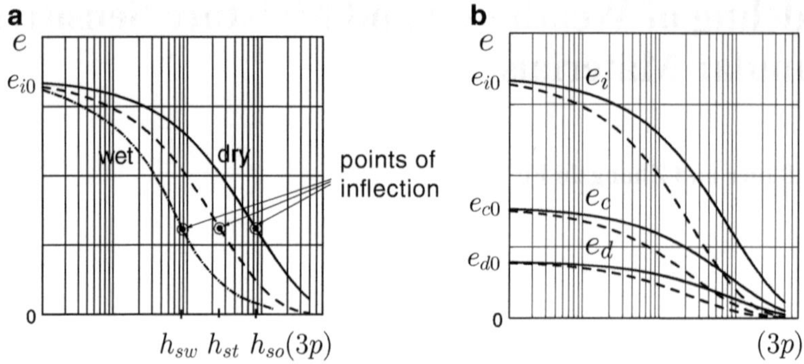

Fig. 1 Influence of the degradation of the solid hardness: (**a**) on the compression behavior; (**b**) on the limit void ratios

$$e_i = e_{i0} \exp\left[-\left(\frac{3p}{h_{st}}\right)^n \right]. \tag{1}$$

Herein the constant e_{i0} denotes the maximum void ratio for $p \approx 0$, h_{st} has the dimension of stress and n is a dimensionless constant. The so-called solid hardness h_{st} is defined for the effective pressure $3\,p$ at which the isotropic compression curve in a semi-logarithmic representation shows the point of inflection and the exponent n is related to the corresponding inclination as illustrated in Fig. 1a. Also for the case that the solid hardness remains unchanged the compression relation of Eq. 1 takes into account the effect of grain crushing as for $p \to \infty$ the void ratio tends towards zero. If the weathered material is moisture sensitive, the course of the compression curve is different for the dry state and the wet state (e.g. Nobari and Duncan 1972; Brauns et al. 1980; Alonso and Oldecop 2000; Oldecop and Alonso 2007). In this case the solid hardness is no longer a material constant. The higher compressibility of the wet material can be explained by a degradation of the stiffness of the solid material due to the reaction with water, which leads to grain abrasion and grain breakage and consequently to a rearrangement of particles into a denser state. The process of degradation of the solid hardness can be repeatedly initiated, for instance by infiltration of water into the micro-cracks and pores that penetrate the surface of the weathered grains. In this paper the degradation of the solid hardness is assumed to be irreversible and only the dry state of the material followed by water saturation is considered. If a reaction with water takes place, the process of degradation of the solid hardness is modeled by the following evolution equation (Bauer 2009):

$$\dot{h}_{st} = -\frac{1}{c}\left(h_{st} - h_{sw}\right). \tag{2}$$

In Eq. 2 the parameter $\dot{h}_{st}$ denotes the rate of the solid hardness, h_{st} is the current solid hardness, h_{sw} denotes the value of the solid hardness when the process of degradation is completed, i.e. $h_{sw} = h_{st}(t \to \infty)$, and c has the dimension of time and scales the velocity of degradation. It follows from Eq. 2 that the solid hardness is a function of the degradation time t, i.e. $h_{st} = h_{st}(t)$, and for an irreversible process of degradation the rate of the solid hardness must be negative, i.e. $\dot{h}_{st} < 0$. Thus the time derivative of Eq. 1 reads:

$$\dot{e} = n\, e \left(\frac{3p}{h_{st}}\right)^n \left[\frac{\dot{h}_{st}}{h_{st}} - \frac{\dot{p}}{p}\right]. \tag{3}$$

Herein $\dot{e}$, $\dot{h}_{st}$ and $\dot{p}$ denote the time derivatives of e, h_{st} and p, respectively. Assuming a constant grain density the following relation for the volume strain rate $\dot{\varepsilon}_v$ under isotropic compression can be derived:

$$\dot{\varepsilon}_v = \frac{\dot{e}}{1+e} = \frac{n\, e}{1+e} \left(\frac{3p}{h_{st}}\right)^n \left[\frac{\dot{h}_{st}}{h_{st}} - \frac{\dot{p}}{p}\right]. \tag{4}$$

From the right-hand side of Eq. 4 it can be concluded that the volume strain rate depends on the current state of e, p and h_{st}, and also on their corresponding rates. For the special case that during degradation of h_{st} the pressure is kept constant, i.e. $\dot{p} = 0$, the material creeps and the volumetric strain rate in Eq. 4 reduces to:

$$\dot{\varepsilon}_v = \frac{n\, e}{1+e} \left(\frac{3p}{h_{st}}\right)^n \frac{\dot{h}_{st}}{h_{st}}. \tag{5}$$

Equation 5 indicates that during creep the volumetric strain rate $\dot{\varepsilon}_v$ is proportional to the rate of the solid hardness, $\dot{h}_{st}$, and it is also influenced by the pressure p and the current void ratio e. Under constant volume, i.e. $\dot{\varepsilon}_v = 0$, Eq. 4 yields for stress relaxation:

$$\dot{p} = p\, \frac{\dot{h}_{st}}{h_{st}}. \tag{6}$$

As a result of the degradation of the solid hardness the value of h_{st} becomes smaller, which also implies a reduction of the minimum void ratio e_d and the critical void ratio e_c as illustrated by the dashed curves in Fig. 1b. In particular it is assumed that the influence of the current solid hardness and the pressure level on e_d and e_c can be modeled as:

$$\frac{e_d}{e_{d0}} = \frac{e_c}{e_{c0}} = \frac{e_i}{e_{i0}} = \exp\left[-\left(\frac{3p}{h_{st}}\right)^n\right]. \tag{7}$$

2 Hypoplastic Constitutive Model

While for unweathered granular materials the solid hardness in the hypoplastic constitutive model by Gudehus (1996) and Bauer (1996) was constant, the influence of degradation of the solid hardness ($\dot{h}_{st} < 0$) is taken into account by the following relation (Bauer 2009):

$$\overset{\circ}{\sigma}_{ij} = f_s\Big[\hat{a}^2\,\dot{\varepsilon}_{ij} + (\hat{\sigma}_{kl}\,\dot{\varepsilon}_{kl})\,\hat{\sigma}_{ij} + f_d\,\hat{a}\,(\hat{\sigma}_{ij} + \hat{\sigma}^*_{ij})\,\sqrt{\dot{\varepsilon}_{kl}\,\dot{\varepsilon}_{kl}}\Big] + \frac{\dot{h}_{st}}{h_{st}}\left(\frac{1}{3}\sigma_{kk}\,\delta_{ij} + \kappa\,\sigma^*_{ij}\right). \tag{8}$$

Herein $\overset{\circ}{\sigma}_{ij}$ is the effective objective stress rate, σ_{ij} is the effective Cauchy stress, $\sigma^*_{ij} = \sigma_{ij} - \sigma_{kk}\delta_{ij}/3$ is the deviatoric part of σ_{ij}, $\hat{\sigma}_{ij} = \sigma_{ij}/\sigma_{kk}$ and $\hat{\sigma}^*_{ij} = \hat{\sigma}_{ij} - \delta_{ij}/3$ are normalised quantities, δ_{ij} is the Kronecker delta and $\dot{\varepsilon}_{ij}$ is the strain rate. As defined in the forgoing section h_{st} and $\dot{h}_{st}$ denote the solid hardness and the rate of the solid hardness, respectively. In contrast to earlier constitutive models for weathered rockfill materials, e.g. Bauer et al. (2008), the present version also takes into account the influence of the deviatoric stress in the last term on the right-hand side of Eq. 8. Investigations by Fu and Bauer (2009) show that the deviatoric stress has a significant influence on the amount of creep deformation. Herein factor κ scales the creep velocity which may be influenced by the stress state and the void ratio. For the present investigations factor κ is assumed to be constant. For states where the degradation of the solid hardness has been completed, i.e. for $\dot{h}_{st} = 0$, the solid hardness is h_{sw}. Then the rate independent hypoplastic model by Gudehus (1996) and Bauer (1996) is recovered from Eq. 8. Function $\hat{a}$ in Eq. 8 is related to critical stress states and depends on the intergranular friction angle φ_c and on the stress deviator σ^*_{ij}. For critical states factor $\hat{a}$ represent the stress limit condition by Matsuoka and Nakai (1977) as outlined in more details by Bauer (2000). The influence of the pressure level and the current void ratio is taken into account in Eq. 8 with the stiffness factor f_s and the density factor f_d. In these factors the current void ratio is related to the limit void ratios and to the critical void ratio (e.g. Gudehus 1996; Bauer 1996, 2009). The limit void ratios and the critical void ratio depend on the current solid hardness and the stress state as defined in Eq. 7.

3 Comparison of Numerical Simulations with Experiments

The performance of the constitutive model is demonstrated by comparing the results obtained from the numerical simulation of triaxial compression tests with the corresponding experiments carried out by Li (1988). All experiments with slightly weathered broken sandstone were performed under drained conditions. In the numerical simulations the influence of the specific weight of the solid grains and of the water is neglected. The effective grain stress is assumed to be equal to the total stress prescribed at the boundary of the specimen. The hypoplastic model proposed in Sect. 2 includes 11 constants. For the initial dry material the following values for the constitutive constants involved were obtained: $\varphi_c = 40°$, $h_{so} = 47\,\text{MPa}$, $n = 0.3$, $e_{io} = 0.59$, $e_{co} = 0.48$, $e_{do} = 0.20$, $\alpha = 0.18$ and $\beta = 2.50$. After water saturation the additional constitutive constants for modeling the degradation of the solid hardness are: $h_{sw} = 11.5\,\text{MPa}$, $\kappa = 0.7$ and $c = 3/\text{day}$.

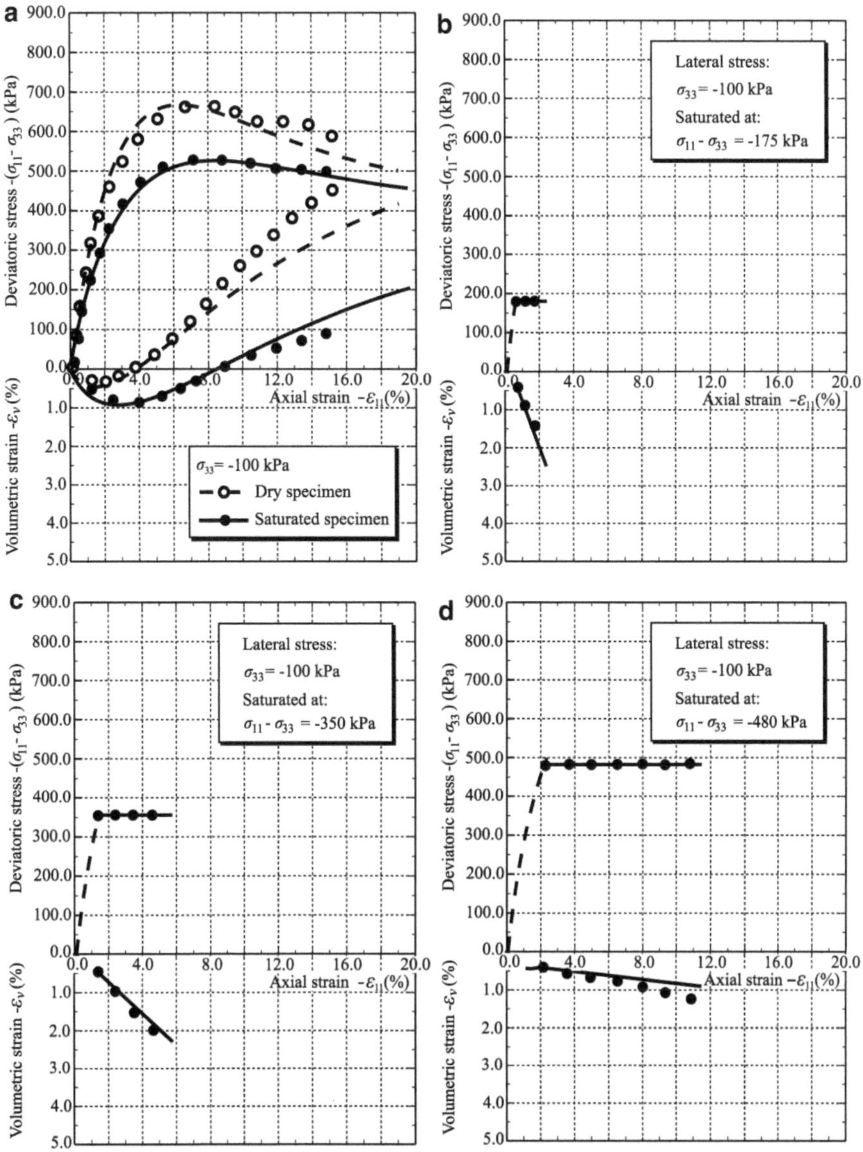

Fig. 2 Triaxial compression tests under constant lateral stress: the shapes indicate experimental data by Li (1988), *dashed curves* and *solid curves* are numerical results for the dry state and the water saturated state, respectively

For a lateral constant stress of $-100\,\mathrm{MPa}$ Fig. 2 shows the evolution of the deviatoric stress and of the volumetric strain versus the axial strain for both the dry material and the water saturated material. As can be seen in Fig. 2a, the deviatoric stress $(\sigma_{11} - \sigma_{33})$ and the volumetric strain ε_v against the axial strain ε_{11} is different for the dry and the water saturated materials. In particular the compaction at the beginning of deviatoric loading is significantly higher for the water saturated material, while the maximum deviatoric stress is higher for the dry material. Moreover, the subsequent dilation is considerably lower for the saturated material, which can be attributed to the more pronounced particle disintegration. In the numerical simulations the solid hardness is the only parameter which differs for the dry material ($h_{so} = 47\,\mathrm{MPa}$) and for the water saturated material ($h_{sw} = 11.5\,\mathrm{MPa}$). For the simulation of creep deformation as a result of wetting, first isotropic compression and deviatoric loading are applied to the dry specimen. Then the specimen is saturated under constant stress, which initiates a time dependent degradation of the solid hardness according to Eq. 2. Figure 2b, c, d shows the behavior of the specimen for three different deviatoric stress states. As can be seen, the axial creep deformation is more pronounced for a higher deviatoric stress while the volume strain curve is steeper under a lower deviatoric stress. The creep paths are almost linear and in good agreement with the experimental results.

References

E. Alonso, L.A. Oldecop, Fundamentals of rockfill collapse, in *Proceedings of the 1st Asian Conference on Unsaturated Soils*, ed. by H. Rahardjo, D.G. Toll, E.C. Leong (Balkema, Paris, 2000), pp. 3–13

E. Bauer, Calibration of a comprehensive hypoplastic model for granular materials. Soils Found. **36**(1), 13–26 (1996)

E. Bauer, Conditions for embedding Casagrande's critical states into hypoplasticity. Mech. Cohes.-Frict. Mater. **5**, 125–148 (2000)

E. Bauer, Hypoplastic modelling of moisture-sensitive weathered rockfill materials. Acta Geotechnica **4**, 261–272 (2009)

E. Bauer, S.F. Tantono, Y. Zhu, L. Sihong, K. Kast, Modeling rheological properties of materials for rockfill dams, in *Proceedings of 1st International Conference on Long Time Effects and Seepage Behavior of Dams*, ed. by Y. Zhu, S. Liu, S. Qiang, A. Chiu (Hohai University Press, Nanjing, 2008), pp. 73–80

J. Brauns, K. Kast, A. Blinde, Compaction effects on the mechanical and saturation behaviour of disintegrated rockfill, in *Proceedings of the International Conference on Compaction*, vol. 1, Paris, 1980, pp. 107–112

Z.Z. Fu, E. Bauer, Hypoplastic constitutive modelling of the long term behaviour and wetting deformation of weathered granular materials, in *Proceedings of the 2nd International Conference on Long Term Behaviour of Dams, Graz, Austria*, ed. by E. Bauer, S. Semprich, G. Zenz (Graz University of Technology, Graz, 2009) pp. 473–478

G. Gudehus, A comprehensive constitutive equation for granular materials. Soils Found. **36**(1), 1–12 (1996)

G.X. Li, Triaxial wetting experiments on rockfill materials used in Xiaolangdi earth dam. Research report from Tsinghua University, 1988

H. Matsuoka, T. Nakai, Stress-strain relationship of soil based on the 'SMP', in *Proceedings of Specialty Session 9, IX International Conference Soil Mechanics and Foundation Engineering*, Tokyo, 1977, pp. 153–162

E.S. Nobari, J.M. Duncan, Effect of reservoir filling on stresses and movements in earth and rockfill dams. Department of Civil Engineering, Report No. TE-72-1, University of California, 1972

L.A. Oldecop, E. Alonso, Theoretical investigation of the time-dependent behaviour of rockfill. Géotechnique **57**(3), 289–301 (2007)

Effect of Different Sample Preparation Methods on the Behavior of Granular Materials Using Bender Elements

M. Aris and N. Benahmed

Abstract This paper discusses results of experimental study. It highlights the influence of modes of reconstitution of samples on structural architecture of granular materials. Undrained triaxial tests were carried out on Hostun sand HN31. Different procedures for the preparation of the samples were used: moist tamping, dry pluviation and water pluviation. Bender elements which transmit and receive shearing waves have been used to study the structural anisotropy induced by each mode of deposition and its evolution during the loading path.

The results obtained showed that the three different methods of preparation induce different structures and anisotropy of soil and lead to different mechanical behavior during the undrained shearing tests.

Keywords Sand fabric • Anisotropy • Bender elements • Triaxial test

1 Introduction

It is well known now that the deposition mode plays a key role on the mechanical behavior of soil in sense that it induces different structures and grain arrangements, thus some degree of anisotropy.

Previous studies have shown that the behavior of sand is strongly affected by the method of reconstitution (Canou 1989; Zlatovic and Ishihara 1997; Vaid et al. 1999; Benahmed 2001; Benahmed et al. 2004; Yamamuro and Wood 2004; Yamamuro et al. 2008). However, most of the published literatures are based on "macroscopic" observations. In fact, a few attempts have been made to investigate the effect of

M. Aris · N. Benahmed (✉)
CEMAGREF d'Aix en Provence, 3275, Route de Cézanne CS, 40061 13182
Aix en Provence Cedex 5, France
e-mail: nadia.benahmed@cemagref.fr

S. Bonelli et al. (eds.), *Advances in Bifurcation and Degradation in Geomaterials*, Springer Series in Geomechanics and Geoengineering 11,
DOI 10.1007/978-94-007-1421-2_45, © Springer Science+Business Media B.V. 2011

method of preparation at micro scale, i.e. on the microstructure of sand. Studies undertaken by Benahmed (2001), Benahmed et al. (2004) have shown, through the use of the Scanning Electron Microscope, that samples of sand prepared by moist tamping present an irregular structure with predominance of aggregates and macropores, and are more unstable than the ones prepared by dry pluviation, which present regular structure without macropores.

Over the years, bender elements have become an interesting tool to assess soil properties in laboratory testing (Dyvik and Madshus 1985; Bellotti et al. 1996; Brignoli et al. 1996; Pennington et al. 2001; Lings and Greening 2001; Leong et al. 2005). They represents a non intrusive method of investigation and offer the possibility of studying shear or compressive waves propagation through the skeleton of soil, hence to obtain information on the anisotropy.

In the present work, this technique was used to study the fabric anisotropy induced by each mode of deposition and its evolution during the followed stress-strain paths.

The paper first presents the mechanical behaviour of Hostun sand at macroscopic scale. Then, some preliminary results of shear wave velocity measurements at initial state and during the undrained shearing stage are presented and discussed.

2 Experimental Device and Bender Elements Equipment

The material used is Hostun sand HN31 which is a fine and uniform sand with angular to sub angular grain shape. It has a medium diameter $D_{50} = 0.34$ mm, and a coefficient of uniformity $Cu = 1.57$.

Three procedures of reconstitution of the samples were used; the moist tamping (MT), the air pluviation (DP) using a specific pluviator, and the water pluviation (WP). All samples were tested under undrained shear on classical triaxial cell equipped with bender elements (Fig. 1). The piezoelectric transducers have been

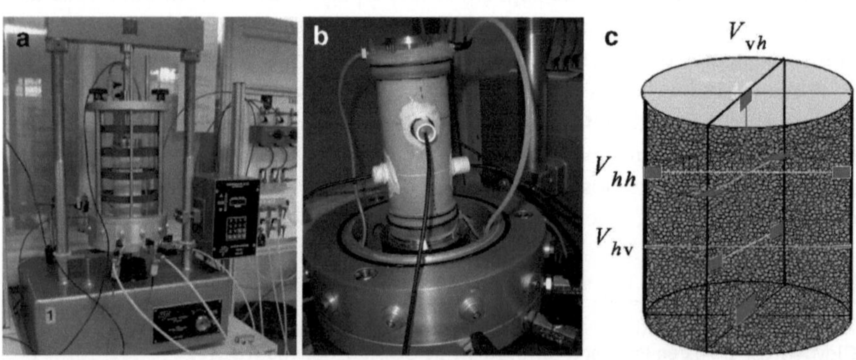

Fig. 1 Photography of the experimental equipment; (**a**) Triaxial apparatus, (**b**) Sample fitted with vertical and lateral bender elements, (**c**) different directions of wave propagation

placed on top cap and pedestal base of the triaxial cell, and on the lateral sides of the sample in order to measure the shear wave velocity in different directions (HH, HV and VH). All determinations were made from the first arrival of the shear wave using the tip-to-tip distance between the bender elements.

3 Test Results and Analysis

3.1 Effect of Mode of Deposition on Mechanical Behavior

Figure 2 presents results of the undrained triaxial tests that were performed on three samples prepared at fixed relative density of about 3% and using the different techniques of reconstitution (MT, DP and WP).

A significant difference on mechanical response is observed. The moist tamping induces a contractant and unstable behavior with strain softening and liquefaction instability, whereas dry and water pluviation lead to a dilative, more stable behavior with strain hardening. It can also be noticed that the sample prepared by sedimentation is slightly more resistant than the one made by dry pluviation.

3.2 Evolution of Vs During Loading

The evolution of shear wave velocities with strain level is plotted in Fig. 3. As can be seen, the values of *Vs* in both directions decrease just after the start of the test and vary throughout the undrained shearing stage. This is partly due to the change of the stresses state and certainly also because of the change of the structure. It is

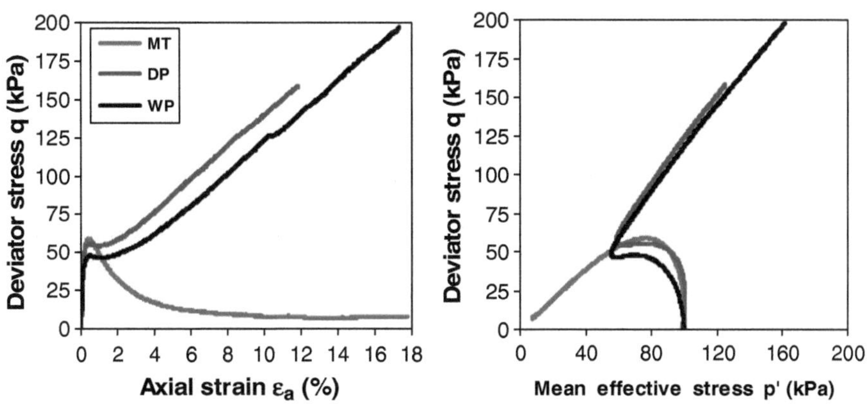

Fig. 2 Effect of method of reconstitution on undrained mechanical behavior of Hostun sand

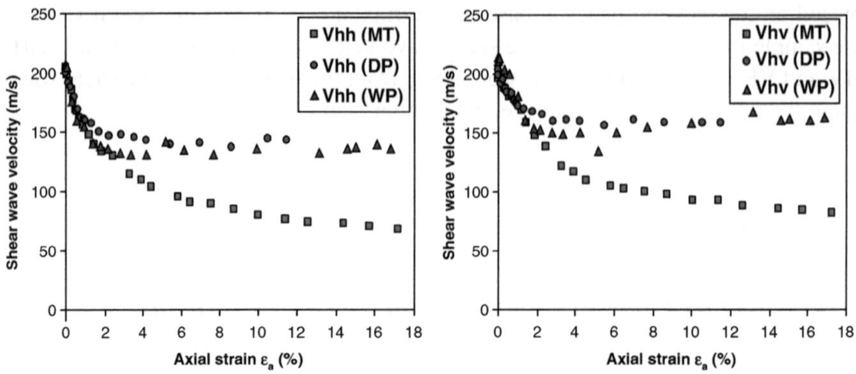

Fig. 3 Evolution of shear wave velocities Vs_{hh} and Vs_{hv} during undrained shear

Table 1 Comparative V_s measurements for different modes of preparation

	Test conditions		Shear wave velocities (m/s)		
Test N°	Preparation mode	D_r(%)	V_{hh}	V_{hv}	V_{hh}/V_{hv}
MT-3a	Moist Tamping	3	203,4	190	1,07
DP-3a	Dry Pluviation	3	213,2	217,4	0,98
WP-3a	Water Pluviation	3	218,5	220,4	0,99

interesting to note that the trend of this variation is in agreement with the mechanical behavior observed (Fig. 2). The moist tamping sample shows a clear reduction of *Vs*, hence of stiffness, according to the occurrence of the liquefaction instability. However, this reduction seems to stabilize when the critical state is reached.

Almost the same trend is observed for the DP and WP samples except that the values of *Vs* reached after the peak strength, i.e. at characteristic state or phase transformation, are much higher. This is consistent with the mechanical behavior exhibited by these two samples (strain hardening).

3.3 *Characterization of Initial Anisotropy*

A set of *Vs* readings taken at initial state of three samples just after their reconstitution is reported on Table 1. The anisotropy is characterized by the ratio Vs_{hh}/Vs_{hv}. It is interesting to notice that in each direction, the values of *Vs* are not the same and the ratio is different from 1. This suggests that the initial state of samples is mainly anisotropic and the level of this anisotropy depends on the mode of reconstitution. The moist tamping procedure seems to be the technique which induces more anisotropic samples. It is important to point out that the initial anisotropy is minimal in all cases.

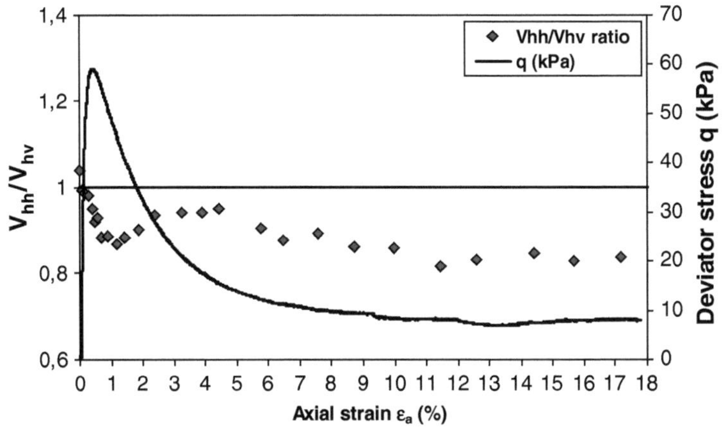

Fig. 4 Variation of anisotropy ratio with the deviator stress

3.4 Characterization of Induced Anisotropy

The evolution of the anisotropy ratio with stress-strain curve is illustrated in Fig. 4. We could see clearly that the anisotropy changes during the test, and this change is more pronounced before and just after the occurrence of the peak strength.

The velocity Vs_{hh} decreases more than Vs_{hv} and this might be associated with the decrease in the horizontal effective stress. At the critical state, the evolution of the velocity Vs_{hh} and Vs_{hv} tend to be similar and the anisotropy ratio to be stable around 0.84.

4 Conclusion

Bender elements were used to characterize the effect of mode of reconstitution on structural anisotropy of granular materials. Three methods of sample reconstitution were used. Shear wave velocity measurements have revealed the existence of inherent and induced anisotropy, which are probably linked to different grain arrangements and organizations, and enabled them to be estimated. The moist tamping method induces more initial anisotropy compared with air and water pluviation method. The initial anisotropy is less important than the anisotropy induced by loading. These new results confirm and expand the observations made by Benahmed et al. (2004) with the scanning electron microscope and triaxial apparatus.

References

N. Benahmed, Comportement mécanique d'un sable sous cisaillement monotone et cy clique: Application aux phénomènes de liquéfaction et de mobilité cyclique, Thèse, ENPC, 2001

N. Benahmed, J. Canou, J.C. Dupla, Structure initiale et propriétés de liquéfaction statique d'un sable. C R Mécanique **332**, 887–894 (2004)

R. Bellotti, M. Jamiolkowski, D.C.F. Lo Presti, D.A. O'Neill, Anisotropy of small strain stiffness in ticino Sand. Geotechnique. **46**(1), 115–131 (1996)

E.G.M. Brignoli, M. Gotti, K.H. Stokoe, Measurement of shear waves in laboratory specimens by means of piezoelectric transducers. Geotech. Test. J. **19**(4), 384–397 (1996)

J. Canou, Contribution à l'étude et à l'évaluation des propriétés de liquéfaction d'un sable, Thèse de doctorat de l'Ecole Nationale des Ponts et Chaussées, Paris, 1989

R. Dyvik, C. Madshus, Lab measurements of Gmax using bender elements, in *Proceedings of the conference on the Advances in the Art of Testing Soil under Cyclic Conditions* (ASCE Geotechnical Engineering Division, New York, 1985), pp. 186–196

E.C. Leong, S.H. Yeo, H. Rahardjo, Measuring shear wave velocity using bender elements. Geotech. Test. J. **28**(5), 488–498 (2005)

M.-L. Lings, P.-D. Greening, A novel Bender/extender element for soil testing, Geotehnique, **51**(8), 713–717 (2001)

D.S. Pennington, D.F.T. Nash, L. Lings, Horizontally mounted bender elements for measuring anisotropic shear moduli in triaxial clay specimens. Geotech. Test. J. **24**(2), 133–144 (2001)

Y.P. Vaid, S. Sivathayalan, D. Stedman, Influence of specimen reconstituting method on the undrained response of sand. Geotech. Test. J. **22**(3), 187–195 (1999)

J.A. Yamamuro, F.M. Wood, Effect of depositional method on the undrained behavior and microstructure of sand with silt. Soil Dyn. Earthq. Eng. 24, 751–760 (2004)

J.A. Yamamuro, F.M. Wood, P.V. Lade, Effect of depositional method on the microstructure of silty sand. Can. Geotech. J. 45, 1538–1555 (2008)

S. Zlatovic, K. Ishihara, Normalized behaviour of very loose non-plastic soils: effects of fabric, Soils and Foundations, **37**(4), 47–56 (1997)

Characterization of the Strain Localization in a Porous Rock in Plane Strain Condition Using a New True-Triaxial Apparatus

P. Bésuelle and S.A. Hall

Abstract We present here first results obtained in a new true-triaxial apparatus that allows observation of the rock specimen under loading (up to a confining pressure equal to 100 MPa). The three principal stresses are independent, however the intermediate stress can also be controlled in order to impose a plane strain condition. Observation of a specimen under load is possible as one surface of the prismatic specimen, is in contact with a hard transparent window. Therefore the evolution of the strain field in the sample can be measured by digital image correlation (DIC) of photographs taken of this surface. We show some results on the evolution of the pattern in the post-localization regime in clay rock specimens.

Keywords Failure • Strain localization • Clay rocks • True-triaxial apparatus • Digital image correlation

1 Introduction

Failure by strain localization is commonly observed in geomaterials. In a previous workshop (IWBI Minneapolis St Paul, 2002), we presented an experimental characterization of strain localization in a porous sandstone (Bésuelle et al. 2000). This study was performed with classical axisymmetric triaxial compression tests. The effect of the confining pressure was observed on several aspects: onset of localization; pattern of localization; porosity evolution inside the localized bands. Complex patterns of localization were observed at high confining pressure in the transition between the brittle and ductile regimes, showing several deformation

P. Bésuelle (✉) · S.A. Hall
CNRS – UJF – Grenoble INP, Laboratoire 3SR, BP53, 38041 Grenoble Cedex 9, France
e-mail: Pierre.Besuelle@hmg.inpg.fr

S. Bonelli et al. (eds.), *Advances in Bifurcation and Degradation in Geomaterials*, Springer Series in Geomechanics and Geoengineering 11,
DOI 10.1007/978-94-007-1421-2_46, © Springer Science+Business Media B.V. 2011

bands in the specimens. However the history (time evolution) of the localization was not accessible because the observations were post-mortem.

Measurements of strain fields and their evolution in time are particularly useful to study strain localization (initiation of deformation bands) and post-localization regimes. Such tools have been developed for soils (e.g., sand specimens in plane strain conditions Desrues and Viggiani 2004 or in triaxial conditions using X-ray tomography Hall et al. 2010). Similar developments for rocks are still difficult, especially because the pertinent confining pressure to reproduce in-situ stresses and material stiffnesses are higher than for soils; only a very few devices exist (e.g., Takemura et al. 2004).

We present here first results obtained in a new true-triaxial apparatus that allows observation of rock specimens under loading. Whilst several triaxial apparatuses exist that allow the application of three different principal stresses, they do not allow observation of specimens under load and such analysis is only possible post-mortem (e.g., Mogi 1967; Atkinson and Ko 1973; Michelis 1985; Wawersik et al. 1997; Haimson and Chang 2000; King 2002; Naumann et al. 2007 and Popp and Salzer 2007). Furthermore, in this new device, as for biaxial apparatuses (e.g., Ord et al. 1991 and Labuz et al. 1996), failure surfaces can develop and propagate in a sample in an unrestricted manner; this can be under true-triaxial or plane-strain (biaxial) conditions as, if required, the intermediate stress can be controlled (with active control) to impose a plane strain condition during a test. The observation of a specimen under load is possible as one surface of the prismatic specimen, which is orthogonal to the plane strain direction, is in contact with a hard transparent window. The deformation of this surface should representative of the deformation in the whole specimen (due to the plane-strain condition), up to and beyond strain localization. Therefore the evolution of the strain field in a sample can be measured by digital image correlation (DIC) of photographs taken of this surface.

The next section describes briefly the new apparatus. We present in a third section preliminary results obtained with a porous clay rock, focusing on strain localisation and crack initiation.

2 Description of the True Triaxial Cell

The true-triaxial apparatus has been developed in Laboratoire 3SR (Grenoble) with the aim to characterize the initiation of localization and the post-localization regime in rocks. With this device three independent stresses can be applied in the three space directions on prismatic rock specimens, with the ability to visualise the specimen under load. The surfaces perpendicular to the major and intermediate stresses (compression) are in contact with rigid platens, which are moved by two perpendicular pistons, while the two surfaces perpendicular to the minor stress are free to deform because the stress is applied by a confining fluid (through a soft membrane). As deformation bands and cracks are generally parallel to the intermediate stress, the specimen has the freedom to deform and fail with no kinematic constraints imposed on the formation of the failure zone. Moreover, one of the two surfaces perpendicular to the intermediate stress is in contact with a hard window to observe the specimen under load. The two pistons can be controlled in stress or displacement. The intermediate stress can be controlled such that there is no deformation in this direction, which allows application of plane strain loading. In such a case, the kinematics over the surface in contact with the window is representative of the kinematics in the whole specimen up to and beyond strain localization.

A simplified schema of the apparatus is presented in Fig. 1. The hydraulic axial piston (1) applies the axial loading on the specimen (2). This is self-compensated with respect to the confining pressure, i.e., it is in equilibrium whatever the confining pressure. The axial loading is controlled in displacement by an external displacement transducer linked to a pressure generator that adapts the pressure applied to the top of the piston to keep a constant displacement rate. The axial piston moves inside a floating axial frame (3), when the piston moves down, the frame and the bottom loading cap move up by about the same value. In such a way, if the specimen deforms homogeneously, the middle of the specimen does not move (or only very slightly). The weight of the floating frame is compensated by an external air piston.

The horizontal piston (4) applies the intermediate stress. This is also self-compensated with respect to the confining pressure inside the confining chamber (5). The piston can be controlled in displacement by the internal displacement transducers (6). One possible mode of operation is to adapt the pressure sent by its generator pressure to keep a zero displacement, i.e., a plane strain condition on the specimen, although all other controls in displacement or stress are possible. The surface of the specimen, opposed to the horizontal piston, is in contact with a thick, transparent sapphire window. This surface can be observed and photographs of the surface can be taken. For a symmetry of the contact, the surface on the side of the horizontal piston is in contact with a thin sapphire platen, to have the same boundaries conditions.

The minor stress is applied by the confining fluid on the two lateral surfaces of the specimen. The specimen is separated from the fluid by a silicone membrane. The membrane wraps around both the specimen and the four loading caps. In such a way, there is a direct contact between the specimen and the window. Note that in the axial direction, a special device of wedges between the specimen and the loading caps ensures that they have the same thickness as the specimen in the direction of intermediate stress. If a compression or extension of specimen in this direction occurs, the set of wedges automatically follow this change. This avoids an extrusion of the membrane during the deformation of the specimen.

A set a three pressure generators (syringe pumps with electronic control) is associated to the apparatus, to apply the loading in the three space directions. Further development is a control of the bottom and top pore pressure inside the specimen to impose a fluid flux. Note also that numerous electrical connectors in the cell allow to put several internal transducers, e.g., for acoustic emissions measurement.

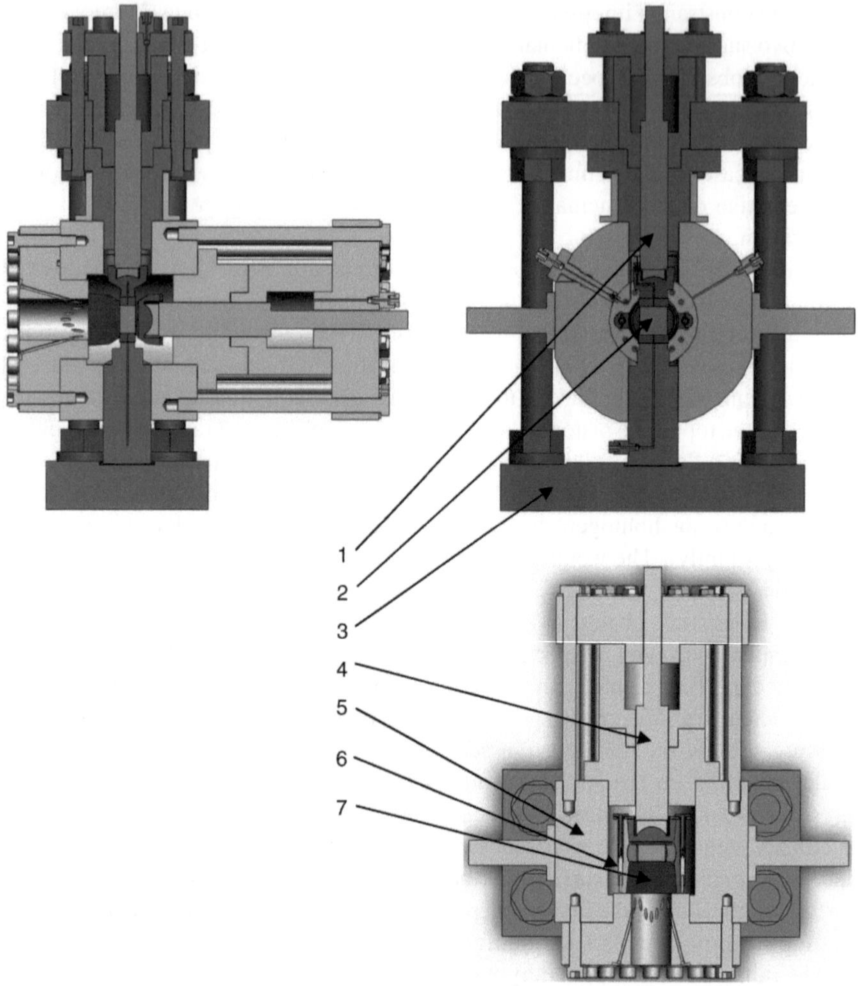

Fig. 1 Scheme of the true triaxial cell with an observation window: (*1*) axial piston, (*2*) specimen, (*3*) floating axial frame, (*4*) horizontal piston, (*5*) confining chamber and fixed frame, (*6*) internal displacement transducers, (*7*) sapphire window

The surface of the specimen is illuminated through the sapphire window by light from a set of LEDs focussed onto the sample surface through optical fibers; this provides a good and homogenous luminosity to take photographs. Photographs of the visible surface of the specimen were taken throughout the loading with a high resolution camera (providing images of 6,080 × 4,044 pixels). DIC analysis can thus be carried out on the resultant images to yield displacement and strain fields over the observed surface; for details on the DIC procedure see Hall et al. (2000). However, it is important to note that for the DIC, it is necessary to have a pattern over the surface of the sample that varies such that different parts of the surface can be uniquely distinguished. Depending on the test specimen, this pattern can be natural (as with some sandstones for example) or artificial (in this case it was necessary to add this pattern as discussed later).

The size of the specimen is 50 mm in the axial direction, 30 mm in the direction of intermediate stress, and 25 or 50 mm in the direction of minor stress, which corresponds to slenderness ratios (the ratio of the height to the width) of two and one, respectively. The capacity of the cell for confining pressure is 100 MPa, the axial piston can apply a force of 500 kN and the horizontal piston a force of 700 kN, which correspond to a differential stress with respect to the confining pressure of 670 and 530 MPa, respectively, for a specimen with a slenderness ratio of two, and half that for a 50 mm width specimen.

3 Selected Results

We present here a test that has been performed on a clay rock specimen, the Callovo-Oxfordian argillite, from the underground research laboratory (URL) at Bure (Eastern France) at approximately 500 m below the ground surface. It is a sedimentary rock composed of particles of calcite and quartz in a clay matrix (Lenoir et al. 2007). The specimen has been prepared with a diamond wire saw and then polished with a fine sandpaper. The surface of the specimen in contact with the window of the triaxial apparatus has been painted with a thin layer of white ink and then a speckle of black ink, using an airbrush. The size of pixels in the photographs correspond to about 10 μm on the sample surface. The test has been performed with an initial isotropic loading to 2 MPa and then an axial loading in plane strain conditions with a displacement rate of 1.25 $\mu m{\cdot}s^{-1}$, i.e., a strain rate of $2.5\ 10^{-5}\ s^{-1}$.

Figure 2 shows the evolution of the differential stress (major stress minus minor stress) with respect to the axial strain (specimen shortening divided by its initial height). A zoom of the full curve (left) is presented (right) for the period when the analyzed photographs were taken. The beginning of the curve at left is quite linear, followed by a small curvature and a first stress peak at 0.02 axial strain, followed by a strong stress drop. Then a slow stress increase is observed, followed by a second stress drop at 0.42 axial strain. After, the stress is quite constant. The two stress drops are associated with major failure by faulting in the specimen. The crack that appeared during the second drop is conjugate to the first crack set, which appeared at the first drop. We focus later on the strain localization at first stress peak.

Figure 3 presents the fields of a few axial displacement increments before and after the stress peak. The specimen deformation during increment 3,388–3,390 seems quite homogeneous, and the displacements are primarily vertical.[1] The increment 3,392–3,393 shows a loss of the homogeneity with a gradient of

[1] The top displacement is approximately the same but opposite to the bottom displacement, due to the conception of the apparatus. The middle of the specimen does not move in the axial direction.

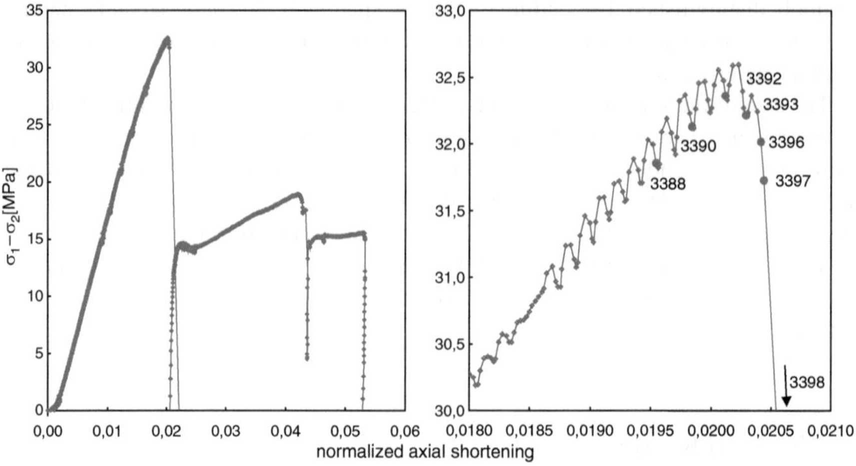

Fig. 2 Evolution of the differential stress (major stress minus minor stress) versus axial strain. At *right*, detail of the full curve at *left*, close to the stress peak. Numbers correspond to the photographs (The small oscillations of the *right curve* have been induced by an imperfect regulation of the axial piston pressure generator)

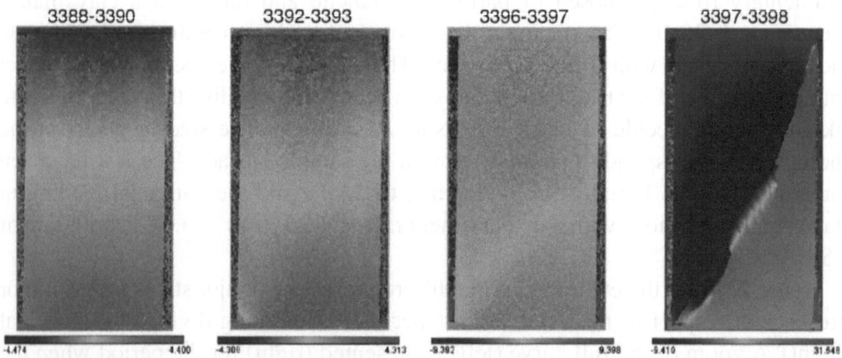

Fig. 3 Fields of axial displacement increments close to the stress peak. Couples of numbers on the *top* of each picture correspond to the photographs numbers used for the DIC. The color scales are expressed in term of pixel size (about 10 μm)

displacement oriented along an inclined line from the bottom left to the top right of the specimen. This corresponds to an incipient strain localization arriving at the stress peak. A shear band is observed in further increments up to photograph 3,397. During increment 3,397–3,398, a strong discontinuity is observed in the displacement field, which corresponds to a crack initiation in the place of the previous shear band. The strain field of this increment is shown in Fig. 4, where the maximum shear strain and volume strain are plotted. A major crack crosses the specimen from the bottom left to the top right. In the central zone of the specimen, there is a set of small conjugate cracks, showing two, quite close, preferential

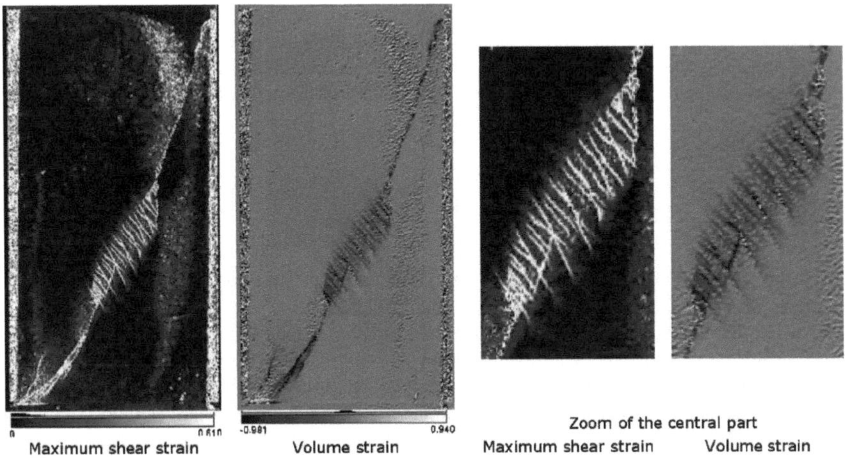

Fig. 4 Fields of shear and volume strain during increment 3,397–3,398

orientations. These small cracks are arranged inside an elongated zone parallel to the major crack. In most of cases, the major and small crack initiation is associated with a compaction combined with shear sliding. The sub-vertical zone of concentration of the shear strain on the right of the specimen is in fact an artefact of the measure due to a change of luminosity of the surface of the specimen in contact with the window (probably a consequence of an initial small default of planarity of the specimen).

A post-mortem analysis of the specimen using our X-ray CT apparatus shows the 3D network of cracks. The major cracks cross the specimen roughly parallel to the direction of intermediate stress, confirming a 2D mechanism of deformation in the specimen. The central zone of conjugate small cracks extends about half of the way into the specimen in the intermediate stress direction, which confirms that the conjugated cracks are not a surface effect but bulk mechanism of deformation.

4 Conclusions

The ability to characterize localized failure in rocks and, in particular, to follow the strain field evolution inside rock specimens during loading using a new true-triaxial apparatus has been demonstrated. Displacement and strain fields results, using DIC, from a plane-strain loading test on a clay rock have been presented, which reveal a complex pattern of localization at failure. Furthermore the evolution of the deformation has been followed from an initially homogeneous deformation through the development of a shear band and subsequent initiation of a set cracks resulting in major faults and small conjugate cracks arranged inside a band parallel to the major fault. Further experiments will focus on the effect of confinement on the processes of failure and localization and also on reproducibility in similar conditions.

Acknowledgments The authors thank the French radioactive waste management agency (ANDRA) who provides the clay rock from its URL at Bure. The development of the true triaxial apparatus has been supported by the 'ACI jeunes chercheurs JC8029' from the French government, by the national research group 'GDR FORPRO' from CNRS and by ANDRA.

References

R.H. Atkinson, H.-Y. Ko, A fluid cushion, multiaxial cell for testing rock specimens. Int. J. Rock Mech. Min. Sci. Geomech. Abstr. **10**, 351–361 (1973)

P. Bésuelle, J. Desrues, S. Raynaud, Experimental characterisation of the localisation phenomenon inside a Vosges sandstone in a triaxial cell. Int. J. Rock Mech. Min. Sci. **37**, 1223–1237 (2000)

J. Desrues, G. Viggiani, Strain localization in sand: an overview of the experimental results obtained in Grenoble using stereophotogrammetry. Int. J. Numer. Anal. Methods Geomech. **28**, 279–321 (2004)

B. Haimson, C. Chang, A new true triaxial cell for testing mechanical properties of rock, and its use to determine rock strength and deformability of Westerly granite. Int. J. Rock Mech. Min. Sci. **37**, 285–296 (2000)

S.A. Hall, D. Muir Wood, E. Ibraim, G. Viggiani, Localised deformation patterning in 2D granular materials revealed by digital image correlation. Granul. Matter (2000). doi:10.1007/s10035-009-0155-1

S.A. Hall, M. Bornert, J. Desrues, Y. Pannier, N. Lenoir, G. Viggiani, P. Bésuelle, Discrete and continuum analysis of localised deformation in sand using X-ray micro CT and volumetric digital image correlation. Géotechnique **60**, 315–322 (2010)

M.S. King, Elastic wave propagation in and permeability for rocks with multiple parallel fracture. Int. J. Rock Mech. Min. Sci. **39**, 1033–1043 (2002)

J.F. Labuz, S.-T. Dai, E. Papamichos, Plane-strain compression of rock-like materials. Int. J. Rock Mech. Min. Sci. Geomech. Abstr. **33**, 573–578 (1996)

N. Lenoir, M. Bornert, J. Desrues, P. Bésuelle, G. Viggiani, Volumetric digital image correlation applied to X-ray microtomography images from triaxial compression tests on argillaceous rock. Strain **43**, 193–205 (2007)

P. Michelis, A true triaxial cell for low and high pressure experiments. Int. J. Rock Mech. Min. Sci. Geomech. Abstr. **22**, 193–188 (1985)

K. Mogi, Effect of the intermediate principal stress on rock failure. J. Geophys. Res. **72**, 5117–5131 (1967)

M. Naumann, U. Hunsche, O. Schulze, Experimental investigations on anisotropy in dilatancy, failure and creep of Opalinus Clay. Phys. Chem. Earth **32**, 889–895 (2007)

A. Ord, I. Vardoulakis, R. Kajewski, Shear band formation in Gosford sandstone. Int. J. Rock Mech. Min. Sci. Geomech. Abstr. **28**, 397–409 (1991)

T. Popp, K. Salzer, Anisotropy of seismic and mechanical properties of Opalinus clay during triaxial deformation in a multi-anvil apparatus. Phys. Chem. Earth **32**, 879–888 (2007)

T. Takemura, M. Oda, M. Takahashi, Microstructure observation in deformed geomaterials using microfocus X-ray computed tomography, in *X-ray CT for Geomaterials, GeoX2004*, ed. by J. Otani, Y. Obara (Balkema, Rotterdam, 2004)

W.R. Wawersik, L.W. Carlson, D.J. Holcomb, R.J. Williams, New method for true-triaxial rock testing. Int. J. Rock Mech. Min. Sci. **34**, 330 (1997)

Granular Materials at Meso and Macro Scale: Photo-Elasticity and Digital Image Correlation

Danuta Lesniewska and David Muir Wood

Abstract Combining integrated photoelasticity and high resolution particle image velocimetry (digital image correlation) provides a tool which can be used to study granular materials at different scales. We have used an experimental arrangement which subjects an assembly of small 3D glass grains to plane strain conditions at the macro-scale. While the quantitative interpretation of individual and incremental photoelastic images in terms of continuum stress quantities presents challenges, some comparisons of strain increment and qualitative stress increment data are presented with concentration on the regions of apparent strain localisation.

Keywords Photoelasticity • Particle image velocimetry • Glass particles • Shear band • Force chains

1 Introduction

Granular materials are assemblies of macroscopic particles of different shapes, origin and surface properties. The traditional continuous approach is typical for engineering applications and adopts all concepts of classical continuum mechanics. However, the continuum properties of granular materials emerge from their

D. Lesniewska
Institute of Hydro-Engineering, Polish Academy of Sciences, Geomechanics Division, 80-328 Gdańsk, Koscierska 7, Poland

Koszalin Technical University, 75-453 Koszalin, Sniadeckich 2, Poland
e-mail: d.lesniewska@ibwpan.gda.pl

D. Muir Wood (✉)
University of Dundee, Fulton Building, Dundee DD1 4HN, UK
e-mail: d.muirwood@dundee.ac.uk

S. Bonelli et al. (eds.), *Advances in Bifurcation and Degradation in Geomaterials*, Springer Series in Geomechanics and Geoengineering 11,
DOI 10.1007/978-94-007-1421-2_47,

particulate nature and a second, discrete, approach attempts to work directly from fundamental laws of physics, applied to contacts within a system of individual grains.

The continuum approach makes use of phenomenological elastic-plastic models whose physical justification is untestable but which have been used predictively, more or less successfully, for engineering problems. Granular physicists admit that the discrete approach is in its infancy (Duran 2000) with a barrier to progress being the lack of experimental knowledge concerning the multi-scale phenomena responsible for load transmission and material deformation in real granular materials. It is extremely difficult to model granular material under laboratory conditions in a way which permits control and knowledge of the state of each individual grain and also of the whole granular assembly. Typically only small collections of artificial 2D grains of regular shape are investigated, using 10^3–10^4 grains, compared with 10^6–10^9 grains for typical soil mechanics tests and billions of grains in any real application.

The greatest mystery regarding granular materials is the way they transmit loads. It is generally accepted that force transmission takes place through discrete chains of particles, surrounded by much less loaded neighbouring particles, but the exact mechanisms ruling this type of behaviour are unknown except from numerical analysis and studies of photelastic systems of 2D circular or elliptical discs (Drescher 1976; Drescher and De Josselin de Jong 1972).

Combining integrated photoelasticity and high resolution PIV appears to provide some possibilities for linkage of stress-related and strain information, at least at the meso-scale. This paper reports the present state of our studies of such linkages. The experimental test arrangement loads an assembly of a large number of 3D glass grains under macro-scale plane strain conditions (Lesniewska and Muir Wood 2009; Muir Wood and Lesniewska 2011).

2 Experimental Observations

The experimental procedure is given in Lesniewska and Muir Wood (2009) and Muir Wood and Lesniewska (2011). The apparatus contains some 10^6 spheroidal glass particles within a series of segmental rigid boundaries some of which are able to move in a controlled way (Fig. 1). The glass particles are immersed in clove oil which has the same refractive index, thus rendering the assembly transparent to light. A photoelastic technique (Drescher 1976; Drescher and De Josselin de Jong 1972) was used to record information concerning the 'stress' fields in the granular material (a typical result is shown in Fig. 2) and particle image velocimetry (PIV; Rechenmacher and Medina-Cetina 2007; White and Take 2002; White et al. 2003; Niedostatkiewicz et al. 2010; Lesniewska and Muir Wood 2011) was used to calculate experimental displacement and strain fields.

Figure 2 shows the result of using polarised light to detect the 'light stripes' which are conventionally assumed to correspond to the average 'force chains' of

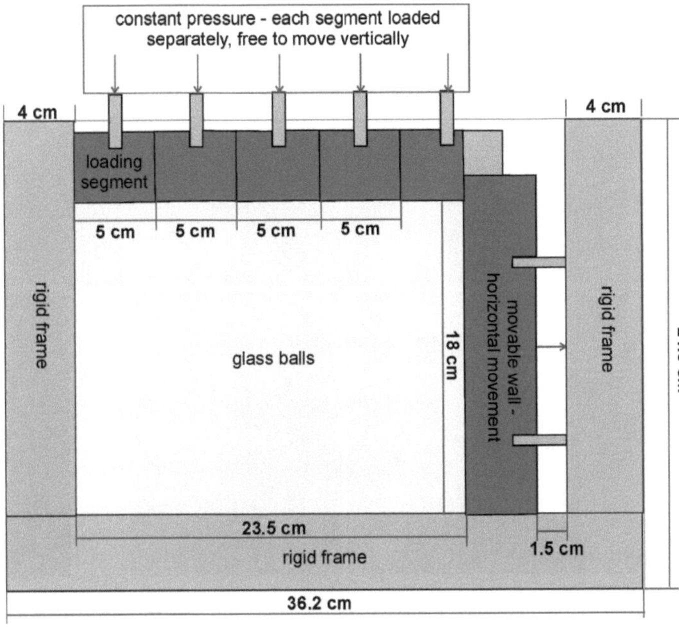

Fig. 1 Apparatus for loading particle assembly with controllable segmental rigid boundaries

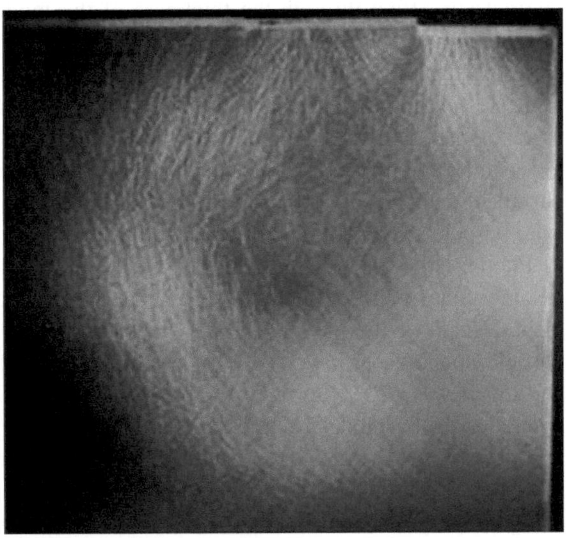

Fig. 2 Assembly of particles under load viewed with monochromatic circularly polarised light

highly loaded particles, somewhat coincident with the directions of major principal stress, carrying the boundary stresses through the granular material. That there is structure to this image is certain; its quantitative interpretation is less clear. It has been shown (Muir Wood and Lesniewska 2011) that the integration of the

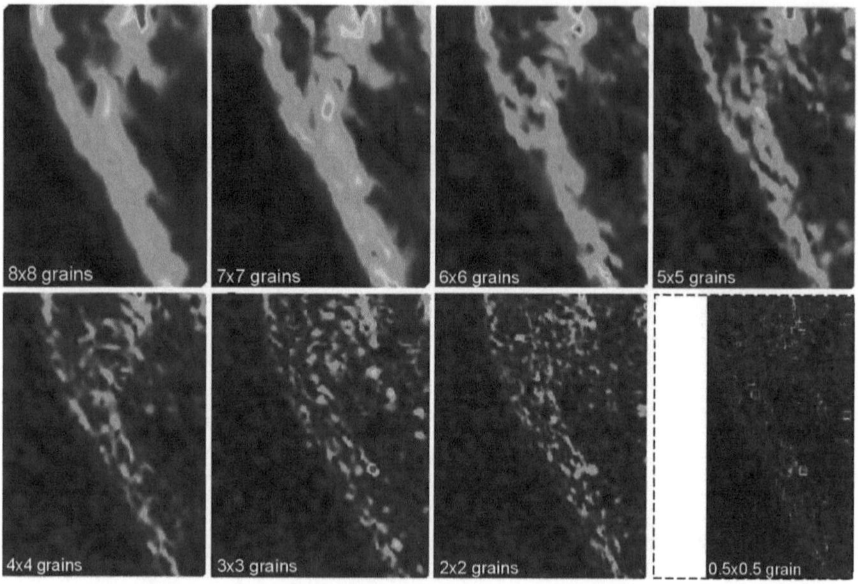

Fig. 3 An example of PIV analysis: fields of shear strains obtained for different dimensions of search patch (quoted as a number of grain diameters)

photoelastic effect through the thickness of the model is equivalent to a product of random matrices – each particle through the thickness of the model forms part of (probably) more than one 3D chain of particles loaded in random directions by multiple neighbours. If the image in Fig. 2 were revealing a continuum view of the assembly it would show smoothly varying light intensity. In fact the structure of the visual information has a scale which, when magnified, can be seen to be of sub-particle dimension and possibly of the order of d_{50}/n where $n = t/d_{50}$ and d_{50} and t are particle size and model thickness respectively.

Photographs taken with white light show sufficient texture of the granular assembly that pairs of such images, taken before and after a small increment of displacement of the movable 'wall' shown in Fig. 1, can be used for analysis by particle image velocimetry (PIV) (Lesniewska and Muir Wood 2009; Muir Wood and Lesniewska 2011) in order to determine the field of displacements, and hence strains, over the model by tracking the location of small regions of visual fabric across the entire image before and after the imposed boundary displacement.

Figure 3 presents fields of shear strains from such a pair of images calculated using the program GeoPIV (White and Take 2002). The use of different patch sizes can capture detailed information about the kinematic conditions within regions of intense shearing at different material scales. An impression of a shear band as a continuous feature is obtained when the size of the correlation region is about 8×8 grain diameters. As the patch size is reduced, at first the presence of the continuous shear band remains evident as some structure within the shear band starts to appear.

In common with the observations reported in Hall et al. (2009), the dimensions of this structure are of the order of a few particle diameters. As the patch size is further reduced, to a size less than the size of a single particle, then the structure and even the existence of the shear band seem to disappear (Fig. 3: 0.5 × 0.5 grain).

3 Combination of Test Results

Although the quantitative interpretation of a photoelastic image (Fig. 2) is uncertain, it is quite feasible to perform arithmetic operations on pairs of such images taken, for example, before and after the wall displacement that produced the fields of strains shown in Fig. 3.

The relationship between the structure of the force chains and the resulting photoelastic image is complex and generally not reversible: calculation of the expected image structure from a distribution of particle contacts and the forces transmitted by these contacts can be made (Muir Wood and Lesniewska 2011), but the inverse calculation, trying to deduce the internal structure of force chains from the recorded image would probably not admit a unique solution.

Figure 4a shows the subtraction of the earlier from the subsequent photoelastic image. Any non-zero intensity pixel in Fig. 4a indicates an increase in average stress (loading) through the width of the model, resulting from the increment of boundary displacement. Comparison between stress increment and strain fields is made easier if the stress field is transformed into a false colour image (Fig. 4a) with the same colour scale as the strain field (Fig. 3). Figure 4b shows an example: more combined images are presented in Lesniewska and Muir Wood (2011).

Precise superposition (Fig. 4b) provides a semi-automatic way to compare experimental stress and strain fields. The main shear band from the coarser patches used for the strain field determination in Fig. 3 matches the brighter region of increasing stress (Fig. 4a): it is deduced that the force chains are strengthening in

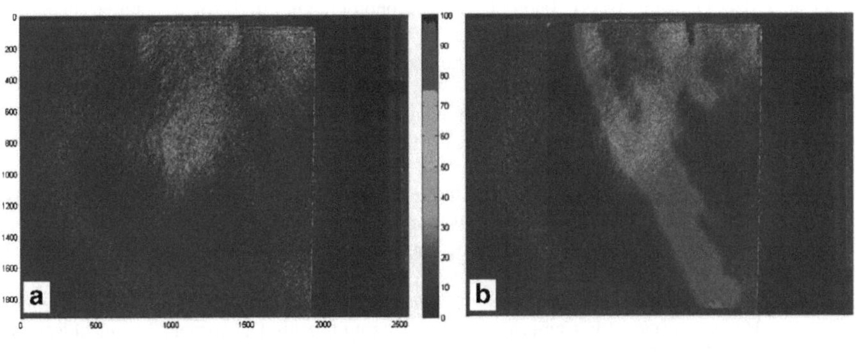

Fig. 4 Combination of digital integrated photoelasticity and PIV: (**a**) shows the difference between successive photoelastic images and (**b**) then superposes this photoelastic difference image on the field of shear strains (Fig. 3, 8 × 8 grains)

this area. The intense shearing immediately below the edge of the right hand loading segment of the top surface lies on a dark region of Fig. 4a, where buckling chains of particles are shedding load (Tordesillas et al. 2009), but immediately to the left of this the bright area in Fig. 4a shows the increased stress resulting from this load transfer.

4 Conclusion

Automatic procedures to compare photoelastic stress fields and strain fields obtained by digital image correlation can help to support development of rational constitutive models. Semi-automatic procedures based on superposition of incremental photoelastic images and corresponding computed strain fields can be used to provide qualitative support for current hypotheses of the behaviour of granular materials in shear bands. The interpretation of changes in photoelastic images is, however, subtle (Muir Wood and Lesniewska 2011).

References

A. Drescher, An experimental investigation of flow rules for granular materials using optically sensitive glass particles. *Géotechnique* **26**(4), 591–601 (1976)

A. Drescher, G. De Josselin de Jong, Photoelastic verification of a mechanical model for the flow of a granular material. *J. Mech. Phys. Solids.* **20**, 337–351 (1972)

J. Duran, *Sands, Powders, and Grains: An Introduction to the Physics of Granular Materials, Partially Ordered Systems* (Springer, New York, 2000)

S. Hall, D. Muir Wood, E. Ibraim, C. Viggiani, Localised deformation patterning in 2D granular materials revealed by digital image correlation. *Granular Matter.* **12**(1), 1–14 (2009). doi:10.1007/s10035-009-0155-1

D. Lesniewska, D. Muir Wood, Observations of stresses and strains in a granular material. ASCE J. Eng. Mech. **135**(9), 1038–1054 (2009)

D. Lesniewska, D. Muir Wood, Photoelastic and photographic study of a granular material. *Géotechnique* (2011). doi: 10.1680/geot.8.T.017

D. Muir Wood, D. Lesniewska, Stresses in granular materials. *Granular Matter.* **13**(4), 395–415 (2011). doi:10.1007/s10035-010-0237-0

M. Niedostatkiewicz, D. Lesniewska, J. Tejchman, Experimental analysis of shear zone patterns in sand for earth pressure problems using Particle Image Velocimetry. *Strain* (2010). doi:10.1111/j.1475-1305.2010.00761.x

A. L. Rechenmacher, Z. Medina-Cetina, Calibration of soil constitutive models with spatially varying parameters. *J. Geotech.Geoenviron. Eng.* **133**(12), 1567–1576 (2007)

A. Tordesillas, J. Zhang, R.P. Behringer, Buckling force chains in dense granular assemblies: physical and numerical experiments. *Geomech. Geoeng.* **4**, 3–16 (2009)

D.J. White, W.A.Take, *GeoPIV: Particle Image Velocimetry (PIV) software for use in geotechnical testing. Manual for GeoPIV.* Cambridge University Engineering Department Technical Report, D-SOILS-TR322 (2002)

D.J. White, W.A. Take, M.D. Bolton, Soil deformation measurement using particle image velocimetry (PIV) and photogrammetry. *Géotechnique* **53**(7), 619–631 (2003)

Index

S. Bonelli et al. (eds.), *Advances in Bifurcation and Degradation in Geomaterials*, Springer Series in Geomechanics and Geoengineering 11,
DOI 10.1007/978-94-007-1421-2, © Springer Science+Business Media B.V. 2011

MIX
Papier aus verantwortungsvollen Quellen
Paper from responsible sources
FSC® C105338

If you have any concerns about our products,
you can contact us on
ProductSafety@springernature.com

In case Publisher is established outside the EU,
the EU authorized representative is:
Springer Nature Customer Service Center GmbH
Europaplatz 3, 69115 Heidelberg, Germany

Printed by Libri Plureos GmbH
in Hamburg, Germany